Lecture Notes in Networks and Systems

Volume 897

The series "Lecture Notes in Networks and Systems" publishes the latest developments in Networks and Systems—quickly, informally and with high quality. Original research reported in proceedings and post-proceedings represents the core of LNNS.

Volumes published in LNNS embrace all aspects and subfields of, as well as new challenges in, Networks and Systems.

The series contains proceedings and edited volumes in systems and networks, spanning the areas of Cyber-Physical Systems, Autonomous Systems, Sensor Networks, Control Systems, Energy Systems, Automotive Systems, Biological Systems, Vehicular Networking and Connected Vehicles, Aerospace Systems, Automation, Manufacturing, Smart Grids, Nonlinear Systems, Power Systems, Robotics, Social Systems, Economic Systems and other. Of particular value to both the contributors and the readership are the short publication timeframe and the world-wide distribution and exposure which enable both a wide and rapid dissemination of research output.

The series covers the theory, applications, and perspectives on the state of the art and future developments relevant to systems and networks, decision making, control, complex processes and related areas, as embedded in the fields of interdisciplinary and applied sciences, engineering, computer science, physics, economics, social, and life sciences, as well as the paradigms and methodologies behind them.

Indexed by SCOPUS, INSPEC, WTI Frankfurt eG, zbMATH, SCImago.

All books published in the series are submitted for consideration in Web of Science.

For proposals from Asia please contact Aninda Bose (aninda.bose@springer.com).

B. Rama Devi · Kishore Kumar · M. Raju ·
K. Srujan Raju · Mathini Sellathurai
Editors

Proceedings of Fifth International Conference on Computer and Communication Technologies

IC3T 2023, Volume 1

Springer

Editors
B. Rama Devi
Department of Electronics
and Communication Engineering
Kakatiya Institute of Technology
and Science
Hanumakonda, Telangana, India

M. Raju
Department of Electronics
and Communication Engineering
Kakatiya Institute of Technology
and Science
Hanamkonda, Telangana, India

Mathini Sellathurai
Department of Science and Engineering
Heriot-Watt University
Edinburgh, UK

Kishore Kumar
Department of Electronics
and Communication Engineering
National Institute of Technology, Warangal
Warangal, Telangana, India

K. Srujan Raju
Department of CSE and IT
CMR Technical Campus
Hyderabad, Telangana, India

ISSN 2367-3370 ISSN 2367-3389 (electronic)
Lecture Notes in Networks and Systems
ISBN 978-981-99-9703-9 ISBN 978-981-99-9704-6 (eBook)
https://doi.org/10.1007/978-981-99-9704-6

This Springer imprint is published by the registered company Springer Nature Singapore Pte Ltd.
The registered company address is: 152 Beach Road, #21-01/04 Gateway East, Singapore 189721, Singapore

Paper in this product is recyclable.

Committees

International Advisory Committee

Dr. Rangaraj M. Rangayyan, Professor, Emeritus of Electrical and Computer Engineering, University of Calgary, Alberta, Canada
Dr. Ajay K. Katangur, Professor and Head, Computer Science, MSU
Dr. Junfeng Que, Professor, Clayton State University, Atlanta, USA
Dr. Anitesh Barua, Professor, UT, USA
Dr. Mathini Sellathurai, Professor, Heriot-Watt University, UK
Dr. Joao Manuel R. S. Tavares, Professor, University of Porto, Portugal
Dr. Kun-Lin Hsieh, NTU, Taiwan
Dr. Ahamad J. Rusumdar, KIT, Germany
Dr. V. R. Chirumamilla, EUT, Netherland
Dr. Halis Altun, MU, Turkey
Dr. P. N. Suganthan, NTU, Singapore
Dr. Mohammad Reza Daliri, Professor, Iran
Dr. Boka Kumsa, Wollega University, Ethiopia

National Advisory Committee

Dr. D. Surya Narayana, Professor, Energy Science Engineering, IIT Mumbai
Dr. N. V. S. Narasimha Sarma, Director, IIIT Trichy
Dr. L. Anjaneyulu, Professor and Head, Department of Head, ECE, NITW
Dr. R. K. Bhagel, Professor, Department of ECE, SV MANIT, Bhopal
Dr. G. Lakshmi Narayanan, Professor and Head, SV ECE, NIT, Trichy
Dr. D. Sriram Kumar, Professor, Department of ECE, NIT, Trichy
Dr. P. Palany Swamy, Professor, Department of Professor, ECE, NIT, Trichy

Dr. S. Anuradha, Professor, Department of ECE, NITW
Dr. J. Ravi Kumar, Professor, Department of ECE, NITW
Dr. N. V. Srikanth, Professor, EED, NITW
Dr. P. Lakshmi Narayana, Director, NERTU, OU, Hyderabad
Dr. P. Malla Reddy, Principal, KUCET, KU, Warangal
Dr. E. Srinivasa Reddy, Principal, UCET, ANU, Guntur
Dr. P. Chandra Shekar, Professor, Department of ECE, UCE, OU, Hyderabad
Dr. M. Asharani, Professor, Department of ECE, JNTU
Dr. Ch. Srinivasa Rao, Professor, Department of ECE, JNTUK
Dr. G. Sashibhushan Rao, Professor, Department of ECE, AU, Visakhapatnam
Dr. P. Rajesh Kumar, Professor and Department of ECE, AU, Visakhapatnam
Dr. B. Anuradha, Professor, Department of ECE, University, Tirupathi
Dr. T. Ramasri, Professor, Department of ECE, University, Tirupathi
Dr. K. Kishan Rao, Professor, Department of SNIST, Hyderabad
Dr. T. Narasa Reddy, Associate SCEE, IIT Mandi
Dr. B. Ravi Kumar, Associate Professor, EED, IIT Hyderabad
Dr. C.V. Rama Rao, Associate Professor, Department of ECE, NITW
Dr. Lalita Gupta, Associate Professor, Department of ECE, MANITB
Dr. P. Suresh Babu, Associate Professor, EED, NITW
Dr. A. Kirubakaran, Associate Professor, EED, NITW
Dr. P. Selvi, Associate Professor, EED, NITW
Dr. P. Subba Reddy, Associate Professor, UCE, OU, Hyderabad
Dr. T. Rama Krishna, Associate Professor, Department of CSE, NITW
Dr. K. Bikshalu, Associate Professor, Department of ECE, KU, Kothagudem
Dr. I. Satish Kumar, Assistant Professor, EED, NITW
Dr. B. K. N. Srinivasa Rao, Assistant Professor, Department of ECE, NITW
Dr. M. Satish, Assistant Professor, Department of ECE, NITW
Dr. A. Subba Rao, Assistant Professor, Department of ECE, MANITB
Dr. Vinay Raj, Assistant Professor, Department of CA, NITT
Dr. K. Sivani, Professor, Department of EIE, KITSW
Dr. T. Anil Kumar, Professor, Department of ECE, CMRIT, Hyderabad
Dr. D. Krishna Reddy, Professor and Head, Department of ECE, CBIT, Hyderabad
Dr. Panyam Narahari Sastry, Professor, Department of ECE, CBIT, Hyderabad
Dr. T. Ranga Babu, Professor and Head, Department of ECE, RVRJC, Guntur
Dr. V. V. K. D. V. Prasad, Professor and Head, Department of ECE, GEC, Andhra Pradesh
Dr. Ramesh Babu Vallabhaneni, Professor, NRIIT
Dr. P. Niranjan Reddy, Professor, Department of CSE, KITSW
Dr. G. Ragotham Reddy, Professor, Department of ECE, KITSW
Dr. P. Kamakshi, Professor, Department of IT, KITSW
Dr. C. Venkatesh, Professor, Department of EEE, KITSW
Dr. K. Venumadhav, Professor, Department of EIE, KITSW

Editorial Board Members

Technical Review Committee

Chairman

Members

Dr. K. Bikshalu, Assistant Professor, Department of ECE, KU Campus, Kothagudem
Dr. E. Hari Krishna, Assistant Professor, Department of ECE, KUCE, Kothagudem
Dr. K. Kishan Rao, Professor, Department of ECE, SIST, Hyderabad
Dr. N. V. Koteswara Rao, Professor, Department of ECE, CBIT
Dr. Panyam Narahari Sastry, Professor, Department of ECE, CBIT
Dr. D. Krishna Reddy, Professor and Head of the Department, Department of ECE, CBIT, Hyderabad
Dr. E. Srinivasa Rao, Professor and Head, Department of ECE, Vasavi Engineering College
Dr. Sunny Dial, Professor, School of Electronics Engineering, VIT, Andhra Pradesh
Dr. M. Sushanth Babu, Professor, Department of ECE, Matrusri Engineering College
Dr. Bhavani Sankar, Professor and Head, Department of ECE, Anjalai Ammal–Mahalingam
Dr. Tummala Rangababu, Professor and Head, Department of ECE, RVRJC, Guntur, Andhra Pradesh
Dr. Shaik Jakeer Hussain, Professor, Head, Vignan University
Dr. Ram DeshMukh, Professor and Head, Department of EEE, SR Engineering College
Dr. B. Ramesh, Professor and Head, Department of ECE, KITS, Huzurabad
Dr. Ravindra Babu Kallam, Professor and Head, Department of CSE, KITS, Huzurabad
Dr. M. N. Tibdewal, Professor, Department of ECE, Shri Sant Gajanan Maharaj College of Engineering, Shegaon
Dr. Ramesh Babu Vallabhaneni, Professor, Department of ECE, NRI Institute of Technology, Vijayawada
Dr. M. Sunil Kumar, Assistant Professor, Department of ECE, SVEC
Humaira Nishat, Professor, Department of ECE, CVR College of Engineering
Dr. S. Sumathi, Professor, Department of EEE, Mahendra Engineering College
Dr. G. Raghotham Reddy, Professor, Department of ECE, Dean SAC, KITSW
Dr. C. Venkatesh, Professor and HoD, Department of EEE, KITSW
Dr. V. Shankar, Professor, Department of CSE, KITSW
Dr. V. Rajagopal, Professor, Department of EEE, KITSW
Dr. J. Tarun Kumar, Professor and Head, ECE Department, SR Engineering College
Dr. P. Prasad Rao, Principal, Vaagdevi Engineering College
Dr. V. Chandra Shekar Rao, Associate Professor, Department of CSE, KITSW
Dr. S. Narasimha Reddy, Associate Professor, Department of CSE, KITSW
Dr. M. Raju, Associate Professor, Department of ECE, KITSW
Dr. V. Venkateshwar Reddy, Associate Professor, Department of ECE, KITSW
Dr. T. Senthil Murugan, Associate Professor and HoD, Department of IT, KITSW
Dr. G. Rajender Naik, Associate Professor, Department of EEE, KITSW
Dr. G. Sudheer Kumar, Associate Professor, Department of EEE, KITSW
Dr. B. Vijay Kumar, Associate Professor, Department of EEE, KITSW
Dr. T. Sai Kumar, Associate Professor, Department of ECE, CMR Tech. Campus
Dr. S. Aruna, Associate professor, Department of ECE, AUCE

Dr. Varun Revuri, Associate Professor, Department of CSE, Vaagdevi Engineering College, Warangal
Dr. Prakash Pareek, Assistant Professor, Department of ECE, Vaagdevi Engineering College, Warangal
Dr. Korlapati Keerti Kumar, Associate Professor, Department of ECE, Vaageswari College of Engineering, Karimnagar
Dr. Guntha Karthik, Associate Professor, Department of ECE, Stanley College of Engineering and Technology for Women
Dr. K. C. T. Swamy, Associate professor, Department of ECE, GPCET
Dr. K. Ramudu, Associate Professor, Department of ECE, KITSW
Dr. Megha, Assistant Professor (Senior Grade), School of Computing and Assistant Dean, Research and Consultancy, DIT University, Dehradun, Uttarakhand
Dr. V. Raju Reddy, Assistant Professor, Department of ECE, KITSW
Dr. M. Chandrasekar, Assistant Professor, Department of ECE, KITSW
Dr. K. Sowjanya, Assistant Professor, Department of ECE, KITSW
Dr. B. Dhanalaxmi, Assistant Professor, Department of ECE, KITSW

Preface

The Springer Scopus indexed 5th International Conference on Computer and Communication Technologies (IC3T) 2023 organized by Department of ECE, KITSW was held in Warangal, Telangana, during October 6–7, 2023. Warangal, the home of a major state university amid pleasant surroundings, was a delightful place for the conference. The 280 scientific participants, 180 of whom were students, had many fruitful discussions and exchanges that contributed to the success of the conference. Participants from 18 countries and 17 states of India made the conference truly international in scope. The 76 papers that were presented on the two days formed the heart of the conference and provided ample opportunity for discussion. This change, allowing the conference to end with invited talks from the industry experts, was a departure from the format used at previous IC3T conferences. The papers were split almost equally between the five main conference areas, i.e., image processing and communications system, VLSI, wireless networks, Internet of Things (IoT), and machine learning (ML), and the posters were distributed across the days of the conference, so that approximately equal numbers of papers in the different areas were scheduled for each day. Of the total number of presented papers, 76 of these are included in this proceedings volume 1 and 2 with each 38 papers, the first time that papers have been published by IC3T. There were six plenary lectures covering the different areas of the conference: **Sri Akula Naresh**, Scientist-F, DRDL-Hyderabad, delivered a keynote address on *Artificial Intelligence in Defence*; **Dr. Kishore Kumar**, Professor in ECE, NITW, delivered a keynote address on *Machine Learning Based Emotion Recognition Techniques*; **Dr. Mathini Sellathurai**, Professor, Heriot-Watt University, UK, on *5G-IoT Noise Cleaning Multi-Model Hearing Aids and Devices*; **Dr. M. Sushanth Babu**, Professor in ECE, CBIT, Hyderabad, delivered a talk on *Research Challenges and Solutions in Vehicular Communications in 5G*; **Dr. K. V. Sridhar**, Associate Professor, NIT Warangal, delivered a talk during valedictory session on *Convex and Non-Convex Optimal Methods for Computer Vision and Image Processing Application.* **Mr. Rohith R. V.**, CEO, Data science AI as a resource person for the two day workshop on AIML, under this workshop total 75 students have trained in this workshop as part of IC3T2023 conference. These public talks were very accessible to a general audience. In addition, notably, this was the 4th

Conference at KITSW, and a formal session was held the first day to honor the event as well as those who were instrumental in initiating the conference.

Generous support for the conference was provided by **Captain V. Lakshmikantha Rao, Honorable EX M.P. (Rajya Sabha)**, and Former Minister, Chairman KITS Warangal. The funds were sizeable, timely, and greatly appreciated and permitted us to support a significant number of young scientists (postdocs and students) and persons from developing/disadvantaged countries. Nevertheless, the number of requests was far greater than the total support available (by about a factor of five!), and we had to turn down many financial requests. We encourage the organizers of the next IC3T to seek a higher level of funding for supporting young scientists and scientists from developing/disadvantaged countries. All in all, the Springer Scopus Indexed 5th IC3T 2023 in Warangal was very successful. The plenary lectures and the progress and special reports bridged the gap between the different fields of Computers and Communication Technology, making it possible for non-experts in a given area to gain insight into new areas. Also included among the speakers were several young scientists, namely postdocs and students, who brought new perspectives to their fields. The next IC3T will take place in 2024, and trend to be continued every year. Given the rapidity with which science is advancing in all of the areas covered by IC3T 2023, we expect that these future conferences will be as stimulating as this most recent one was, as indicated by the contributions presented in this proceedings volume.

We would also like to thank the authors and participants of this conference, who have considered the conference above all hardships. Finally, we would like to thank all the reviewers, session chairs, and volunteers who spent tireless efforts in meeting the deadlines and arranging every detail to make sure that the conference runs smoothly.

Hanumakonda, India — Dr. B. Rama Devi
Warangal, India — Dr. Kishore Kumar
Hanumakonda, India — Dr. M. Raju
Hyderabad, India — Dr. K. Srujan Raju
Edinburgh, UK — Dr. Mathini Sellathurai
October, 2023

Acknowledgements

A big vote of thanks must go to our Management, Ekashila Education Society, especially to Captain V. Lakshmikantha Rao, Honorable EX M.P. (Rajya Sabha) and Former Minister, Chairman, KITS Warangal, for generous support in organizing the conference.

The organizing committee wishes to acknowledge the assistance and encouragement that we have received from our organization and the many other individuals, who helped to prepare this event.

The organizing committee also wishes to acknowledge the assistance that we had received from chairmen of technical review committee of this conference Dr. K. Srujan Raju, Professor, CMR Technical Campus, Hyderabad, for all his hard work with the preparation of the conference proceedings.

Contents

Square Shaped SIW Based Slotted MIMO Antenna 1
Ajith Sai Visal, Darisi Abhinay, and P. Rahul Lal

Food Recognition and Calorie Measurement Using Machine Learning 9
Maganti Vasudha, D. Rashmi, and B. A. Mahalakshmi Jain

Breast Cancer Prediction with Gradient Boost and XGBoost 19
Avantika Mahadik, Prashant Sharma, and Vaibhav Narawade

Data Cube Technology for Accessing of Large Database 39
Vipin Saxena, Vishal Verma, Versha Verma, and Karm Veer Singh

Handwritten Character Recognition for South Indian Languages Using Deep Learning 49
C. S. Arjun, N. Shobha Rani, and Akshatha Prabhu

CavFind: A Novel Algorithm to Detect Cavities on Protein Structure ... 65
Swati Adhikari and Parthajit Roy

Tire Imprint Identification and Classification using VGG19 73
Disha N. Girish and M. Priyanka

Uncovering Insights in Agroforestry: A Text Mining Analysis Using Web Scraping and Topic Modeling 95
Parisa Monika, Desu Devi Sri, and M. Suneetha

Generative Adversarial Networks: A Comprehensive Review 105
Pavani Kotha, V. Janardhan Babu, and Sreejyothsna Ankam

System Modification Process to Reduce Calculation Complexity in Backward Forward Sweep Algorithm 115
T. S. N. G. Sarada Devi and T. V. L. N. Pavan Phani Kumar

Exploit Reversible Gates to Implement Fast Quantum Sorting Algorithm 125
Subhasree Bhattacharjee, Kunal Das, Arindam Sadhu, Soumyadip Sarkar, and Bikramjit Sarkar

Analyzing the Performance Variations of Naive Bayes, Linear SVM, and Random Forest for Spam Detection: A Comprehensive Study on the &Quot; Spam or Ham" Dataset 135
Bhawna Ojha, Pradeep Yadav, Rakhi Arora, Nitin Dixit, Gaurav Dubey, and Khemchand Shakyawar

A Novel Encryption Framework to Improve the Security of Medical Images 145
M. Senthilkumar, K. Suthendran, S. V. Suji Aparna, Mahesh Kotha, S. Kirubakaran, Srinivasarao Dharmireddi, and Voruganti Naresh Kumar

Enhancing DDoS Attack Detection in SDN: A Novel Approach with IG-RFFI Feature Selection 161
Konda Srikar Goud and Srinivasa Rao Giduturi

A Novel Fusion Study on Disease Detection in Cotton Plants Using Embedded Approaches of Neural Networks 171
Samuel Chepuri and Y. Ramadevi

A Security Framework for the Detection of Targeted Attacks Using Honeypot 183
P. Subhash, Mohammed Qayyum, C. Likhitha Varsha, K. Mehernadh, J. Sruthi, and A. Nithin

SNMP Watcher: A Model to Detect DoS and DDoS Attacks Using SNMP Management Information Base Analysis 193
V. Surya Narayana Reddy, P. Subhash, Ch. Vinay Kumar, M. Samuel, G. Vishal Reddy, and G. Tharun

Comparative Analysis for Loan Approval Prediction System Using Machine Learning Algorithms 205
Ashok Reddy Kandula, K. Divya, Sri Rama Krishna Datta Movva, Mohan Sai Motineni, Harika Pappala, and Karthika Venkata Satish Reddy Jakkireddy

Classification of Credit Card Delinquent Customers Using Random Forest Algorithm 217
Vadapally Praveen Kumar, Satyanarayana Nimmala, Burri Naresh, and Ch Ravikumar

Evaluating E-learning Engagement Through EEG Signal Analysis with Convolutional Neural Networks 225
Dharmendra Pathak and Ramgopal Kashyap

Food Item Recognition and Calories Estimation Using YOLOv5 241
Mahmoud Y. Shams, Ahmed Hussien, Aya Atiya, Logyn Medhat, and Roheet Bhatnagar

Fuzzy Cognitive Security Techniques for Data Aggregation in Capillary Networks 253
Jayshree D. Mallapur, Kirankumar B. Balavalad, Kirankumar Y. Bendigeri, Santosh B. Kumbalavati, and Nandini Sidnal

Performance Analysis on Optimal Data Integrity and Security-Constrained Cloud Storage Framework with Meta-heuristic-Based Mechanism 263
K. Brindha, K. Karuppasamy, and K. Kalaivani

Computer Vision-Based Model for Classification of Citrus Fruits Diseases with Pertinent Image Preprocessing Method 275
Ashok Kumar Saini, Roheet Bhatnagar, and Devesh Kumar Srivastava

FLAML-HDPS Model: An Efficient and Intelligent AutoML Approach for Heart Disease Prediction 287
P. Deepan, R. Vidhya, B. Rajalingam, R. Santhoshkumar, and N. Arul

An Efficient Selective Features Approach to Detect Hypothyroid Using Machine Learning 297
N. Subhash Chandra, Srinivasa Rao Dhanikonda, Dhanamma Jagli, and Nalla Siddhartha

Kullback–Leibler Divergence-Based Feature Selection Method for Image Texture Classification 309
M. Subba Rao, Guntoju Kalpana Devi, Suraya Mubeen, Badam Prashanth, Tazzeen Fatima, K. Reddy Madhavi, Voruganti Naresh Kumar, and Charan Yadav Chintalacheri

Generative Adversarial Networks: Overview 319
Shivani Pachika, A. Brahmananda Reddy, Bhavishya Pachika, and Akhil Karnam

Horse Race Prediction Using Machine Learning Algorithms 329
Puja Chavan, Pankaj Kunekar, Prajakta Musale, Mrunal Shinde, Atharva Shinde, and Nikhil Shinde

An Approach to Blockchain-Based E-Voting System Using Peer-To-Peer Protocol 337
Divya Boreda, Ashok Kumar Nanda, Chaitra Sai Jalda, V. Pradeep Kumar, and Kumar Gautam

Artificial Cognitive Intelligence and Information Technology in Cybersecurity 347
Tamanna Jena Singhdeo, Adyesha Singhdeo, J. R. Mohanty, and Suresh Satapathy

Analyzing the Counterfeit Medicines Based on Classification Using Machine Learning Techniques 355
Binitha S. Thomson and W. Rose Varuna

Water Quality Classification Using Multi-cell RNN in Aquaculture Ponds for Catla Fish 363
Peda Gopi Arepalli and Jairam Naik Khetavath

Advanced Techniques for Color Image Blind Deconvolution to Restore Blurred Images 371
Jonnadula Narasimharao, P. Deepthi, Bolla Aditya, C. R. Shruthi Reddy, Avala Raji Reddy, and Ganpat Joshi

CRNN-Based Eye Behavior Analysis for Drowsiness Detection 391
J. Sasi Kiran, Gurram Sunitha, Marri Sireesha, U. Mahender, K. Reddy Madhavi, Swathi Rudra, and Voruganti Naresh Kumar

Sentiment Analysis Principle Technical Approach on Online Social Network Data Using CNN for Detection of Stress 401
K. Srujan Raju, Nagu Chandra Sekhar Reddy, Matoori Praveen, V. S. Manoj Kumar Chenna, Sandhyarani, and Bollampelly Chandana

Performance Study of Three Different Types of Data Center Network Architectures, NovaCube, BCube, and FatTree Using NS-3 Simulator 411
A. Ranjith, P. Subhash, G. Nagaraju, M. Srinivas, and A. Prashanth

Author Index 423

Editors and Contributors

About the Editors

Dr. B. Rama Devi received the Ph.D. from the JNTUH College of Engineering, Hyderabad, Telangana, India, in April 2016. She completed her M.Tech. in Digital Communication Engineering in 2007 from Kakatiya University, Warangal (KITSW). She joined the faculty of electronics and communication engineering at KITSW in 2007. From January 19, 2019, to December 16, 2021, she was the head of the Department of Electronics and Communication Engineering at KITSW. At present, she is working as a professor in the Department of ECE. She published more than 40 papers in various journals and conferences. She published three books and had two Indian patents granted to her. She acted as an editor and program chair for the Springer International conferences IC3T 2022 and ICDECT 2021. She acted as a session chair for various international conferences.

Dr. Kishore Kumar earned his Ph.D. in signal processing in 2004. He headed the Department of Electronics and Communication Engineering at NIT Warangal from 2015 to 2017. He was also the head of the Institute Computer Center, NIT Warangal, from 2018 to 2020. He was a visiting professor at ESIGELEC University in France in 2009, a visiting professor at TALLINN University, Estonia, for a semester for an MS in Computer Science, and a visiting professor at the Asian Institute of Technology, Bangkok, in 2019 under the Secondment of Indian Faculty by MHRD, Government of India. He worked as a technical officer in the Cabinet Secretariat, Prime Minister's Office, Government of India. He is a member of the Central Council Board (CCB) and a member of the Andhra Pradesh State Council for Higher Education (APSCHE).

Dr. M. Raju received his Ph.D. from Kakatiya University, Telangana, India, in 2017. In 2009, he received his M.Tech. in Instrumentation and Control Systems from JNTU Kakinada. He has 23 years of teaching experience. He published more than 35 papers in various journals and conferences. She published four books. He has been guiding two research scholars. His areas of interest include wireless communication and signal processing for communications. He is a member of IETE, a member of ISTE, a member of IAENG, and a senior member of IRED.

Dr. K. Srujan Raju obtained his Doctorate in Computer Science in the area of network security. He has more than 20 years of experience in academics and research. His research interest areas include computer networks, information security, data mining, cognitive radio networks and image processing, and other programming languages. He has undergone specific training conducted by Wipro Mission 10X and NITTTR, Chennai, which helped his involvement with students and is very conducive for solving their day-to-day problems. He has guided various student clubs for activities ranging from photography to Hackathon. He mentored more than 100 students for incubating cutting-edge solutions. He has organized many conferences, FDPs, workshops, and symposiums. He has established the Centre of Excellence in IoT and Data Analytics.

Dr. Mathini Sellathurai is currently the dean of Science and Engineering and the head of the Signal Processing for Intelligent Systems and Communications Research Group, Heriot-Watt University, Edinburgh, UK and leading research in signal processing for Radar and wireless communications networks. Professor Sellathurai has five years of industrial research experience. She held positions with Bell-Laboratories, New Jersey, USA and with the Canadian (Government) Communications Research Centre, Ottawa, Canada. She was an associate editorship for the IEEE Transactions on Signal Processing (2005–2018) and IEEE Signal Processing for Communications Technical Committee member (2013–2018). She was an organizer for the IEEE International Workshop on Cognitive Wireless Systems, IIT Delhi, India, 2009, 2010, and 2013; and the general chair of the 2016 IEEE Workshop on Signal Processing Advances in Wireless Communications (SPAWC), Edinburgh, UK.

Contributors

Darisi Abhinay Department of Electronics and Communication Engineering, Amrita Vishwa Vidyapeetham, Amritapuri, India

Swati Adhikari Department of Computer Science, The University of Burdwan, Golapbag, Purba Bardhaman, West Bengal, India

Bolla Aditya Department of IT, CMR Technical Campus, Hyderabad, Telangana, India

Sreejyothsna Ankam Department of AIML, GRIET, Bachupally, Hyderabad, India

S. V. Suji Aparna Department of CSE, CMR Technical Campus, Hyderabad, Telangana, India

Peda Gopi Arepalli Department of Computer Science and Engineering, National Institute of Technology Raipur, Raipur, India

C. S. Arjun Department of Computer Science, School of Computing, Mysuru Campus, Amrita Vishwa Vidyapeetham, Mysuru, India

Rakhi Arora ITM-Institute of Technology and Management, Gwalior, India

N. Arul St. Peter's Engineering College, Hyderabad, India

Aya Atiya Faculty of Artificial Intelligence, Kafrelsheikh University, Kafrelsheikh, Egypt

Kirankumar B. Balavalad Department of Electronics and Communication Engineering, Basaveshwar Engineering College, Bagalkot, Karnataka, India

Kirankumar Y. Bendigeri Department of Electronics and Communication Engineering, Basaveshwar Engineering College, Bagalkot, Karnataka, India

Roheet Bhatnagar Department of CSE, School of Computer Science & Engineering, Manipal University Jaipur, Rajasthan, India

Subhasree Bhattacharjee Department of Computer Application, Narula Institute of Technology, Kolkata, West Bengal, India

Divya Boreda B V Raju Institute of Technology Vishnupur, Narsapur, Telangana, India

K. Brindha CSE, RVS College of Engineering and Technology, Coimbatore, India

Bollampelly Chandana Department of CSE (Data Science), CMR Technical Campus, Hyderabad, Telangana, India

N. Subhash Chandra Department of Computer Science and Engineering, CVR College of Engineering, Hyderabad, India

Puja Chavan Vishwakarma Institute of Technology, Pune, India

V. S. Manoj Kumar Chenna Department of CSE, CMR Technical Campus, Hyderabad, Telangana, India

Samuel Chepuri Department of Computer Science and Engineering, Osmania University, Hyderabad, Telangana State, India

Charan Yadav Chintalacheri Masters in Computer Science, University of Kansas, Lawrence, USA

Kunal Das Department of Computer Science, Acharya Prafulla Chandra College, Kolkata, West Bengal, India

P. Deepan St. Peter's Engineering College, Hyderabad, India

P. Deepthi Department of CSE (AI&ML), CMR Institute of Technology, Hyderabad, Telangana, India

Guntoju Kalpana Devi Department of CSE, CMR Technical Campus, Hyderabad, Telangana, India

Desu Devi Sri Department of Information Technology, VR Siddhartha Engineering College, Vijayawada, India

Srinivasa Rao Dhanikonda Department of Information Technology, BVRIT HYDERABAD College of Engineering for Women, Hyderabad, India

Srinivasarao Dharmireddi Department of Cybersecurity, MasterCard, St. Louis, Missouri, USA

K. Divya Department of AI&DS, Seshadri Rao Gudlavalleru Engineering College, Gudlavalleru, Andhra Pradesh, India

Nitin Dixit ITM-Institute of Technology and Management, Gwalior, India

Gaurav Dubey ITM-Institute of Technology and Management, Gwalior, India

Tazzeen Fatima Department of ECE, Shadan Women's College of Engineering and Technology, Hyderabad, Telangana, India

Kumar Gautam Gwangju Institute of Science and Technology (GIST), Gwangju, South Korea

Srinivasa Rao Giduturi Department of CSE, GITAM School of Technology, GITAM University, Visakhapatnam, Andhra Pradesh, India

Disha N. Girish Department of Computer Science, School of Computing, Amrita Vishwa Vidyapeetham, Mysore, India

Konda Srikar Goud Department of CSE, GITAM School of Technology, GITAM University, Visakhapatnam, Andhra Pradesh, India;
Department of IT, BVRIT HYDERABAD College of Engineering for Women, Hyderabad, India

Ahmed Hussien Faculty of Artificial Intelligence, Kafrelsheikh University, Kafrelsheikh, Egypt

Dhanamma Jagli V.E.S.Institute of Technology, University of Mumbai, Mumbai, India

Karthika Venkata Satish Reddy Jakkireddy Department of AI&DS, Seshadri Rao Gudlavalleru Engineering College, Gudlavalleru, Andhra Pradesh, India

Chaitra Sai Jalda B V Raju Institute of Technology Vishnupur, Narsapur, Telangana, India

V. Janardhan Babu Department of AI, SVPCET, Puttur, India

Ganpat Joshi Department of Computer Science and Engineering, Madhav University, Rajasthan, India

K. Kalaivani AIML, Coimbatore Institute of Technology, Coimbatore, India

Ashok Reddy Kandula Department of AI&DS, Seshadri Rao Gudlavalleru Engineering College, Gudlavalleru, Andhra Pradesh, India

Akhil Karnam VNR Vignana Jyothi Institute of Engineering and Technology, Hyderabad, India

K. Karuppasamy CSE, RVS College of Engineering and Technology, Coimbatore, India

Ramgopal Kashyap Amity University, Chhattisgarh, India

Jairam Naik Khetavath Department of Computer Science and Engineering, National Institute of Technology Raipur, Raipur, India

J. Sasi Kiran Department of CSE, Lords Institute of Engineering and Technology (Autonomous), Himayathsagar, Hyderabad, Telangana, India

S. Kirubakaran CMR College of Engineering and Technology, Hyderabad, Telangana, India

Mahesh Kotha Department of CSE (AI&ML), CMR Technical Campus, Hyderabad, Telangana, India

Pavani Kotha Department of CSE, SVPCET, Puttur, India

Voruganti Naresh Kumar Department of CSE, CMR Technical Campus, Hyderabad, Telangana, India

V. Pradeep Kumar B V Raju Institute of Technology Vishnupur, Narsapur, Telangana, India

Santosh B. Kumbalavati Department of Electronics and Communication Engineering, Basaveshwar Engineering College, Bagalkot, Karnataka, India

Pankaj Kunekar Vishwakarma Institute of Technology, Pune, India

C. Likhitha Varsha CSE-(CyS, DS) and AI&DS, VNR VJIET, Hyderabad, Telangana, India

K. Reddy Madhavi Department of AI & ML, School of Computing, Mohan Babu University, Tirupati, Andhra Pradesh, India

Avantika Mahadik Pacific (PAHER) University, Udaipur, India

B. A. Mahalakshmi Jain Department of Information Science and Engineering, CMR Institute of Technology, Bengaluru, India

U. Mahender Department of CSE, CMR ENGINEERING COLLEGE, Hyderabad, Telangana, India

Jayshree D. Mallapur Department of Electronics and Communication Engineering, Basaveshwar Engineering College, Bagalkot, Karnataka, India

Logyn Medhat Faculty of Artificial Intelligence, Kafrelsheikh University, Kafrelsheikh, Egypt

K. Mehernadh CSE-(CyS, DS) and AI&DS, VNR VJIET, Hyderabad, Telangana, India

J. R. Mohanty KIIT University, Bhubaneswar, India

Parisa Monika Department of Information Technology, VR Siddhartha Engineering College, Vijayawada, India

Mohan Sai Motineni Department of AI&DS, Seshadri Rao Gudlavalleru Engineering College, Gudlavalleru, Andhra Pradesh, India

Sri Rama Krishna Datta Movva Department of AI&DS, Seshadri Rao Gudlavalleru Engineering College, Gudlavalleru, Andhra Pradesh, India

Suraya Mubeen Department of ECE, CMR Technical Campus, Hyderabad, Telangana, India

Prajakta Musale Vishwakarma Institute of Technology, Pune, India

G. Nagaraju Department of CSE (AIML & IOT), VNRVJET, Hyderabad, India

Ashok Kumar Nanda B V Raju Institute of Technology Vishnupur, Narsapur, Telangana, India

Jonnadula Narasimharao Department of CSE, CMR Technical Campus, Hyderabad, Telangana, India

Vaibhav Narawade Department of Computer Engineering, Ramrao Adik Institute of Technology, D. Y. Patil Deemed to be University, Nerul, India

Burri Naresh Department of CSE (CS), CVR College of Engineering, Hyderabad, India

Satyanarayana Nimmala Department of CSE (Data Science), CVR College of Engineering, Hyderabad, India

A. Nithin CSE-(CyS, DS) and AI&DS, VNR VJIET, Hyderabad, Telangana, India

Bhawna Ojha ITM-Institute of Technology and Management, Gwalior, India

Bhavishya Pachika NIT Kurukshetra, Kurukshetra, India

Shivani Pachika Output Media, Windsor, Canada

Harika Pappala Department of AI&DS, Seshadri Rao Gudlavalleru Engineering College, Gudlavalleru, Andhra Pradesh, India

Dharmendra Pathak Amity University, Chhattisgarh, India

T. V. L. N. Pavan Phani Kumar Tata Advanced Systems Limited, Bangalore, India

Akshatha Prabhu Department of Computer Science, School of Computing, Mysuru Campus, Amrita Vishwa Vidyapeetham, Mysuru, India

A. Prashanth Bandenburg University of Technology (BTU), Cottbus, Senftenberg, Germany

Badam Prashanth Department of CSE (AI&ML), CMR Technical Campus, Hyderabad, Telangana, India

Matoori Praveen Department of CSE (Data Science), CMR Technical Campus, Hyderabad, Telangana, India

Vadapally Praveen Kumar Department of CSE (Data Science), CVR College of Engineering, Hyderabad, India

M. Priyanka Department of Computer Science, School of Computing, Amrita Vishwa Vidyapeetham, Mysore, India

Mohammed Qayyum Computer Science Engineering, King Khalid University, Abha, Saudi Arabia

P. Rahul Lal Department of Electronics and Communication Engineering, Amrita Vishwa Vidyapeetham, Amritapuri, India

B. Rajalingam St. Martin's Engineering College, Hyderabad, India

K. Srujan Raju Department of CSE, CMR Technical Campus, Hyderabad, Telangana, India

Y. Ramadevi Department of AIML, CBIT, Hyderabad, Telangana State, India

N. Shobha Rani Department of Computer Science, School of Computing, Mysuru Campus, Amrita Vishwa Vidyapeetham, Mysuru, India

A. Ranjith School of Computer Engineering, KIIT University, Bhubaneswar, India

M. Subba Rao Department of Computer Science and Engineering, Annamacharya Institute of Technology and Sciences (Autonomous), Rajampet, Andhra Pradesh, India

D. Rashmi Department of Information Science and Engineering, CMR Institute of Technology, Bengaluru, India

Ch Ravikumar Department of Artificial Intelligence and Data Science, Chaitanya Bharathi Institute of Technology, Hyderabad, India

A. Brahmananda Reddy VNR Vignana Jyothi Institute of Engineering and Technology, Hyderabad, India

Avala Raji Reddy Department of Mechanical Engineering, CMR Technical Campus, Hyderabad, Telangana, India

C. R. Shruthi Reddy Department of CSE, CMR Technical Campus, Hyderabad, Telangana, India

Nagu Chandra Sekhar Reddy Department of CSE, MLR Institution of Technology, Hyderabad, Telangana, India

Parthajit Roy Department of Computer Science, The University of Burdwan, Golapbag, Purba Bardhaman, West Bengal, India

Swathi Rudra Department of CSE(AI & ML), CMR Technical Campus, Hyderabad, Telangana, India

Arindam Sadhu Department of Electronics and Communication Engineering, Greater Kolkata College of Engineering and Management, Kolkata, West Bengal, India

Ajith Sai Visal Department of Electronics and Communication Engineering, Amrita Vishwa Vidyapeetham, Amritapuri, India

Ashok Kumar Saini Department of CSE, School of Computer Science & Engineering, Manipal University Jaipur, Rajasthan, India

M. Samuel CSE-(CyS, DS) and AI&DS, VNR VJIET, Hyderabad, Telangana, India

Sandhyarani Department of CSE (Data Science), CMR Technical Campus, Hyderabad, Telangana, India

R. Santhoshkumar St. Martin's Engineering College, Hyderabad, India

T. S. N. G. Sarada Devi Department of Electrical Engineering, Malla Reddy Engineering College for Women (Autonomous), Hyderabad, India

Bikramjit Sarkar Department of Computer Science and Engineering, JIS College of Engineering, Kalyani, West Bengal, India

Soumyadip Sarkar Department of Computer Application, Narula Institute of Technology, Kolkata, West Bengal, India

Suresh Satapathy KIIT University, Bhubaneswar, India

Vipin Saxena Department of Computer Science, Babasaheb Bhimrao Ambedkar University, Lucknow, India

M. Senthilkumar Department of Artificial Intelligence and Data Science, Sethu Institute of Technology, Virudhunagar, India

Khemchand Shakyawar ITM-Institute of Technology and Management, Gwalior, India

Mahmoud Y. Shams Faculty of Artificial Intelligence, Kafrelsheikh University, Kafrelsheikh, Egypt

Prashant Sharma Department of Computer Science and Engineering, Pacific (PAHER) University, Udaipur, India

Atharva Shinde Vishwakarma Institute of Technology, Pune, India

Mrunal Shinde Vishwakarma Institute of Technology, Pune, India

Nikhil Shinde Vishwakarma Institute of Technology, Pune, India

Nalla Siddhartha Indian Institute of Technology Tirupati, Tirupati, India

Nandini Sidnal Software Engineering and IT, Torrens University, Ultimo, Sydney, Australia

Karm Veer Singh Department of Computer Science, Babasaheb Bhimrao Ambedkar University, Lucknow, India

Adyesha Singhdeo University of British Columbia, Vancouver, Canada

Tamanna Jena Singhdeo Fairleigh Dickinson University, Vancouver, Canada

Marri Sireesha Department of CSE, CMR Technical Campus, Hyderabad, Telangana, India

M. Srinivas Department of CSE, Vaageswari College of Engineering, Karimnagar, India

Devesh Kumar Srivastava Department of IT, School of Information Technology, Manipal University Jaipur, Rajasthan, India

J. Sruthi CSE-(CyS, DS) and AI&DS, VNR VJIET, Hyderabad, Telangana, India

P. Subhash Department of CSE (CyS, DS) and AI&DS, VNRVJIET, Hyderabad, Telangana, India

M. Suneetha Department of Information Technology, VR Siddhartha Engineering College, Vijayawada, India

Gurram Sunitha Department of AI & ML, School of Computing, Mohan Babu University, Tirupati, Andhra Pradesh, India

V. Surya Narayana Reddy CSE-(CyS, DS) and AI&DS, VNR VJIET, Hyderabad, Telangana, India

K. Suthendran Department of Information Technology, Kalasalingam Academy of Research and Education, Srivilliputhur, India

G. Tharun CSE-(CyS, DS) and AI&DS, VNR VJIET, Hyderabad, Telangana, India

Binitha S. Thomson Department of Information Technology, Bharathiar University, Coimbatore, India

W. Rose Varuna Department of Information Technology, Bharathiar University, Coimbatore, India

Maganti Vasudha Department of Information Science and Engineering, CMR Institute of Technology, Bengaluru, India

Versha Verma Department of Computer Science, Babasaheb Bhimrao Ambedkar University, Lucknow, India

Vishal Verma Department of Computer Applications and Science, School of Management and Science, Lucknow, India

R. Vidhya B.S. Abdur Rahman Crescent Institute of Science and Technology, Chennai, India

Ch. Vinay Kumar CSE-(CyS, DS) and AI&DS, VNR VJIET, Hyderabad, Telangana, India

G. Vishal Reddy CSE-(CyS, DS) and AI&DS, VNR VJIET, Hyderabad, Telangana, India

Pradeep Yadav ITM-Institute of Technology and Management, Gwalior, India

Square Shaped SIW Based Slotted MIMO Antenna

Ajith Sai Visal, Darisi Abhinay, and P. Rahul Lal

Abstract The paper deals with the design and performance analysis of a cavity-backed substrate integrated waveguide (SIW)-based slotted antenna operating at 5.8 GHz. The antenna accommodates four ports oriented in four different directions which make it a multiple input multiple output (MIMO) antenna. The four ports are designed in such a way that it enables the antenna to radiate in multiple directions with linear and circular polarisation. The return loss of the antenna is 36.59 dB and has a maximum gain of 9.55 dB. The proposed research aims to contribute to the advancement in high-performance antennas for wireless communication like 5G, radar communications, etc.

Keywords Circular polarisation · Multiple input multiple output (MIMO) · Substrate integrated waveguide (SIW) · Slotted antenna

1 Introduction

The ever-increasing demand for high-speed wireless communication systems has necessitated the development of advanced antenna technologies capable of efficient and precise beam steering. In this paper, we focus on a substrate integrated waveguide (SIW)-based slotted antenna which is operated at a 5.8 GHz frequency band. This unlicensed operating frequency range has found extensive applications in various wireless communication systems, including Wi-Fi, Bluetooth, point-to-point microwave links, etc. Achieving beam agility and beam scanning at this

A. Sai Visal (✉) · D. Abhinay · P. Rahul Lal
Department of Electronics and Communication Engineering, Amrita Vishwa Vidyapeetham, Amritapuri, India
e-mail: amenu4ece20105@am.students.amrita.edu

D. Abhinay
e-mail: amenu4ece20114@am.students.amrita.edu

P. Rahul Lal
e-mail: rahullal@am.amrita.edu

B. R. Devi et al. (eds.), *Proceedings of Fifth International Conference on Computer and Communication Technologies*, Lecture Notes in Networks and Systems 897,
https://doi.org/10.1007/978-981-99-9704-6_1

frequency band can significantly enhance the performance of these systems. These slotted antennas utilise the phenomenon of wave leakage to generate radiation. By introducing periodic perturbations or discontinuities along a waveguide structure, a portion of the guided wave energy is allowed to radiate into free space. This controlled leakage results in a beam of electromagnetic energy propagating along a specific direction, offering beam steering capabilities. The SIW technology provides a compact and low-cost solution for slotted antennas.

Several slotted antenna designs utilising SIW technology are available in the literature. In [1], four planar SIW-based horn antennas are combined circularly which makes the antenna excite in multiple directions. This antenna offers a wide bandwidth of the order of 5 GHz at an operating frequency of 26 GHz. A SIW-based pill box-shaped array of slots is presented in [2] to reduce the aperture blockage by additional parabolic reflectors. This antenna operates at 29 GHz with a side lobe level as low as − 12 dB. Both of the above-mentioned antenna configurations lack polarisation reconfigurability.

A multi-beam antenna presented in [3] utilises a 90° shifted two-dimensional structure to radiate at a peak gain of 8.5 dBi. This antenna is capable of radiating in sixteen different directions based on the excitation scheme employed to the ports. A SIW-based multi-beam antenna presented in [4] utilises an array of monopoles printed on Rogers R04003 substrate. A high permittivity dielectric is positioned to control beam aiming and stop surface currents. This antenna operates at 30 GHz and has a peak gain of 12.3 dBi. Polarisation reconfigurability is lacking for the above-mentioned designs.

A hybrid SIW cavity-based antenna in [5] can generate six major beams. This antenna is linearly polarised and operates at 5 GHz with a return loss of 21.34 dB and a gain of 6.4 dB. A rectangular SIW cavity is engraved with rectangle radiating slots on the top and bottom face is dealt in [6]. This design incorporates slots in specific orientations on the top and bottom face of the SIW cavity. Bidirectional radiation is achieved from the antenna with a gain of 5 dBi. In both of these studies, circular polarisation is not achieved. In [7] to attain a dual circularly polarised antenna a hybrid coupler is combined with a linearly polarised leaky wave antenna. This antenna operates in the range of 27–32 GHz and has an axial ratio below 3 dB.

The antenna under the current study presents a cavity-backed SIW-based antenna that has the ability to radiate in several directions. By varying the input amplitude and phase to the four coaxial ports, the antenna is also capable of switching its polarisation states from linear to circular. All the results are verified with the help of HFSS available in the Ansys Electronics Desktop software. The detailed design and characteristics of the antenna are presented in the preceding sections.

2 Antenna Design

The antenna's construction ensures that it is centred around the four SIW cavities utilising the equations mentioned in [8–11]. This design shall take advantage of a Rogers RT duroid substrate with a dielectric constant of 2.2 and thickness of 1.57 mm. For the antenna to operate at 5.8 GHz, the optimised values utilising the parametric analysis in Ansys HFSS are as depicted in Fig. 1. As shown in Fig. 1, four SIW cavities are arranged to form a square-shaped structure. The metallic vias with a diameter of 14.49 mm are spaced at a distance of 23 mm. The spacing between the periodic vias on either end is 54.51 mm. The total length of a single SIW cavity is 76.245 mm. Four coaxial ports are positioned at the start of each cavity. Five radiating slots are positioned perpendicular to the excitation ports. The slot measures 6.9 mm in width and 46 mm in length. The spacing between the radiating slots is 13.4 mm.

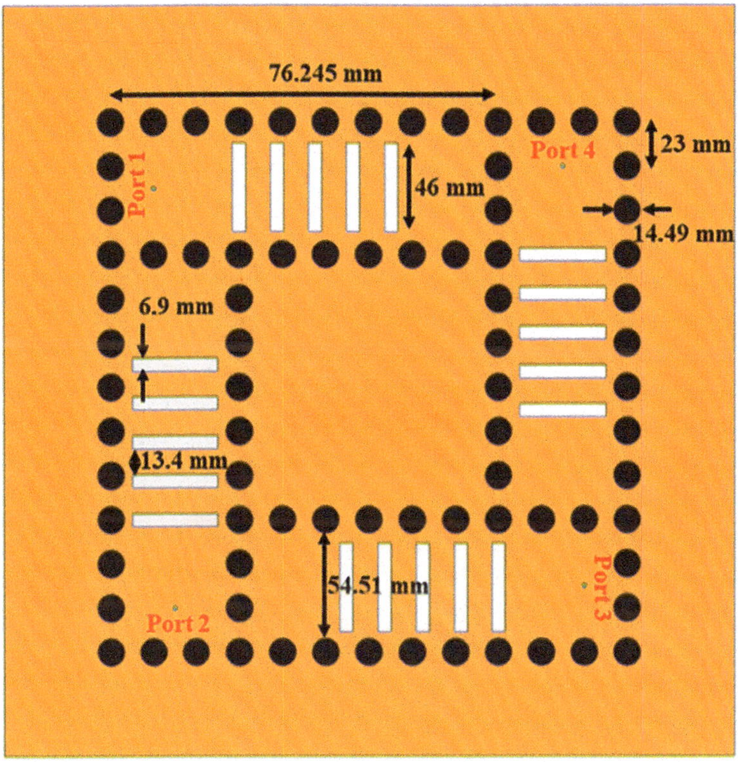

Fig. 1 Plan view of SIW cavity with the dimensions

3 S-parameter and Gain Characteristics of the Antenna

The antenna operation at 5.8 GHz is verified with respect to the S-parameter plot depicted in Fig. 2. Return loss of the antenna is 36.59 dB which ensures significant matching at the operating frequency. The isolation with the adjacent ports is in excess of -30 dB which minimises the cross-talk between the ports. A slender BW of 30 MHz is attained for this configuration. When one of the four ports is excited, the current antenna radiates at a peak gain of 9.55 dB (Fig. 3).

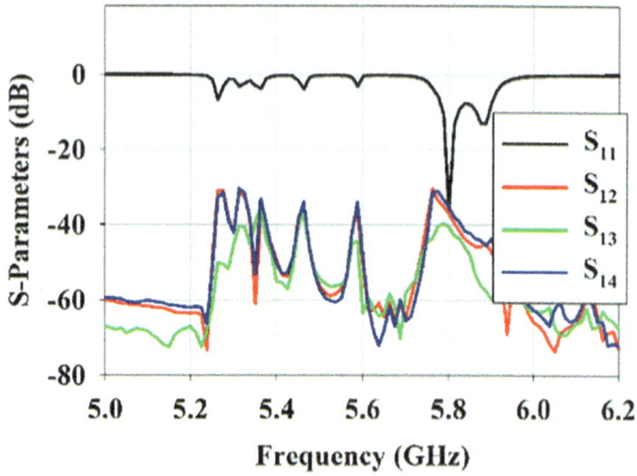

Fig. 2 S-parameter plot of the current antenna

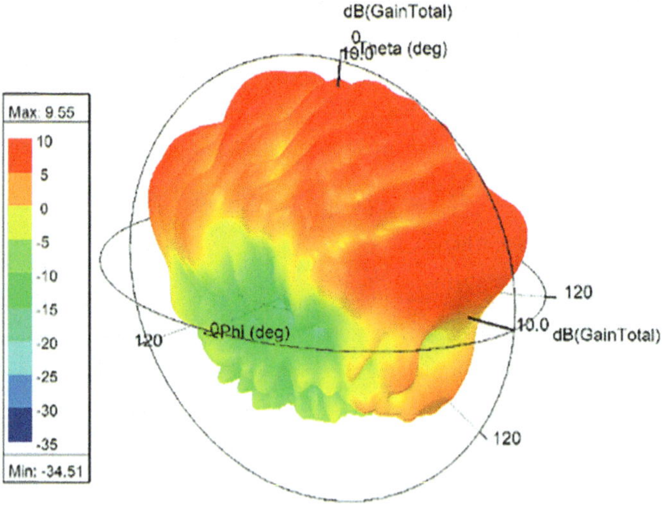

Fig. 3 3D gain plot of the current antenna with a peak gain of 9.55 dB

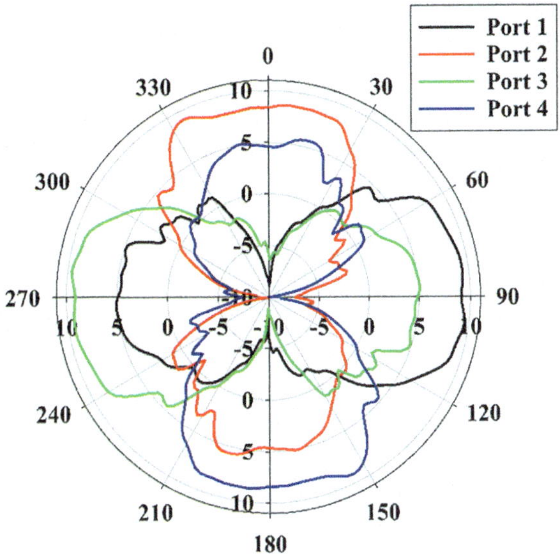

Fig. 4 Radiation pattern of the antenna when the ports are excited individually

4 Multi-beam and Polarisation Characteristics of the Antenna

Utilising the four available input ports, the antenna in the current literature efficiently radiates in multiple directions with different polarisation states. A detailed analysis of the same is presented in the preceding sections.

4.1 Individual Excitation of the Ports with Zero Phase

The excitation of the four individual ports allows the antenna to orient its maximum radiation in four different directions. When the ports are excited in the sequence of 1, 2, 3, and 4, the radiation maximum is oriented along 90°, 0°, 270°, and 180°, respectively. Peak gain under this configuration is measured to be 9.55 dB as shown in Fig. 4. Under this mode of excitation the antenna is linearly polarised. The same is verified utilising the axial ratio plot depicted in Fig. 5. The peak value of the axial ratio at the operating frequency is 25 dB.

4.2 Excitation of Two Ports with a Phase Difference of 90°

Under this configuration, the antenna is capable of radiating in multiple directions, namely two. The structure's symmetry causes the maximum radiation to coincide.

Fig. 5 Axial ratio plot when the ports are excited individually

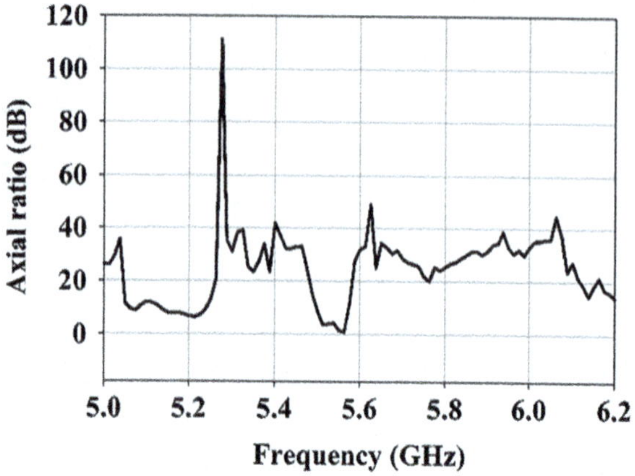

The ports are grouped in the counterclockwise direction by combining ports {1(0°), 2(90°)}, {2(0°), 3(90°)}, {3(0°), 4(90°)}, and {4(0°), 1(90°)}. A phase shift of 90° is maintained to generate a circularly polarised wave. A peak gain of 8.1 dB is generated by the antenna. The multi-beam characteristics are shown in Fig. 6. The axial ratio plot shown in Fig. 7. with a peak value of 1.1 dB at the operating frequency confirms that the antenna is generating a circularly polarised wave under this condition.

Fig. 6 Radiation pattern with a phase difference of 90° when two ports of the antenna are parallelly excited

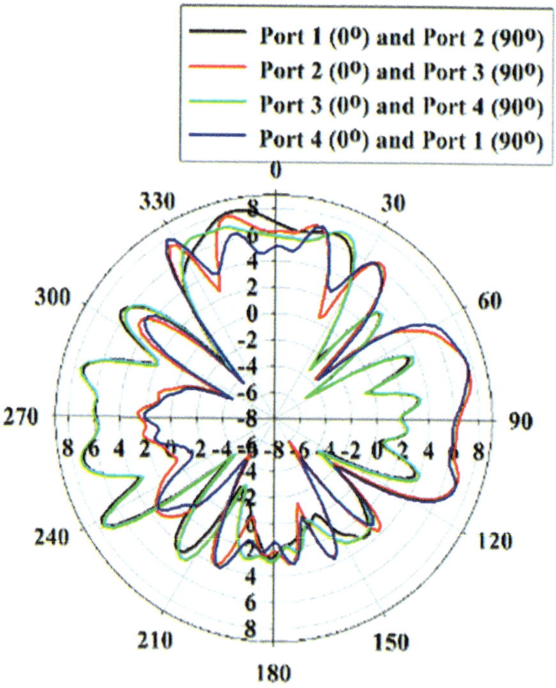

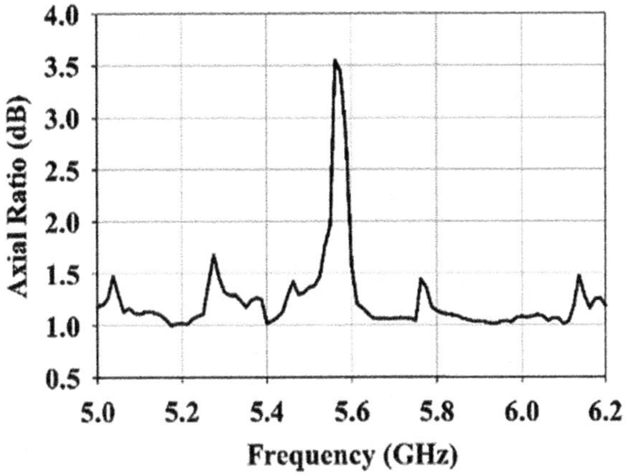

Fig. 7 Plot of axial ratio when the antenna's two ports are excited parallelly and have a phase difference of 90

Table 1 Comparison of various parameters of various antennas from literature survey with the antenna in this work

Reference	Operating frequency (GHz)	Peak gain (dB)	No. of radiation maximums	Polarisation
[2]	28	14	5	NA
[4]	30	12.3	2	NA
[6]	23.4	5	2	NA
[11]	90	8.7	4	Linear
Current work	5.8	9.55	4	Linear and circular

5 Conclusion

An antenna that is capable of radiating in multiple directions along with polarisation reconfigurability between linear and circular is presented in this work. The antenna is having an excellent return loss of 36.59 dB at an operating bandwidth of 30 MHz. This antenna is capable of radiating at six different azimuth angles under different excitation configurations. Table 1 provides a detailed comparison of several antenna parameters used in this work to various antennas found in the literature. This antenna is extremely well suited for MIMO systems because of its isolation at more than 30 dB from the adjacent ports. Being a circularly polarised antenna, this antenna can find applications in 5G communication systems, radars, sensors, and radio frequency identification tags.

References

1. Chaudhari AD, Kumar A, Ghosh S, Mukherjee S (2022) Design and implementation of multi-beam multi-panel antenna array for cellular mm-wave 5G vehicle-to-everything (V2X) communications. In: 2022 IEEE microwaves, antennas, and propagation conference (MAPCON), Bangalore, India, pp 410–414. https://doi.org/10.1109/MAPCON56011.2022.10047071
2. Yan S-P, Zhao M-H, Ban Y-L, Lian J-W, Nie Z (2019) Dual-layer SIW multibeam pillbox antenna with reduced sidelobe level. IEEE Antennas Wirel Propag Lett 18(3):541–545. https://doi.org/10.1109/LAWP.2019.2896183
3. Cheng YJ, Xuan ZJ (2017) Two-dimensional beam scanning antenna array with 90-degree SIW twist. In: 2017 international workshop on antenna technology: small antennas, innovative structures, and applications (iWAT), Athens, Greece, pp 264–266. https://doi.org/10.1109/IWAT.2017.7915375
4. Di Paola C, Zhao K, Zhang S, Pedersen GF (2019) SIW multibeam antenna array at 30 GHz for 5G mobile devices. IEEE Access 7:73157–73164. https://doi.org/10.1109/ACCESS.2019.2919579
5. Rahul Lal P, Rajeswari P (2022) A hybrid substrate integrated waveguide cavity based leaky wave antenna capable of generating six major beams. In: 2022 international conference on wireless communications signal processing and networking (WiSPNET), Chennai, India, 2022, pp 258–261. https://doi.org/10.1109/WiSPNET54241.2022.9767108
6. Joseph S, Rahul Lal P, Menon SK (2016) Substrate integrated waveguide based antenna with bi-directional radiation for radar applications. In: 2016 international conference on wireless communications, signal processing and networking (WiSPNET), Chennai, India, 2016, pp 95–98. https://doi.org/10.1109/WiSPNET.2016.7566098
7. Elsaadany M, Ali MMM, Shams SI, Gagnon G (2021) On the design of SIW dual circularly-polarized leaky wave antenna for mm-wave applications. In: 2021 15th european conference on antennas and propagation (EuCAP), Dusseldorf, Germany, pp 1–4. https://doi.org/10.23919/EuCAP51087.2021.9411113
8. Gopinath A, Rahul Lal P (2017) Substrate integrated waveguide based hybrid cavity filter for Ku band applications. In: 2017 international conference on communication and signal processing (ICCSP), Chennai, India, pp 1563–1566. https://doi.org/10.1109/ICCSP.2017.8286651
9. Gopinath A, Rahul Lal P (2017) Substrate integrated waveguide based circular cavity filter in Ku band. In: 2017 international conference on wireless communications, signal processing and networking (WiSPNET), Chennai, India, pp 911–913. https://doi.org/10.1109/WiSPNET.2017.8299894
10. Devika C, Karthika K, Raj N, Swetha K, Rahul Lal P (2018) Substrate integrated waveguide based rectangular cavity resonator filter for WiFi applications. In: 2018 9th international conference on computing, communication and networking technologies (ICCCNT), Bengaluru, India, pp 1–4. https://doi.org/10.1109/ICCCNT.2018.8494083
11. Rahul Lal P, Purushothaman A (2022) Design of a leaky wave multi-beam antenna with a cross shaped pattern. In: 2022 first international conference on electrical, electronics, information and communication technologies (ICEEICT), Trichy, India, pp 1–6. https://doi.org/10.1109/ICEEICT53079.2022.9768413

Food Recognition and Calorie Measurement Using Machine Learning

Maganti Vasudha, D. Rashmi, and B. A. Mahalakshmi Jain

Abstract This study is one of many that investigate the relationship between determining the nutritional ingredients in food and calculating the calories using data analysis utilizing machine learning techniques. Due to the availability of multifood photos, which must be cropped before processing, the Indian food recipes database is used for the research. The study uses a large dataset of various food photos to train state-of-the-art deep convolutional neural networks (CNNs) to recognize and categorize distinct food items with an amazing 99.89% accuracy. This study's applicability spans several sectors in addition to food recognition, including calorie measurement, meal planning services, and nutritional monitoring systems. The solution is widely available to a wide range of users thanks to a user-friendly web interface. The system's 99.89% accuracy in food detection and calorie measurement demonstrates its dependability and distinguishes it from competing options. Its ability to improve individual health, fight obesity, and encourage healthy eating habits makes it a vital tool in today's health-conscious culture.

Keywords Food recognition · Calorie measurement · Machine learning · CNN · Dietary monitoring

M. Vasudha (✉) · D. Rashmi · B. A. Mahalakshmi Jain
Department of Information Science and Engineering, CMR Institute of Technology, Bengaluru, India
e-mail: maval9is@cmrit.ac.in

D. Rashmi
e-mail: rash.premnath@gmail.com

B. A. Mahalakshmi Jain
e-mail: majal9is@cmrit.ac.in

B. R. Devi et al. (eds.), *Proceedings of Fifth International Conference on Computer and Communication Technologies*, Lecture Notes in Networks and Systems 897,
https://doi.org/10.1007/978-981-99-9704-6_2

1 Introduction

The study of nutrition and health depends heavily on the accurate food identification that is essential for monitoring dietary consumption and ensuring that a person meets their daily calorie and nutritional needs. Being able to finish this duty swiftly and accurately is more crucial than ever right now, especially in view of the rising obesity rates and the risks they represent to the general public health. As a result, researchers have been creating numerous techniques for food recognition in an effort to increase the process' accuracy and effectiveness.

Calorie meter-based food recognition is one among the approaches. This method utilizes a calorimeter, a tool that calculates the estimated caloric content of food items according to their weight, ingredients [1], and nutritional makeup. The effectiveness of food recognition using calorie meters has been the subject of numerous research studies. For instance, Kusuma et al. (2018) created a food identification system that made use of a portable calorimeter to determine the amount of calories in food products in real time. A dataset of 20 food products was utilized in testing the system, and it had an accuracy rate of 86.7%. The same method was employed by Kim et al. (2019) when they created a smartphone app, which had a comparable accuracy rate of 77.8% for a dataset of 12 food items. These experiments show that calorie meter-based food recognition holds promise as a reliable technique for precisely calculating the caloric content of food products. Our work is able to identify the nutrition that we may get affected by lacking certain nutritional ingredients in our body and recommends the food that can benefit the rehabilitation of different age groups.

2 Problem Statement

The food recognition and calorie measurement system seeks to overcome the lack of an accurate and user-friendly method for detecting and quantifying food products' nutritional value. Existing methods frequently suffer from low accuracy, limited food item coverage, and insufficient portion size calculation. The study's goal is to create a cutting-edge system that uses cutting-edge computer vision and deep learning techniques to recognize various food products with greater accuracy and minimum time complexity. Furthermore, the system intends to give real-time calorie monitoring and personalized nutritional advice, allowing users to make educated food choices and simply promote better lives.

3 Existing Work

Researchers in the fields of nutrition and health have already completed a number of works on calorie meter-based meal recognition. These below are a few illustrations of recent works on this subject.

1. "Calorie estimation for food images using deep learning techniques" by Chen et al. (2017)—The method used convolutional neural networks to extract data from the image and ascertain the number of calories in the food item. The system had an accuracy rating of 84.9% on a dataset comprising 33 food items.
2. "Food volume estimation according to a single image using a depth-enhanced convolutional neural network" was published by Matsuda et al. in 2021. In this study, a depth-enhanced convolutional neural network was used to build a method for determining the volume of food items from a single photograph, from which the food item's caloric value could be calculated. On a dataset of 43 food products, the system's accuracy rate was 90.5%.
3. The article "A machine learning-based approach for estimating nutritional information from food images" by Rajaraman et al. (2018)—The caloric content and macronutrient composition of food products were estimated by the system using an amalgam of deep learning and computer vision techniques. On a dataset of 100 dishes [2], the system's accuracy rate was 92.3%.
4. Kim and colleagues' "Real-time food calorie estimation using a compact vision sensor" (2020)—In this work, a portable vision sensor was created that instantly estimates the calories in various food items. The method uses a regression model to calculate the caloric content and a network of convolutional neurons to categorize photos of food. On a dataset of ten dishes, the system had a 90.6% accuracy rate.

4 Study and Assessment of the Indian Food Recipe Images Dataset

The dataset [3] includes a variety of dishes throughout India. There are 80 classes in the dataset, and each class has between 45 and 60 photos:

i. It contains 4308 images each with a size of 224 × 224.
ii. A comma-separated values (.csv) file with the whole set of calories and class labels.

5 Methodology

5.1 Implementation of CNN

The convolutional neural network (CNN) architecture depicted in Fig. 1 is built for this study to analyze and reliably categorize food images [4]. The CNN receives the food picture as input and performs numerous important operations such as rescaling, Conv2D, max pooling, flattening, and dense layers. Convolutional layers are in charge of feature extraction. Small filters or kernels move across the picture, recognizing relevant patterns and characteristics like edges, textures, and forms. Max pooling is used after each Conv2D layer to downsample the feature maps, lowering their spatial dimensions. This procedure keeps the most significant characteristics while decreasing computational complexity. The flatten layer converts the 2D feature maps into a 1D vector, preparing them for the fully linked layers [5] that follow. The flattened features are passed through dense (fully connected) layers. These layers learn high-level representations and make decisions based on the extracted features, ultimately classifying the food image into different categories.

5.2 Plan of Implementation

The following are the phases involved in implementing the food recognition and calorie measurement system.

- **Download the Indian Recipe Dataset**: Downloading the Indian recipe dataset, which consists of 80 various foods and 1000 photos for each category, is required in this phase [6]. The dataset, which covers a diversity of cuisine from many nations and cuisines, is frequently used in studies on food recognition.
- **Preprocessing of images**: Cleaning the dataset, uniformly scaling the photos, and separating it into training and testing sets are all included in this phase. It is essential to preprocess the dataset because it improves the accuracy of the food recognition model by removing noise and pointless data.
- **Classification of food products**: After preprocessing the dataset, the food products should be categorized. To be able to accurately categorize the food products in the dataset considering their attributes and ingredients, this stage entails creating a deep learning system, that is a network of convolutional neurons [7]. The preprocessed dataset serves as the algorithm's training set, and the testing set serves as its accuracy test.
- **Front-end development**: Creating the application's user interface is a component of front-end development. In this stage, the application's layout is designed, HTML, CSS, and JavaScript are used to create the web pages, and the acknowledgement of the food algorithm is integrated into the program.

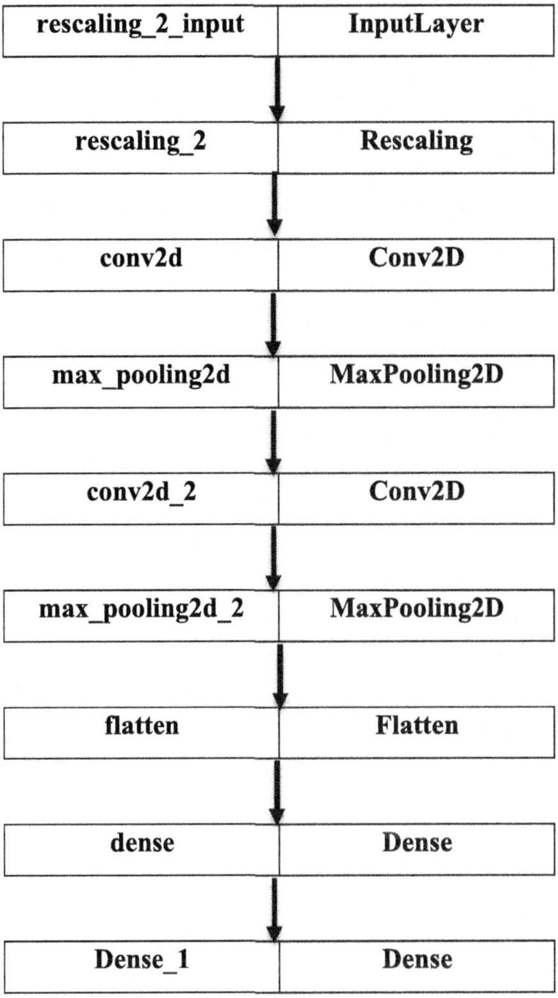

Fig. 1 CNN architecture

- **Algorithm management**: Managing and improving the algorithm for food acknowledgement [8] is a step of the algorithm management stage. This step entails fine-tuning the procedure to increase accuracy, optimizing the model's hyperparameters to increase performance, and incorporating new features into the algorithm.
- **Back-end development**: Creating the server side of the programme is a component of back-end development. Creating the database, creating the server-side scripts, and fusing the front-end and back-end components of the programme are all pieces of this process.

6 Working of the Proposed System

When a user submits an image through the web app, the app will predict what food is in the picture using the pre-trained Keras model and return the predicted dish along with its probability. The model predicts the food recipe chosen in the Web site and provides along with the calories of that particular recipe with the corresponding accuracy rate. The model has attained a maximum of 99.89% accuracy.

This web application offers a method for calculating the suggested daily calorie intake based on a person's height, weight, age, gender, and degree of exercise. These Web sites use formulas based on research to estimate how many calories an individual needs to consume each day to maintain their current weight, lose weight, or gain weight [9]. The Harris–Benedict equation, which considers an individual's basal metabolic rate (BMI) and activity level to predict their daily calorie demands, is employed for this purpose and activity level to estimate their daily calorie needs.

6.1 System Architecture

Figure 2 represents the food recognition and calorie measurement system architecture. The user uploads a photo from a local host to the Web site once the model has undergone training, provided the picture is part of a dataset. The CNN system predicts the image and delivers the dish titles, accurate nutritional information. For the recognition, after the training of CNN model, the food image's features will be extracted. By doing this, the image is subjected to the CNN model, and one or more layers' outputs are recorded [10]. The output of feature extraction is then checked against a database of well-known recipes.

The CNN is trained using a big dataset of food photos and their corresponding labels, which identify the type of food and its related calorie information. The input image is subjected to a sequence of convolutional layers in a CNN, each of which

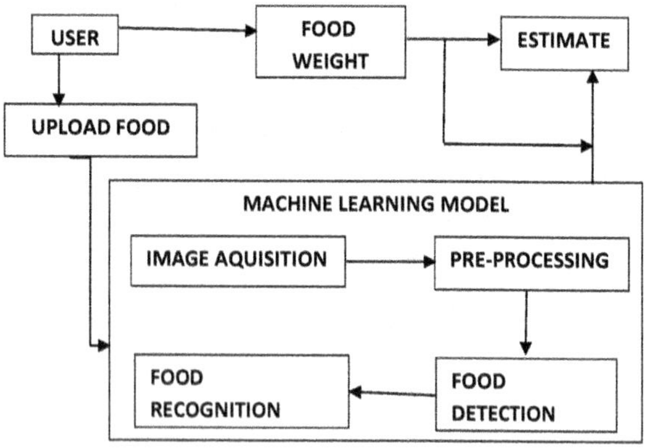

Fig. 2 System architecture

Table 1 Comparing accuracies

Method	Dataset used	Accuracy rate (%)
Resnet-50	256 food images	86.5
DenseNet-121	500 food images	87.1
Ensemble (ResNext-101, DenseNet-161)	UEC food-100 dataset	90.02
CNN	**Indian food dataset**	**98.89**

Note The specific row in bold conveys that the Method is implemented in our project and have obtained the accuracy of 98.89% when compared to the remaining methods

applies a unique set of filters [11]. These filters are tiny matrices that move across the image and compute a dot product between the weights of the pixels they overlap and the weights of the pixels in the image. This process creates a feature map that draws attention to patterns in the image, like edges and corners..

7 Results and Discussion

The Food Recognition and Calorie Measurement System achieves an impressive 99.89% accuracy. To analyze and validate the performance of the models for various food recipes prediction using CNN model, the dataset was partitioned into training and testing sets [12]. The remaining 20% of the data was used for testing, while the remaining 80% was used for training. To train the models on the training set, cross-validation with a batch size of 32 and 10 epochs was utilized. Accuracy, precision, recall, and $F1$-score were the evaluation metrics. In comparison with the previous works, this model performs better in terms of accuracy (Table 1).

8 Future Enhancements

The food recognition and calorie measurement system that employs CNN has several potential areas for development. Enhanced Food Database: Continuously expand the database with a wide variety of food images from different cuisines and cultures with real-time updates to improve the system's accuracy and recognition capabilities. Allergen Detection: Develop a feature to detect common food allergens, alerting users about potential allergens in their meals. Integration with Health Apps: Integrate the system with existing health and fitness apps for a seamless user experience and a holistic approach to health tracking. Offline Mode: Implement an offline mode to enable users to access basic food recognition and calorie measurement functionalities without an Internet connection.

9 Conclusion

Food recognition based on caloric meters using state-of-the-art deep CNN has been built and analyzed with a phenomenal accuracy of 99.89%. The system has the potential to increase the precision and effectiveness of tracking calorie consumption by identifying food photos and estimating their calorie content. By combining this technology with their past history [13], it is possible to create a personalized nutrition plan that takes into consideration the person's nutritional goals and preferences. Despite the encouraging outcomes produced by calorie meter-based food recognition systems, some restrictions and difficulties still need to be resolved. For instance, variables like lighting conditions, how food is presented, and a person's unique metabolism might have an impact on how accurate these systems are. Furthermore, the accuracy of calorie meter-based food recognition is limited to predicting caloric intake and provides no details about other significant nutritional aspects including macronutrient composition, vitamin and mineral content, and overall meal quality [14]. The variety in portion sizes and meal presentation are two of the obstacles that must yet be overcome.

References

1. Katiyar B et al (2019) Food calorie estimation using deep learning and mobile images. In: Research gate conference
2. Memis S, Arslan B, Batur OZ, Sonmez EB (2020) A comparative study of deep learning methods on food classification problem. In: Proceedings of the 2020 innovations in intelligent systems and applications conference (ASYU), pp 1–4
3. Indian food dataset from "Kaggle". https://www.kaggle.com/datasets/nehaprabhavalkar/indian-food-101
4. Matsuda Y, Hoashi H, Yanai K (2012) Recognition of multiple-food images by detecting candidate regions. In: Proceedings of the 2012 IEEE international conference on multimedia and expo, pp 25–30
5. Szegedy C, Vanhoucke V, Ioffe S, Shlens J, Wojna Z (2016) Rethinking the inception architecture for computer vision. In: Proceedings of the IEEE conference on computer vision and pattern recognition (CVPR), June 2016
6. Kawano Y, Yanai K (2014) Automatic expansion of a food image dataset leveraging existing categories with domain adaptation. In: Proceedings of the European conference on computer vision. Springer, pp 3–17
7. Jia Y, Shelhamer E, Donahue J, Karayev S, Long J, Girshick R, Guadarrama S, Darrell T (2014) Caffe: convolutional architecture for fast feature embedding. Association for Computing Machinery, New York, NY, USA
8. Hassannejad H, Matrella G, Ciampolini P, De Munari I, Mordonini M, Cagnoni S (2016) Food image recognition using very deep convolutional networks. In: Proceedings of the 2nd international workshop on multimedia assisted dietary management, pp 41–49
9. Martinel N, Piciarelli C, Micheloni C (2016) A supervised extreme learning committee for food recognition. Comput Vis Image Underst 148:67–86
10. Zagoruyko S, Komodakis N (2016) Wide residual networks. In: Proceedings of the British machine vision conference (BMVC), pp 87.1–87.12

11. Krizhevsky A, Sutskever I, Hinton GE (2012) Imagenet classification with deep convolutional neural networks. In: Proceedings of the 25th international conference on neural information processing systems, vol 1. Ser. NIPS 12. Curran Associates Inc., pp 1097–1105
12. Jiang S, Min W, Liu L, Luo Z (2020) Multi-scale multi view deep feature aggregation for food recognition. IEEE Trans Image Process 29:265–276
13. Yanai K, Kawano Y (2015) Food image recognition using deep convolutional networks with pre-training and fine-tuning. In: Proceedings of the IEEE international conference on multimedia expo workshops (ICMEW), pp 1–6
14. Rodriguez P, Gonfaus JM, Cucurull G, XavierRoca F, Gonzalez J (2018) Attend and rectify: gated attention mechanism for fine grained recovery. In: Proceedings of the European conference on computer vision (ECCV)

Breast Cancer Prediction with Gradient Boost and XGBoost

Avantika Mahadik, Prashant Sharma, and Vaibhav Narawade

Abstract In the arena of cancer research, it has been able to discover cancer patients utilizing genetic factor information grouped together with artificial intelligence innovations. To strengthen the foretelling paradigms, it is essential to integrate biological, social, and demographic data. In our research article, the gradient boost technique, the (SVM) support vector machine classier, the (LR) logistic regression technique, random forest, the (DT) decision tree classifier, *k* nearest neighbour, XGBoost, and adaptive linear neuron classifiers have been separately evaluated. In the suggested model, primary objective is to choose the most suitable features. By using correlation matrix and principal component analysis (PCA), we scaled down features for the better enhancement of the model's efficacy. It gave us 10 important features from the dataset. We have successfully achieved 100% accuracy in the prediction through gradient boost and XGBoost classifier with a 97.50% cross-validation score.

Keywords Breast cancer · Machine learning · Gradient boost · Correlation matrix · Principal component analysis · Grid search · Cross validation score

A. Mahadik (✉)
Pacific (PAHER) University, Udaipur, India
e-mail: avantika_mahadik@rediffmail.com

P. Sharma
Department of Computer Science and Engineering, Pacific (PAHER) University, Udaipur, India
e-mail: prashant.sharma@pacific-it.ac.in

V. Narawade
Department of Computer Engineering, Ramrao Adik Institute of Technology, D. Y. Patil Deemed to be University, Nerul, India
e-mail: vaibhav.narawade@rait.ac.in

B. R. Devi et al. (eds.), *Proceedings of Fifth International Conference on Computer and Communication Technologies*, Lecture Notes in Networks and Systems 897,
https://doi.org/10.1007/978-981-99-9704-6_3

1 Introduction

Cancer prevention is one of the most significant medical concerns of this century. The fight against cancer is crucial. The most recent available scientific information indicates that effectual core preventive measures may avoid a minimum of forty percent of cancer instances and prompt identification of tumours can further decrease lifespan. Bladder cancer, cancer of the breast, cerebral and the nervous system cancer, cancer of the cervical area, malignancy in children, cancer of the colon, cancer of the endometrium, tumours of the gallbladder, leukaemia, carcinoma of the kidneys, liver as well as lung melanoma, stomach malignance, skin tumours, cancer of the pancreatic gland, and thyroid cancer are just a few of the cancers that the IARC (which stands for the International Agency for Research on Cancer) classified. Research shows the probability of breast cancer has recently been rapidly increasing. The frequency among women may be indorsed to lifestyle choices and inheritance patterns. According to WHO statistics, breast cancer is a foremost threat to the health of women among all other types of cancers [1–3]. However, different precautions, including tests, scans, and routine clinical check-ups, can reduce the risk and significantly increase the probability of survival. World Health Organisation (WHO) stated that in the year 2020 highest females in Asia suffer from cancer [3], which is presented in Fig. 1. The frequency of new cancer diagnosis and the percentage of female cancer mortality are highlighted in Fig. 2.

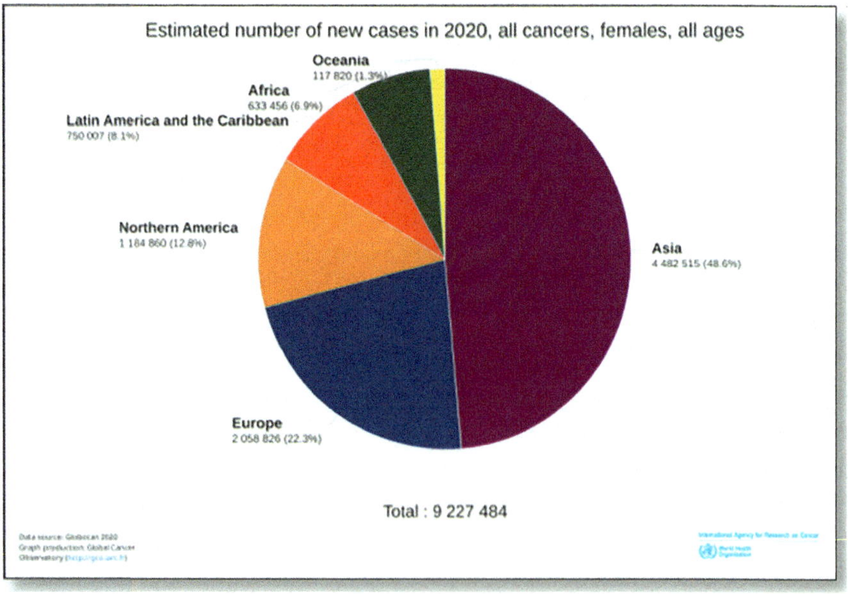

Fig. 1 Showing the estimated new cancer cases in females according to world health organization in 2020 [2]

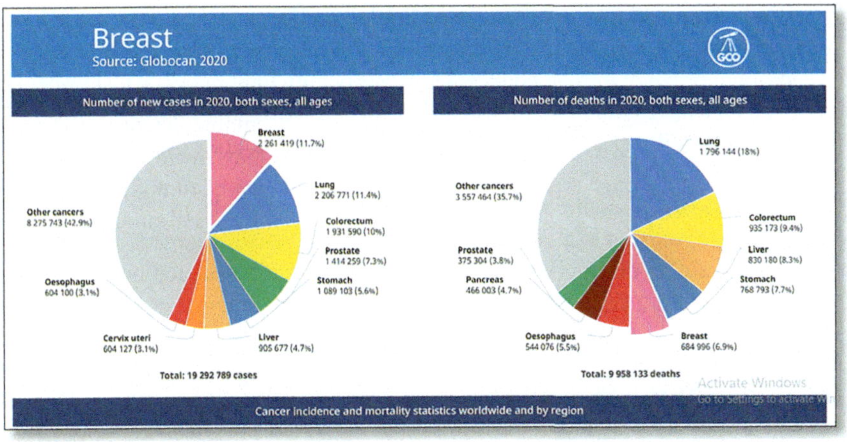

Fig. 2 Showing detailed statistics of various types of cancers found in males and females [2]

1.1 Breast Cancer

Throughout our lifespan, billions of cells will typically grow and divide as needed. Most cells die when they become aberrant or aged. When something goes wrong during this process, our cells continue to divide, preventing the old or abnormal cells from dying as they should. This is how cancer develops [4, 5]. The uncontrolled growth of cancer cells might push away healthy cells. In simple words, when an abnormal cell comes together and forms a lump, it is called cancer [6, 7]. Figure 3 shows the information about breast cancer tumours. Cancer initiates in the cells. New cells can occasionally arise in the body when the process of proliferation of cells goes wrong when the body really does not require them. In this kind of situation, the unwanted new cells, which are also called muted cells, form a lump or tumour in the breasts and that is called breast cancer.

The cancer tumour categorized into benign and malignant. Malignant tumours are those that have cancer, while benign that does not. Malignant tumours have irregular

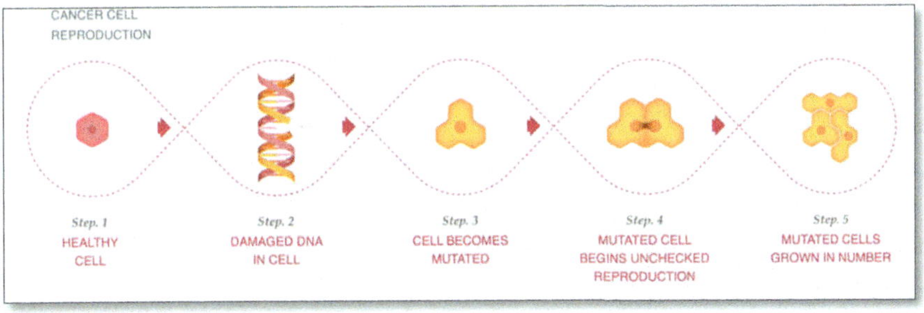

Fig. 3 Cancer cell reproduction [8]

borders which spread fast and damage additional bodily organs. On the other hand, the tumour, which is large in size, grows in regular borders, and fails to extend to the additional organ of human body, is known as benign tumours [9, 10]. The benign tumours which are round or oval in shape and also have regular borders showed into Fig. 4a, b [11]. Whereas (c) and (d) show malignant tumours which have irregular borders and are not defined clearly.

Figure 5 shows the count of malignant cells and benign cells in our experiments. Where we are given 0 to benign and 1 to malignant. Extracting the records which are relevant for training our model is essential. We performed that step before training the model. The exact counts are shown in Fig. 5.

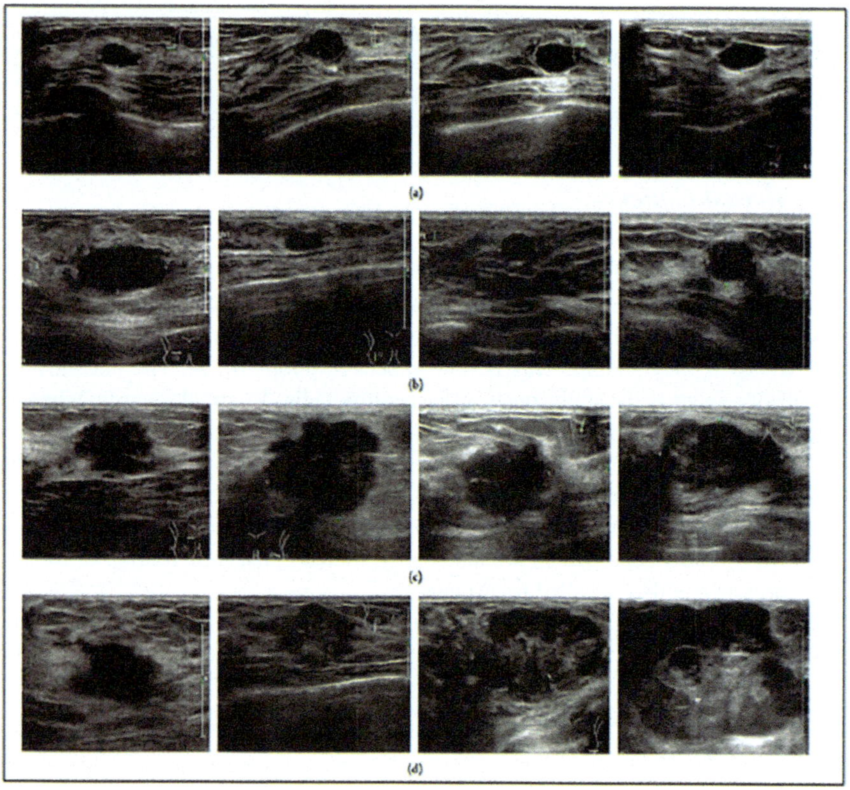

Fig. 4 Samples of ultrasound images [11]

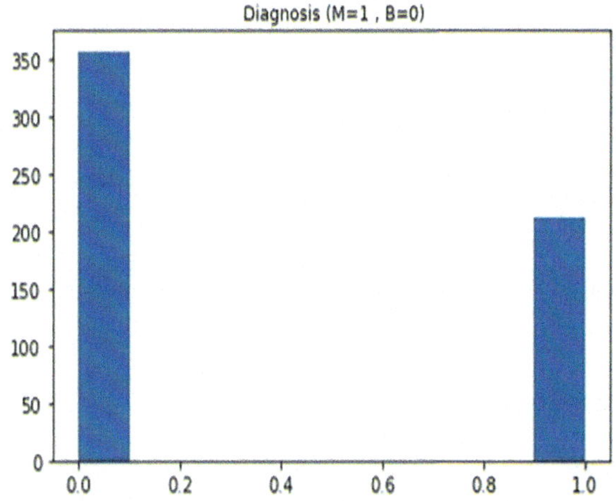

Fig. 5 Benign and malignant count

2 Related Work

The major goal of Elsadig et al. in [12] was to create the model based on five feature selection approaches. Methods for feature selection which are used to prioritize the critical features include gain ratio, gini index, chi square, ANOVA, and ReliefF [12]. The best machine learning technique overall, according to the authors, is SVM which has a 97.7% prediction accuracy.

Rabiei et al. in [13] pointed out that mammography reports are insufficient for the early recognition of breast cancer. Images from mammography are not more effectively working. While predicting the disease age, family history, personal history, and breast density are four important features stated by the authors. Gradient boosting is given the highest prediction accuracy among random forest and SVM [13].

Hosni et al. [14] founded in their assessment that artificial neural networks (ANN) classifier along with decision tree and SVM classifier usually build the ensemble model for predicting breast cancer. Additionally, authors observed that researchers frequently used trusted breast cancer datasets which are available on WBCD for their experiments.

3 Proposed System

We observed, many researchers faced three major problems: data cleaning, feature selection, and model training. Our research is assisting to figure out the ideal features to train the model with in order to attain the maximum degree of accuracy.

3.1 Breast Cancer Dataset

In our experiment, 569 breast cancer records which were collected from a kaggal Website were tested. The model's performance is based on data which we are provided. Unstructured, irrelevant data causes poor performance of the model. We conducted our experiments by using the Wisconsin Diagnostic Breast Cancer (WDBC) datasets [15]. Although working with datasets, data cleaning is a major task. After removing noisy data, we focused on feature selection.

3.2 Data Preprocessing

Checking the missing values, removing noisy data is our primary phase. We removed unstructured data from the WDBC dataset as well as, and by using the is_null command, we removed the missing values records from the collected dataset. We designed our proposed model which is shown in Fig. 5.

3.3 Architecture of Proposed Model

Our proposed prediction system is based on four key approaches. Correlation matrix, principal component analysis (PCA), grid search, and cross-validation are applied in our system. Figure 6 shows the architectural design of the suggested model.

3.4 Selection of Features

Picking up best features from 32 common features to train our model is the greatest step. Table 1 shows all 32 common features of breast cancer prediction. Out of these we picked ten features. We used correlation matrix and principal component analysis (PCA) for this task, which worked enormously and returned ten important features.

3.4.1 Correlation Matrix

The correlation matrix, which evaluated whether the two variables in the dataset were significantly correlated or not, also showed that the relationship between the two variables was neutral. Heatmaps are used to show how variables are correlated with one another. With the help of a correlation matrix, we summarized the features. In Fig. 7, we are shown the 15 related features. And also Table 2 shows the detailed names of all 15 features which we received after implementing the correlation matrix on

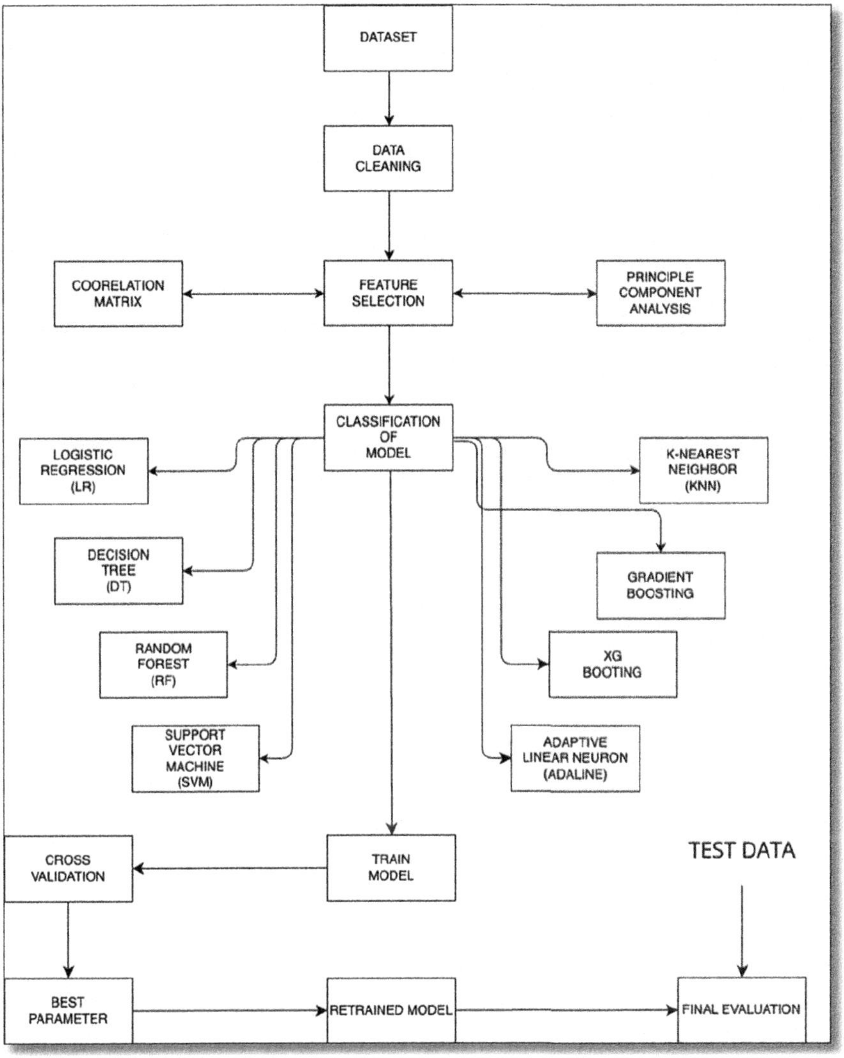

Fig. 6 Proposed breast cancer prediction model

the breast cancer dataset. We focused on correlation matrix and principal component analysis together in the process of feature selection.

Table 1 Breast cancer features [16–18]

No.	Feature name	Feature details
i.	Patient_Identity_No	Patient identity number
ii.	Type_Diagnosis	Type of breast cancer tumour where m indicating malignant and benign indicated by b
iii.	Radius_Of_Mean	Perimeter to centre distance indicated
iv.	Texture_Of_Mean	It shows grayscale values through standard deviation
v.	Perimeter_Of_Mean	It shows actual perimeter size of tumour
vi.	Area_Of_Mean	
vii.	Smoothness_Of_Mean	It shows mean of local variation in radius lengths
viii.	Compactness_Of_Mean	It shows mean of perimeter^2/area − 1.0
ix.	Concavity_Of_Mean	It shows mean of severity of concave portions of the contour
x.	Concave_Points_Mean	It shows mean for number of concave portions of the contour
xi.	Symmetry_Mean	
xii.	Fractal_Dimension_Mean	It shows mean for "coastline approximation" − 1
xiii.	Radius_Standard_Error	It shows standard error for the mean of distances from centre to points on the perimeter
xiv.	Texture_Standard_Error	It shows standard error for standard deviation of grayscale values
xv.	Perimeter_Standard_Error	
xvi.	Area_Standard_Error	
xvii.	Smoothness_Standard_Error	It shows standard error for local variation in radius lengths
xviii.	Compactness_Standard_Error	It shows standard error for perimeter^2/area − 1.0
xix.	Concavity_Standard_Error	It shows standard error for severity of concave portions of the contour
xx.	Concave_Points_Standard_Error	It shows standard error for number of concave portions of the contour
xxi.	Symmetry_ Standard_Error	
xxii.	Fractal_Dimension_ Standard_Error	It shows standard error for "coastline approximation" − 1
xxiii.	Radius_Worst	It shows worst/largest value of mean distances from centre to perimeter
xxiv.	Texture_Worst	It shows worst/largest value of mean for standard deviation of grayscale value

(continued)

Table 1 (continued)

No.	Feature name	Feature details
xxv.	Perimeter_Worst	
xxvi.	Area_Worst	
xxvii.	Smoothness_Worst	It shows worst/largest value of mean for local variation in radius lengths
xxviii.	Compactness_Worst	It shows worst/largest value of mean for perimeter^2/area − 1.0
xxix.	Concavity_Worst	It shows worst/largest value of mean for severity of concave portions of the contour
xxx.	Concave_Points_Worst	It shows worst/largest value of mean for number of concave portions of the contour
xxxi.	Symmetry_Worst	
xxxii.	Fractal_Dimension_Worst	

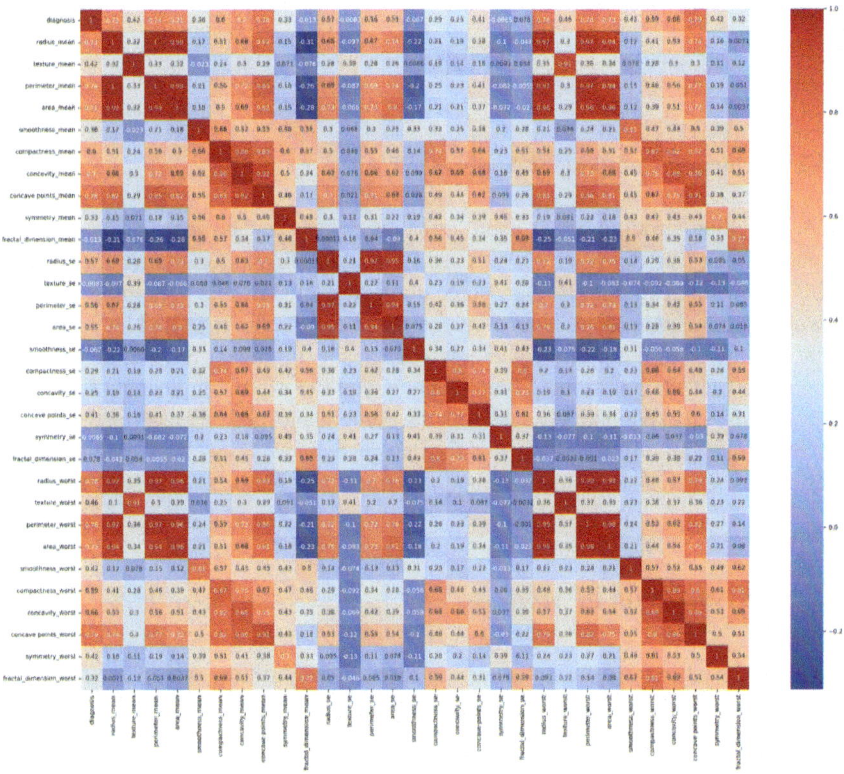

Fig. 7 Correlation matrix of breast cancer features

Table 2 Shows 15 selected features received after correlation matrix [17–19]

Feature no.	Feature name
I.	Radius_Of_Mean
II.	Perimeter_Of_Mean
III.	Area_Of_Mean
IV.	Compactness_Of_Mean
V.	Concavity_Of_Mean
VI.	Concave_Points_Of_Mean
VII.	Radius_Standard_Error
VIII.	Perimeter_Standard_Error
IX.	Area_Standard_Error
X.	Radius_Worst
XI.	Perimeter_Worst
XII.	Area_Worst
XIII.	Compactness_Worst
XIV.	Concavity_Worst
XV.	Concave_Points_Worst

3.4.2 Principal Component Analysis (PCA)

PCA is another strong practice used in feature selection. The second key approach in our proposed model is PCA. We used PCA after correlation matrix for the lessening the features amount further. Scree plot is depicted in Fig. 8, helping to decide how many principal components to maintain. The main aim is to find an "elbow" or point of nuance in the plot. This elbow specifies a significant drop in eigenvalues, suggesting that the additional components beyond that point contribute less to the overall variance explained. By selecting the components before the elbow, we maintain those that capture the majority of the data discrepancy as well as remove the less useful ones. Table 3 shows the ten final and important features (Fig. 9).

3.5 Classification of Models

We have tested eight different machine learning algorithms which are shown in Table 4.

3.5.1 Decision Tree Classifier (DT Classifier)

Both classification and regression can be performed via decision tree techniques. In the decision tree technique, three key nodes, namely leaf, root, and decision. Possible

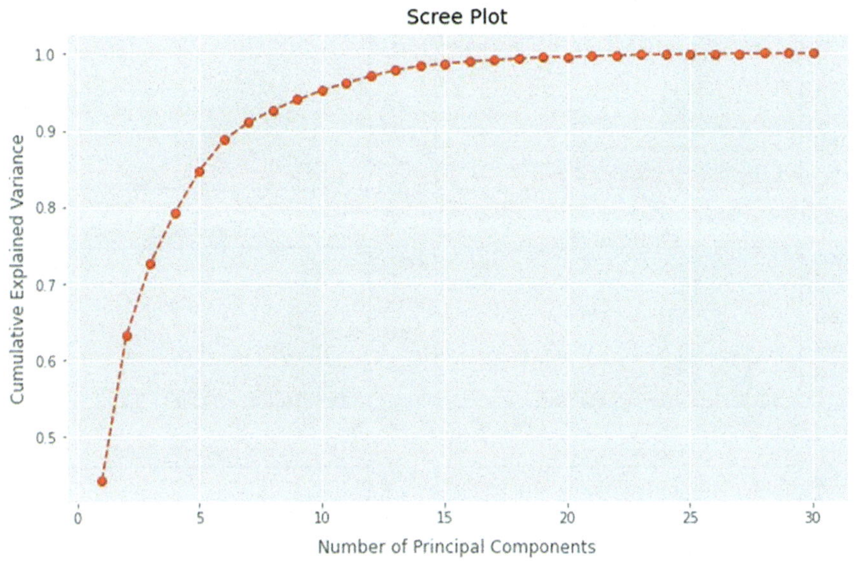

Fig. 8 Principal component analysis (PCA)

Table 3 Final ten features [18]

Feature no.	Feature name [18]
a.	Radius_Mean
b.	Texture_Mean
c.	Perimeter_Mean
d.	Area_Mean
e.	Smoothness_Mean
f.	Compactness_Mean
g.	Concavity_Mean
h.	Concave_Points_Mean
i.	Symmetry_Mean
j.	Fractal_Dimension_Mean

outcomes are shown by leaf nodes. Decision nodes are split into several branches which are involved in the process of generating conclusions [20, 21].

3.5.2 Random Forest Classifier (RF Classifier)

Number of decision trees has been generated in this classifier using different subsets of the input data. It is combining different classifiers to solve a difficult hitch and enhance model's efficacy. This merging of different models is referred to as an

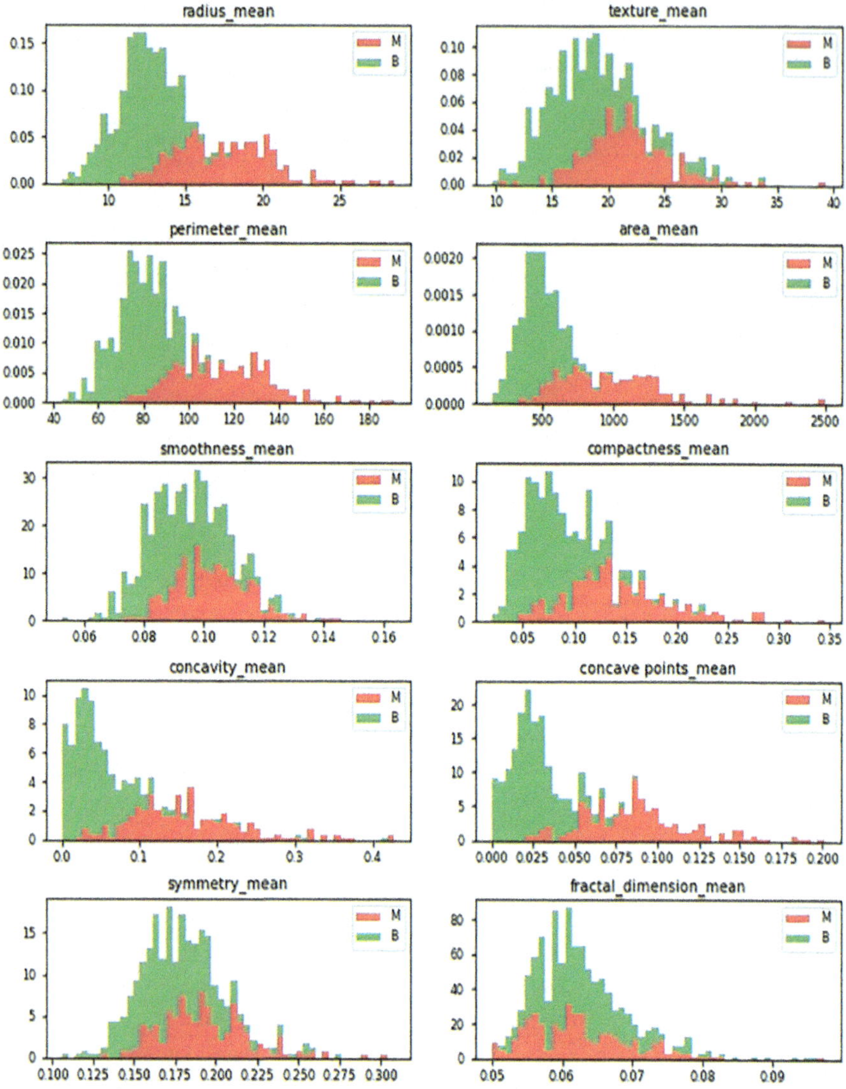

Fig. 9 Final ten features Source: Author

ensemble, and it is made possible through bagging and boosting. A distinct training subset is created by the process of bagging using a sample training dataset. A majority vote is required to determine the result. Boosting is the process of transforming weak learners into strong ones by developing successive models to reach the highest level of accuracy [20, 21].

Table 4 Names of models which are being tested in our proposed model

No.	Machine learning algorithms
(a)	Decision tree classifier (DT)
(b)	Random forest classifier (DT)
(c)	*K* nearest neighbour classifier (DT)
(d)	Support vector machine (SVM)
(e)	Logistic regression
(f)	Gradient boost
(g)	XGBoost
(h)	Adaptive linear neuron (Adaline)

3.5.3 *K*-Nearest Neighbour Classifier (*K*NN Classifier)

This classifier mainly applied for categorization. The classification of a new sample point using data point is the primary aim of this classifier. It classifies items based on distance and according to the classes of their immediate neighbours. The majority of *K*NN applications are in classification jobs. The parameter k in *K*NN indicates how many labelled points (neighbours) are taken into account for classification. The value of k indicates how many of these points are required to calculate the result. Our tasks are to calculate the distance and find the categories that are closest to our unidentified entity [20, 21].

3.5.4 SVM

This algorithm was created by Vladimir Vapnik. The key goal of this technique is to identify the "hyperplane". The precise plane, line, or decision boundary known as a hyperplane can group many features (sometimes referred to as n-dimensional space) into sets. For the hyperplane construction, the SVM identifies the most intense endpoints and vectors. Sever situation named by support vectors and support vector machine algorithm is named after them. This classifier works in classification as well as regression [20, 21].

3.5.5 Logistic Regression Classifier (LR Classifier)

The purpose of this classifier is to figure out likelihood instances which are going to a specific style [15, 20, 21]. A sigmoid function is used to determine the probability for the given class. The logistic regression predicts the likelihood that an instance will belong to a specific class or not, whereas the output of linear regression is a continuous value that can be anything. [20, 21].

3.5.6 Gradient Boosting

This is also called ensemble approach in machine learning which constructed on the integration of the combination of boosting and optimization algorithm [20, 21].

3.5.7 XGBoosting

This is a leading machine library for regression and classification. This is an ensemble learning method which combines a weak model's prediction to give a strong prediction. XGBoost handles the missing values very well.

3.5.8 Adaptive Linear Neuron (Adaline)

Finding the line or hyperplane that minimizes the vertical offsets is the goal of the ordinary least squares (OLS). Linear regression approach has a strongly connected with Adaline classifier [22, 23].

3.6 Model Training

Grid search is the third powerful technique we used in the proposed model. Finding the best parameter is the key feature of grid search. Grid search applied in all eight models, including the above ten important features. The best parameter helped to achieve the highest accuracy in the prediction. Grid search we used for every model. While training our model, we did cross-validation of each machine learning model. Through those best parameters, we retrained the model and tested the data. Finally, our model predicted breast cancer with the highest prediction accuracy.

The fourth key concept of our model is cross-validation score. Cross-validation score helped in the process of finalizing a prediction model among eight models. Gradient boost, XGBoost, random forest, and decision tree gave 100% accuracy in the prediction. We further checked cross-validation score. Accuracy and cross-validation score together are the criteria for finalizing the best model. So, the gradient boost and XGBoost are given 100% highest accuracy along with a 97.50% cross-validation score.

4 Result Analysis

After feature selection was completed with correlation matrix and PCA, we received ten prominent features, those the expended in all eight models. For each model, we used a grid search. We cross-validated each machine learning model as we trained our own. We retrained the model using those optimal parameters and finally tested the findings. Finally, our model achieved the highest prediction accuracy in breast cancer. All eight models were cross-validated with a cross-validation score. So, our result analysis is based on the accuracy percentage along with their cross-validation score. Figures 10, 11, 12, 13, 14, 15, 16 and 17 show accuracy and cross-validation score of each model.

Once we obtained the percentage of accuracy and cross-validation score of every model, we compared the result and depicted in Table 5. After applying our four key conceptions of the planned model, we achieved 100% percentage accuracy of gradient boosting, XGBoosting, DT classifier, and RF classifier. Gradient boosting and XGBoosting are equally, achieving a cross-validation score that is 97.50%.

Fig. 10 Cross-validation score of gradient boost Source: Author

```
Accuracy : 100.000%
Cross-Validation Score : 97.500%
Cross-Validation Score : 93.750%
Cross-Validation Score : 93.750%
Cross-Validation Score : 94.363%
Cross-Validation Score : 93.972%
```

Fig. 11 Cross-validation score of XGBoost Source: Author

```
predictor_var = features_mean
xgb = XGBClassifier(objective = 'binary:log

classification_model(xgb, traindf,predictor

Accuracy : 100.000%
Cross-Validation Score : 97.500%
Cross-Validation Score : 94.375%
Cross-Validation Score : 93.750%
Cross-Validation Score : 94.047%
Cross-Validation Score : 93.718%
```

Fig. 12 Cross-validation score of random forest Source: Author

```
Accuracy : 100.000%
              precision    recall  f1-score   support

           0       1.00      1.00      1.00       254
           1       1.00      1.00      1.00       144

    accuracy                           1.00       398
   macro avg       1.00      1.00      1.00       398
weighted avg       1.00      1.00      1.00       398

Cross-Validation Score : 93.750%
Cross-Validation Score : 93.125%
Cross-Validation Score : 94.583%
Cross-Validation Score : 93.722%
Cross-Validation Score : 93.206%
```

Fig. 13 Cross-validation score of decision tree Source: Author

```
Accuracy : 100.000%
              precision    recall  f1-score   support

           0       1.00      1.00      1.00       254
           1       1.00      1.00      1.00       144

    accuracy                           1.00       398
   macro avg       1.00      1.00      1.00       398
weighted avg       1.00      1.00      1.00       398

Cross-Validation Score : 92.500%
Cross-Validation Score : 90.000%
Cross-Validation Score : 90.000%
Cross-Validation Score : 89.968%
Cross-Validation Score : 89.190%
```

Fig. 14 Cross-validation score of SVM Source: Author

```
Accuracy : 97.990%
              precision    recall  f1-score   support

           0       1.00      0.97      0.98       260
           1       0.95      0.99      0.97       138

    accuracy                           0.98       398
   macro avg       0.97      0.98      0.98       398
weighted avg       0.98      0.98      0.98       398

Cross-Validation Score : 88.750%
Cross-Validation Score : 88.125%
Cross-Validation Score : 86.250%
Cross-Validation Score : 85.890%
Cross-Validation Score : 85.168%
```

Fig. 15 Cross-validation score of logistic regression Source: Author

```
Accuracy : 90.452%
              precision    recall  f1-score   support

           0       0.94      0.91      0.93       262
           1       0.84      0.89      0.86       136

    accuracy                           0.90       398
   macro avg       0.89      0.90      0.90       398
weighted avg       0.91      0.90      0.91       398

Cross-Validation Score : 92.500%
Cross-Validation Score : 90.625%
Cross-Validation Score : 89.583%
Cross-Validation Score : 89.656%
Cross-Validation Score : 89.699%
```

Random forest and decision tree were cross-validated with 94.58% and 92.50%, respectively. SVM showed 97.99% accuracy, but after cross-validation the validated score was decreased to 88.75%. We observed that *K*NN, logistic regression, and Adaline accuracy percentage are less compared to cross-validation percentages. Table 5 shows the comparison of eight models. The final evaluation was based on

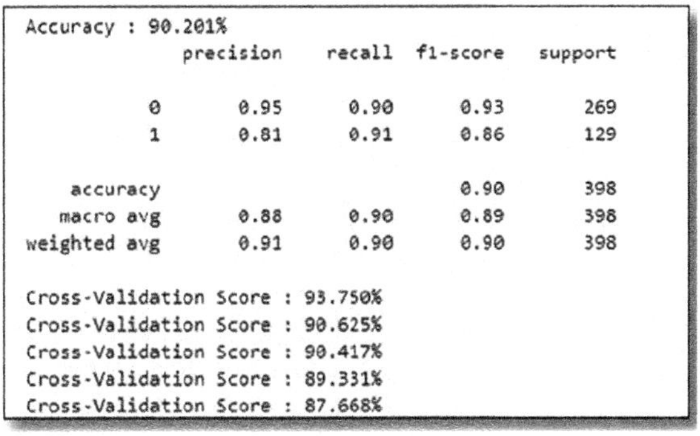

Fig. 16 Cross-validation score of *K*NN Source: Author

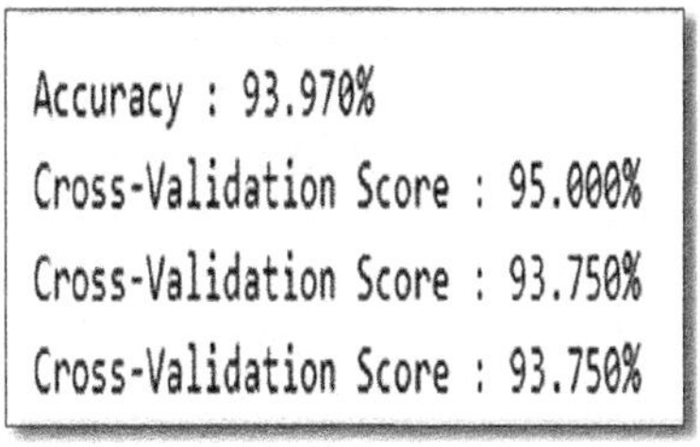

Fig. 17 Cross-validation score of Adaline Source: Author

accuracy along with cross-validation score, so gradient boost and XGBoost are two models we are suggesting for final prediction.

Table 5 Comparison of eight models for final evaluation

Model name	Accuracy in percentage	Cross-validation score in percentage
Gradient boost	**100**	**97.50**
XGBoost	**100**	**97.50**
Random forest	100	94.58
Decision tree	100	92.50
SVM	97.99	88.75
*K*NN	90.20	93.75
Logistic regression	90.45	92.50
Adaline	93.97	95.00

Bold value represents highest accuracy and highest cross validation score in percentage

5 Conclusion

The Wisconsin diagnostic breast cancer (WDBC) dataset was used in our research. We observed that feature selection is the crucial task. Elimination of features to received appropriate features for achieving the highest accuracy in breast cancer prediction possible through correlation function and principle component analysis (PCA). We reduced the size of features. Table 3 shows the ten major features of the model. Grid search and cross-validation score are given accuracy after selecting the best hyperparameter. Eight models are trained and evaluated on the basis of accuracy and cross-verification score, which are summarized in Table 5. After comparing all eight models, we summarized that gradient boost and XGBoost are the best models, which provided 100% accurateness in the predication of breast cancer along with the 97.50% cross-validation score.

We focused on numeric data. Our model can be enhanced through convolutional neural network (CNN) technique which are used for image data. There is a scope for further research, where researcher can combine different types of data such as numeric, categorical, image data, and data which are collected from IoT's device. Also, it is necessary to see the secrecy and authenticity of data. We used the K-fold cross-validation to retrained our model. Further, researchers can use different types of cross-validation techniques and other reliable dataset for greater achievement.

References

1. https://www.who.int/news-room/fact-sheets/detail/cancer6
2. https://www.iarc.who.int/
3. El-Toukhy S, El-Daly S, Kamel M, Hanafi H (2022) The diagnostic significance of circulating miRNAs and metabolite profiling in early prediction of breast cancer in Egyptian women. J Cancer Res Clin Oncol 149(8):5437–5451
4. Shaikh FJ, Rao DS (2022) Prediction of cancer disease using machine learning approach. Mater Today Proc 50:40–47
5. Islam MM, Haque MR, Iqbal H, Hasan MM, Hasan M, Kabir MN (2020) Breast cancer prediction: a comparative study using machine learning techniques. SN Comput Sci 1:1–14
6. https://www.indiancancersociety.org
7. https://www.indiancancersociety.org/cancer-information/
8. Siddiqui SY, Haider A, Ghazal TM, Khan MA, Naseer I, Abbas S, Ateeq K (2021) IoMT cloud-based intelligent prediction of breast cancer stages empowered with deep learning. IEEE Access 9:146478–214649
9. Fatima N, Liu L, Hong S, Ahmed H (2020) Prediction of breast cancer, comparative review of machine learning techniques, and their analysis. IEEE Access 8:150360–150376
10. Thomas T, Pradhan N, Dhaka VS (2020) Comparative analysis to predict breast cancer using machine learning algorithms: a survey. In: 2020 international conference on inventive computation technologies (ICICT). IEEE, pp 192–196
11. Gupta P, Garg S (2020) Breast cancer prediction using varying parameters of machine learning models. Proc Comput Sci 171:593–601
12. Elsadig MA, Altigani A, Elshoush HT (2023) Breast cancer detection using machine learning approaches: a comparative study. Int J Electr Comput Eng 13(1). ISSN 2088-8708

13. Naji MA, El Filali S, Aarika K, Benlahmar EH, Abdelouhahid RA, Debauche O (2021) Machine learning algorithms for breast cancer prediction and diagnosis. Proc Comput Sci 191:487–492
14. Hosni M, Abnane I, Idri A, de Gea JMC, Alemán JLF (2019) Reviewing ensemble classification methods in breast cancer. Comput Methods Program Biomed 177:89–112
15. https://www.kaggle.com
16. Asri H, Mousannif H, Al Moatassime H, Noel T (2016) Using machine learning algorithms for breast cancer risk prediction and diagnosis. Proc Comput Sci 83:1064–1069
17. https://docs.mindsdb.com
18. https://manikanthp.github.io/
19. Singh P, Mohan P, Rajput R (2023) Combining K-means and Gaussian mixture model for better accuracy in prediction of ductal carcinoma in situ (DCIS)-breast cancer. In: IEEE international conference on integrated circuits and communication systems (ICICACS)
20. Sharma P, Narawade V, Mahadik A (2023) A comparison and analysis of supervised machine learning algorithms towards accurate predicting of heart diseases. Biogecko 12(1):1–8
21. Mahadik A, Sharma P, Narawade V (2023) Prediction and analysis of diabetes using machine learning. J Data Acquisit Process 38(2):1330–1341
22. Singh G (2020) Breast cancer prediction using machine learning. Int J Sci Res Comput Sci Eng Inf Technol 8(4):278–284
23. Bayrak EA, Kırcı P, Ensari T (2019) Comparison of machine learning methods for breast cancer diagnosis. In: 2019 scientific meeting on electrical-electronics and biomedical engineering and computer science (EBBT). IEEE, pp 1–3

Data Cube Technology for Accessing of Large Database

Vipin Saxena, Vishal Verma, Versha Verma, and Karm Veer Singh

Abstract During the past years and due to coronavirus pandemics, the online business of the companies has been increased, and people are working from the home through high-speed Internet connectivity. Daily, large amount of database is getting space over the cloud servers, but when one is accessing the desired information from the various search engines, customers are not getting accurate information within minimum time frame. Therefore, present work is an attempt to use of the data cube technology for representation of large amount of the database and faster accessing of desired database as per the customer's requirements. A sample of database of the Life Insurance of India (LIC), which is a largest leading company working in the insurance sector, is selected for validation of the proposed system model which is designed through the unified modeling language (UML), and the computed results are depicted in the form of the tables and graphs.

Keywords Business · Database · Data cube technology · Insurance and correlation

1 Introduction

Many of the companies are having the complex database related to various tasks performed by the companies, and data analysist divides the data into various segments for analysis purpose. From the literature, it is seen that various online analytical methods are described to analyse the data stored by the companies in the warehouses called as data warehouses. The concept of online analytical processing (OLAP) is gaining popularity for faster accessing of database with high efficiency and accuracy.

V. Saxena (✉) · V. Verma · K. V. Singh
Department of Computer Science, Babasaheb Bhimrao Ambedkar University, Lucknow 226025, India
e-mail: profvipinsaxena@gmail.com

V. Verma
Department of Computer Applications and Science, School of Management and Science, Lucknow 226027, India

B. R. Devi et al. (eds.), *Proceedings of Fifth International Conference on Computer and Communication Technologies*, Lecture Notes in Networks and Systems 897,
https://doi.org/10.1007/978-981-99-9704-6_4

The data can be represented in the form of multidimensional arrays represented as data cube for using the OLAP technology. It is a tool for performing multidimensional analysis of complex data stored in data warehouse or in a centralized store. On the other hand, data mining is defined as a process for extracting the relevant information in the minimum time frame from large database, and it can be done using various tools and methods. It is a computational process which analyses the data from different perspective, dimension, and angles.

In the present work, a database of leading LIC of India is considered which is having huge database and can be represented in the form of multidimensional data related to various policies of different plans taken by the various customers. LIC Company offers the different plans to the customers based on the annual income, age, etc., and customers may pay the premium amount monthly, quarterly, or yearly, and after the maturity of the policy, the customer gets the maturity amount, bonus, and other benefits. Initially, the database is stored in form of data warehouse; thereafter a platform-independent unified modeling language (UML) model is created for accessing of the database by the desired customers. By the proposed method, the data can be collected from multiple sources and arranged in the form of data cube. Data is categorized by dimensions which is stored in dimension table in each OLAP cube. Thereafter, a computing method is used for representing the data through data cube technology, and customers may select, extract, and perform the various queries over the data for analysis purpose with different angles of view. It may be useful for creating financial report, budget, forecasting, etc., and customers or analysts may perform operations like roll-up, drill-down, slice, dice, etc.

2 Related Work

In the present work, extraction of relevant data is performed from a large set of databases through data cube technology. Therefore, important research papers available in the literature and related to the present work are explained in brief. Pratheebha et al. [1] have explained the concept of data mining, the process of knowledge discovery database (KDD), and different techniques like clustering, association, classification, prediction, and outlier detection. Palacios et al. [2] proposed the models through the concepts of machine learning algorithms for extracting the meaningful information related to the students with various levels of the education from large database of higher education. The model allows to compute the drop rate of the students from various courses of studies. In the year 2021, Alshareef et al. [3] have introduced a decision-making structure that extracts the knowledge and visualization, and the structure proposed by the authors uses the concepts of data mining techniques to extract the useful patterns for the purpose of analysis and decision-making. Rutkowski et al. [4] have described, how to adapt static decision tree to accommodate the data stream, and for this purpose, new decision trees are designed, non-parametric techniques and non-stationary regression are employed. Jiawei et al.

[5] explored the concept of KDD and data mining for the issues related to the feasibility, usefulness, effectiveness, and scalability of techniques for the discovery of hidden patterns from large data sets.

In the present paper, a system model is developed through UML, and hence, a brief history of UML is described in brief. In the year 1997, OMG [6] described all UML specifications and further developed the UML as a standardization, general prospects modelling language in the field of software engineering [7]. It contains the graphical notations which are used to create visual model of software. In the year 2002, Booch [8] has described the facilities available in the UML which are used for communication purpose and reduces the confusion among project stakeholders. It allows to user to model everything from enterprise information system. Further, OMG [9] has described UML as a super structure specification in the year 1999 and released the various versions of UML. UML has given significant contributions in the history of the object-oriented programming language that is why the modelling language is still widely used by the researchers. In the year 2018, Abdelhedi et al. [10] have proposed a logical model and developed Object2NoSQL in UML for creating a big data warehouse. Ciccozzi et al. [11] have conducted a systematic review through UML by considering 63 research articles and 19 tools among 5400 entries through systematic search and selection process. Koc et al. [12] have given systematic review on the UML diagram and conducted a comprehensive review from 2000 to 2019. The scientists focused on the utilization of UML diagram in software engineering research. This identified the research questions and keywords that how the UML diagrams are utilized. Semenova et al. [13] have presented a review of the UML which helps to the scientists that how the UML diagrams with application are used, the merits and demerits of the UML language.

Again, in the year 2005, Booch [14] has updated the various kinds of UML diagrams which may be useful for development of the software designs. Dennis et al. [15] helped to the scientists and researchers who are able to learn new techniques and approaches to develop the system more effectively and efficiently. Booch et al. [16] have also illustrated the application of the UML to complex modelling problems across a variety of application domain. Foster et al. [17] adopted a methodical approach to solve the software engineering problems and have given fundamentals of software engineering, software investigation and analysis, software implementation, design, and software tools. Shao et al. [18] have proposed a model named Internet of things (IoT)-based efficient data visualization framework (EDVF) for business intelligence.

The present work is showing how the complex data is represented using multidimensional model through OLAP. The important research papers related to the OLAP technology are also described in brief. In the year 2019, Kumar et al. [19] described the methodology of designing the OLAP data cube. Lamani et al. [20] have introduced the data mining techniques for improvement of OLAP cubes which are useful for decision-making in the fully automatic environment. In the year 2019, Djiroun et al. [21] have discussed the problem of designing and creation of cube as the various business analysts need the implementation of complex database for accessing the data in very less time. In the year 2020, Xio et al. [22] have applied the technique which is based on probability, for creating the cube obtained through cube aggregation. Queiroz-Sousa and Salgado [23] defined various criteria along with a critical review of applications of the OLAP technology. In 2021, Mathur et al. [24] have presented the tool which may be used for design of OLAP cube. Further, Francia et al. [25] have contributed intelligence analytics model (IAM). Chaudhary and Saxena [26] have also created the data cube of debit cards system and computed the correlation coefficient for analysis purpose of the data.

3 Methodology

Let us consider the database of LIC of India for insurance of the customers with a sample size of 25 customers for demonstration purpose. When the size of data is increased, then it is necessary to extract the desired information of customer in the minimum time frame. The following Table 1 gives the information related to the eight fields like Policy_no, Cust_id, Cust_name, Plan_id, Plan_no, Plan_name, Due_date, and Premium represented in the different columns. A customer may have more than one policy of different branches or multiple policies of single branch of LIC.

In the above table, the names of the customer may be same which represents that a customer may select more than one policy. Later on, a system model is designed through UML and is represented in Fig. 1 with offline and online processes for issuing the policy.

When the customer goes through the offline process by visiting the nearest LIC branch and fills and submits the registration form. All the details of the customer are loaded along with supporting documents over the server of the branches connected to the LIC branch. Thereafter, agent will process the LIC policy and provides the details regarding policy to the customer. The company checks eligibility of taking the policy by one of the officers controlled by Officer class. When it is properly verified, then officer informed to the customer for payment of the applicable premium of the policy. After successful deposit of the premium of the desired policy, the customer receives the documents of the policy through online system controlled through server class. The OLAP cube contains three fields as Plan_id, Cust_id, and Policy_no through data source view and is created after designing the pivot table. Before inserting the database into a cube, a star schema of the data source view is also designed from LIC_Branch to policy, customer, and plan. The three fields like Plan_id, Cust_id, and Policy_no of the database are summarized and bonded in the form of the data cube which can be expanded for large volume of database; however, 25 important fields are selected and represented below in Fig. 2.

From the above figure, one can perform several queries to find the desired policy details of the customer, which is due for depositing the payment of the premium from the large database of policy. The information is retrieved by performing the various queries discussed in the next section.

Table 1 Data set of policy of customers of Life Insurance Corporation of India

Policy_no	Cust_id	Cust_name	Plan_id	Plan_no	Plan_name	Due_date	Premium
247,700,298	1011	Vishnu	229	936/25	New Jeevan Labh	13-12-2021	6335
247,700,567	1012	Amit	229	936/25	New Jeevan Labh	13-12-2021	7413
208,333,018	1013	Vijay	221	836/21	Jeevan Labh	10-07-2018	5716
248,351,433	1014	Ram	232	943/20	Adhaar Stambh	13-03-2022	735
248,351,468	1015	Arvind	232	943/20	Adhaar Stambh	13-03-2022	732
229,228,999	1016	Devendra	229	936/25	New Jeevan Labh	23-07-2020	12,157
236,428,518	1017	Keshav	224	165/16	Jeevan Saral	26-10-2011	919
229,227,436	1018	Vineet	229	936/25	New Jeevan Labh	27-02-2020	2415
226,429,038	1019	Goldee	231	815/20	New Jeevan Anand	28-09-2015	1533
248,388,442	1020	Vibha	230	945/52	Jeevan Umang	13-05-2022	5012
225,546,858	1021	Ravi	225	165/20	Jeevan Saral	19-10-2012	3064
225,546,859	1022	Raj	225	165/20	Jeevan Saral	19-10-2012	3064
209,445,105	1023	Shrawan	222	836/25	Jeevan Labh	23-11-2018	2303
225,970,469	1024	Mahendra	223	165/15	Jeevan Saral	23-12-2013	510
226,582,071	1025	Yashika	226	832/25	New Children Plan	23-04-2016	1078
247,712,203	1026	Ashish	229	936/25	New Jeevan Labh	25-01-2022	2010
226,891,069	1027	Sushant	222	836/25	Jeevan Labh	26-07-2019	32,618
206,591,021	1028	Rahul	221	836/21	Jeevan Labh	28-12-2017	2326
244,839,786	1029	Abhishek	233	943/17	Adhaar Stambh	27-10-2021	1028
244,839,787	1030	Ganesh	234	943/18	Adhaar Stambh	28-10-2021	954
244,839,789	1031	Deepa	235	943/19	Adhaar Stambh	28-10-2021	888
244,839,838	1032	Praveen	229	936/25	New Jeevan Labh	28-10-2021	7265
244,839,875	1033	Doli	232	943/20	Adhaar Stambh	28-10-2021	830
244,134,724	1034	Sharafat	229	936/25	New Jeevan Labh	30-07-2021	2423
208,371,071	1035	Shyam	222	836/25	Jeevan Labh	30-10-2018	3631

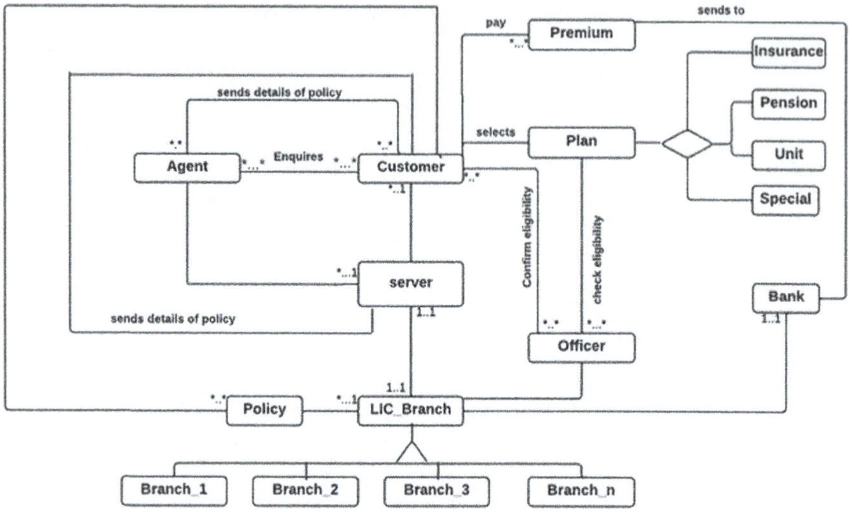

Fig. 1 UML model for accessing of database for LIC

Fig. 2 Data cube representation for database of policy

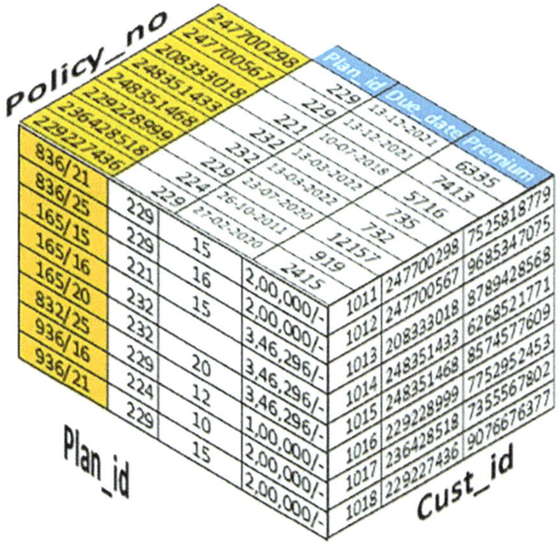

4 Results and Discussion

By the use of database of Life Insurance of Policy, millions of customers are enjoying the benefits of the insurance. The present work shall be very helpful for finding the best policy among the existing schemes of the policy. On the basis of data cube, various sequential queries have been performed for validation of the system model and explained below in brief.

Sample Query 1

The queries have been executed in the environment of Visual Studio 2022, and OLAP data cube is designed named as MYLICcube.cube. Data source view is designed with the fact table named LIC_Branch, final Star Schema is obtained, and using this Star Schema data cube is designed with three dimensions as Plan_no, Policy_no, and Cust_id. The query is performed on the SQL Server Management Studio by connecting analysis services engine, the following MDX query command is running on the environment, and executing the following query gives the following result.

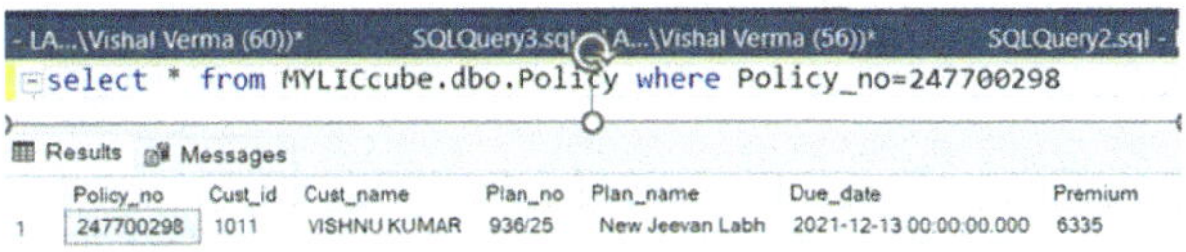
- LA...\Vishal Verma (60))*   SQLQuery3.sql - A...\Vishal Verma (56))*   SQLQuery2.sql -

```
select * from MYLICcube.dbo.Policy where Policy_no=247700298
```

Results   Messages

	Policy_no	Cust_id	Cust_name	Plan_no	Plan_name	Due_date	Premium
1	247700298	1011	VISHNU KUMAR	936/25	New Jeevan Labh	2021-12-13 00:00:00.000	6335

The Policy-no = 247,700,298 gives result related to the Vishnu Kumar with Cust_id as 1011. In the sample Query1 when the command is executed, the details of Policy_no = 247,700,298 are fetched from the dimension table, and it is run in the environment of SQL Server Management Studio when connecting to analysis services package.

Sample Query 2

In this case, when the command is executed, the details of Plan_name = New Jeevan Labh are fetched from the plan table. It is represented below:

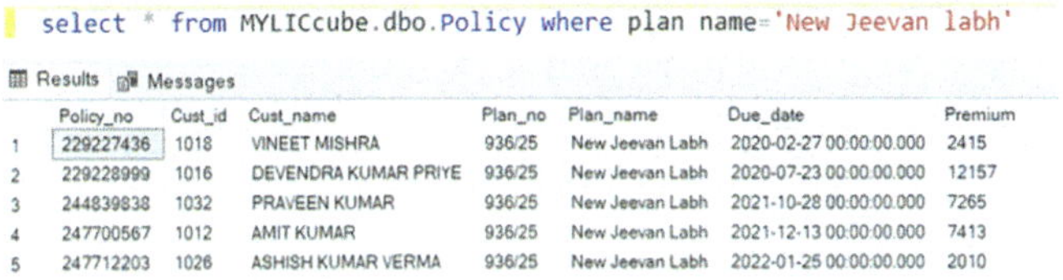

```
select * from MYLICcube.dbo.Policy where plan name='New Jeevan labh'
```

Results   Messages

	Policy_no	Cust_id	Cust_name	Plan_no	Plan_name	Due_date	Premium
1	229227436	1018	VINEET MISHRA	936/25	New Jeevan Labh	2020-02-27 00:00:00.000	2415
2	229228999	1016	DEVENDRA KUMAR PRIYE	936/25	New Jeevan Labh	2020-07-23 00:00:00.000	12157
3	244839838	1032	PRAVEEN KUMAR	936/25	New Jeevan Labh	2021-10-28 00:00:00.000	7265
4	247700567	1012	AMIT KUMAR	936/25	New Jeevan Labh	2021-12-13 00:00:00.000	7413
5	247712203	1026	ASHISH KUMAR VERMA	936/25	New Jeevan Labh	2022-01-25 00:00:00.000	2010

Sample Query 3

When the command is executed for the details of all the customer, then it is fetched from the customer table which is represented below:

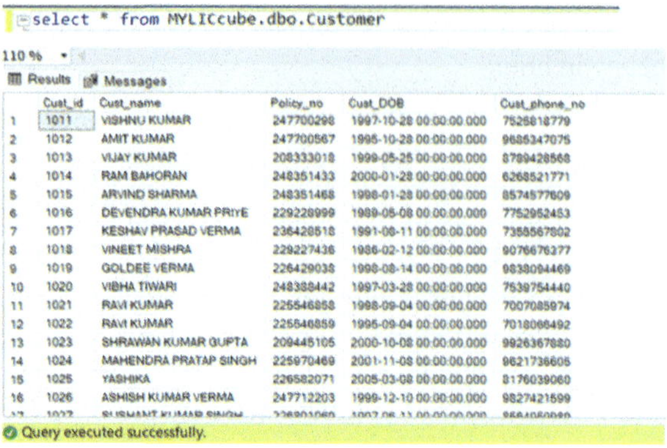

```
select * from MYLICcube.dbo.Customer
```

110 %

Results | Messages

	Cust_id	Cust_name	Policy_no	Cust_DOB	Cust_phone_no
1	1011	VISHNU KUMAR	247700298	1997-10-28 00:00:00.000	7526818779
2	1012	AMIT KUMAR	247700567	1995-10-28 00:00:00.000	9685347075
3	1013	VIJAY KUMAR	208333018	1999-05-25 00:00:00.000	8789428568
4	1014	RAM BAHORAN	248351433	2000-01-28 00:00:00.000	6268521771
5	1015	ARVIND SHARMA	248351468	1998-01-28 00:00:00.000	8574577609
6	1016	DEVENDRA KUMAR PRIYE	229228999	1989-05-08 00:00:00.000	7752952453
7	1017	KESHAV PRASAD VERMA	236428518	1991-08-11 00:00:00.000	7358567802
8	1018	VINEET MISHRA	229227436	1986-02-12 00:00:00.000	9076676377
9	1019	GOLDEE VERMA	226429038	1998-08-14 00:00:00.000	9838094469
10	1020	VIBHA TIWARI	248388442	1997-03-28 00:00:00.000	7539754440
11	1021	RAVI KUMAR	225546858	1998-09-04 00:00:00.000	7007085974
12	1022	RAVI KUMAR	225546859	1995-09-04 00:00:00.000	7018066492
13	1023	SHRAWAN KUMAR GUPTA	209445105	2000-10-08 00:00:00.000	9926367880
14	1024	MAHENDRA PRATAP SINGH	225970469	2001-11-08 00:00:00.000	9621736605
15	1025	YASHIKA	226582071	2005-03-08 00:00:00.000	8176039060
16	1026	ASHISH KUMAR VERMA	247712203	1999-12-10 00:00:00.000	9827421599

Query executed successfully.

Sample Query 4

When the command is executed for one record of the customer which is inserted into the customer table when browsing the data cube named MYLIC.cube, the resultant is shown below:

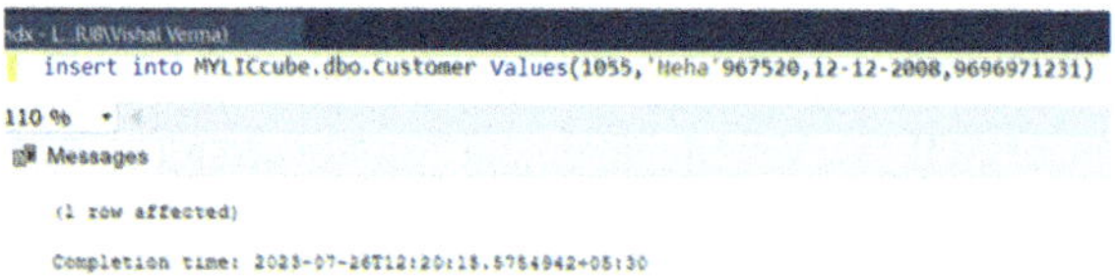

```
insert into MYLICcube.dbo.Customer Values(1055,'Neha'967520,12-12-2008,9696971231)
```

110 %

Messages

```
(1 row affected)

Completion time: 2023-07-26T12:20:15.5754942+05:30
```

For best interactions among the various fields of the database, a concept of correlation coefficient is used between the data set which may be positive, zero, or negative and used to define the relationship between two data sets. The formula is given below:

$$r = \frac{n(\sum xy) - (\sum x)(\sum y)}{\sqrt{\left[n\sum x^2 - (\sum x)^2\right]\left[n\sum y^2 - (\sum y)^2\right]}} \tag{1}$$

The range of correlation coefficient is depicted through r which lies between 0 and 1including both the values and r signifies that when r is 0 then two data sets have no relation, whereas the value 1 represents the perfect relationship between data sets. For the selected data, Cust_id and No_of_policies are taken as x and y, respectively, and the correlation coefficient (r) is computed as 0.437337, which shows the positive correlation and indicates that there is positive relationship between the policy and plan data sets issued to the customers by the company.

5 Conclusions

A platform-independent UML is used by the software industries for the development of the software designs. To tackle the large database, a system model is proposed through UML which has been validated through the concepts of various sample queries which are successfully executed and the relationship among the data set is valid. Further the present work may be extended for finding the hidden relationship among the large data set of LIC of India which can help to the customers for finding the relationships among the policy and the plan.

References

1. Pratheebha T, Indumathi V, Megala SS (2021) An empirical study on data mining techniques and its applications. Int J Softw Hardw Res Eng (IJSHRE) 9:23–31
2. Palacios CA, Reyes-Suárez JA, Bearzotti LA, Leiva V, Marchant C (2021) Knowledge discovery for higher education student retention based on data mining. Machine learning algorithms and case study in Chile. Entropy 23(4):1–23
3. Alshareef HN, Mazrashi A, Helal M, Tahir M (2021) Knowledge extraction and data visualization. A proposed framework for secure decision-making using data mining. Int J Adv Comput Sci Appl 12(8):483–489
4. Rutkowski L, Jaworski M, Duda P (2020) Stream data mining: algorithms and their probabilistic properties. Springer, Cham, p 330
5. Jiawei H, Pei J, Tong H (2022) Data mining concept and techniques. Morgan Kaufmann, p 752
6. OMG (1997) Unified modelling language specification. http://www.omg.org
7. OMG (1999) Unified modelling language specification (UML) version 1.3, June 1999
8. Booch G (2002) Growing the UML. Softw Syst Model 1(2):157–160
9. OMG (2003) Unified modelling language superstructure specification (UML 2.0)
10. Abdelhedi F, Brahim AA, Zurfluh G (2018) Formalizing the mapping of UML conceptual schemas to column-oriented databases. Int J Data Warehouse Min 14(3):44–68
11. Ciccozzi F, Malavolta I, Selic B (2019) Execution of UML models. A systematic review of research and practice. Springer, pp 2313–2360
12. Koc H, Erdoğan AM, Barjakly Y, Pekeret S (2021) UML diagrams in software engineering research. A systematic literature review. Multidiscip Digit Publ Inst Proc 74(1):1–13
13. Semenova E, Tynchenko V, Chashchina S, Suetin V, Stashkevich A (2022) Using UML to describe the development of software products using an object approach. In: IEEE international IOT, electronics and mechatronics conference (IEMTRONICS), pp 1–4
14. Booch G, Rambaugh J, Jacobson I (2005) The unified modelling language user guide. Pearson Education, p 496
15. Dennis A, Wixom BH, Tegarden D (2012) System analysis & design with UML version 2.0. An object-oriented approach, 4th edn. Wiley, Hoboken, p 253
16. Booch G, Rambaugh J, Jacobson I (2017) The Unified modelling language user guide. China Machine Press, Beijing, p 504
17. Foster EC, Towle BA Jr (2021) Software engineering: a methodical approach. CRC Press, Boca Raton
18. Shao C, Yang Y, Juneja S, Seetharam TG (2022) IoT data visualization for business intelligence in corporate finance. Inf Process Manage 59(1):1–14
19. Kumar N, Verma V, Saxena V (2012) Data cube representation for vehicle insurance policy system. Int J Comput Appl 975:8887

20. Lamani A, Erraha B, Elkyal M, Sair A (2019) Data mining techniques application for prediction in OLAP cube. Int J Electr Comput Eng (IJECE) 9(3):2094–2102
21. Djiroun R, Boukhalfa K, Alimazighi Z (2019) Data cubes retrieval and design in OLAP systems: from query analysis to visualization tool. Int J Bus Intell Data Min 14(1/2):267
22. Xie X, Zou K, Hao X, Pedersen TB, Jin P, Yang W (2020) OLAP over probabilistic data cubes II: parallel materialization and extended aggregates. IEEE Trans Knowl Data Eng 32(10)
23. Queiroz-Sousa PO, Salgado AC (2020) A review on OLAP technologies applied to information networks. ACM Trans Knowl Discov Data 14(1):1–25
24. Mathur S, Gupta SL, Pahwa P (2021) Optimizing OLAP cube for supporting business intelligence and forecasting in banking sector. J Inf Technol Manage 13(1):80026
25. Francia M, Marcel P, Peralta V, Rizzi S (2021) Enhancing cubes with models to describe multidimensional data. Inf Syst Front
26. Choudhary B, Saxena V (2021) Faster access of credit cards through data cube. technology. Webology 18(60). ISSN: 1735-188X

Handwritten Character Recognition for South Indian Languages Using Deep Learning

C. S. Arjun, N. Shobha Rani, and Akshatha Prabhu

Abstract Character recognition is a crucial component of many applications, including document processing, handwriting recognition, and optical character recognition (OCR). We present an OCR model-based character recognition workflow in this research that involves feature extraction from the input images, model comparison, and performance assessment. We compared the performance of four OCR models, including FAST R-CNN, Efficient Net B7, VGG16, and VGG19, on Kannada and Malayalam handwritten character datasets. The results showed that Efficient Net B7 outperformed the other models with the highest accuracy. We also analysed the performance of each model in terms of precision, recall, and $F1$-score. Our results showed that FAST R-CNN was second with accuracy of 94.55%, Efficient Net B7 was first with accuracy of 96.66%, VGG19 was third with accuracy of 92.23%, and VGG16 was last with accuracy of 91.38%. The workflow provides a structured approach for developing and evaluating OCR models for character recognition tasks in Indian scripts. The findings can be useful for various applications that require accurate and efficient character recognition. Future work could focus on exploring other deep learning architectures and datasets for character recognition in Indian scripts.

Keywords Malayalam · Deep learning · OCR · Fast R-CNN

C. S. Arjun · N. S. Rani (✉) · A. Prabhu
Department of Computer Science, School of Computing, Mysuru Campus, Amrita Vishwa Vidyapeetham, Mysuru, India
e-mail: n_shobharani@my.amrita.edu

A. Prabhu
e-mail: p_akshatha@my.amrita.edu

B. R. Devi et al. (eds.), *Proceedings of Fifth International Conference on Computer and Communication Technologies*, Lecture Notes in Networks and Systems 897,
https://doi.org/10.1007/978-981-99-9704-6_5

1 Introduction

Optical character recognition (OCR) is an essential technology that has revolutionized the way we interact with printed and handwritten documents. It involves using computer algorithms to analyse digital images and extract text, which can then be converted into machine-encoded text. OCR has numerous applications in various industries, including finance, health care, and legal, among others. However, while OCR technology has been successful in recognizing text from images in several languages, it has traditionally faced challenges in recognizing non-Latin scripts, such as Kannada and Malayalam. These languages use complex scripts with unique characters and diacritical marks, making it challenging for OCR software to accurately recognize them. This project aims to develop an OCR text recognition system that is specifically designed to recognize and extract text from images written in the Kannada and Malayalam languages. The system will utilize computer algorithms and deep learning techniques to accurately recognize and extract text from digital images, regardless of the font, size, or orientation of the text. The project is essential as it will enhance the accessibility and usability of documents written in Kannada and Malayalam, improving the efficiency and effectiveness of businesses and organizations that work with these languages. Additionally, it will provide a valuable tool for individuals who are visually impaired by allowing them to convert digital images into machine-encoded text, which can then be read aloud using text-to-speech technology. Overall, this project has the potential to significantly improve the accuracy and efficiency of OCR text recognition in Kannada and Malayalam languages, enabling more efficient and accessible processing of documents written in these scripts. Optical character recognition (OCR) is a technology that enables the recognition of printed or handwritten characters in an image and then converts them into machine-readable text. Deep learning-based OCR systems have achieved remarkable success in recent years due to their ability to learn and recognize complex patterns from large amounts of data. Kannada and Malayalam are two of the major languages spoken in India. Despite their significance, OCR for these languages has not been extensively explored compared to other languages like English or Hindi. This is mainly due to the complexity of the script and the limited availability of datasets. Motivation for developing an OCR system for Kannada and Malayalam is to provide a more accessible way for people who speak these languages to digitize and store their documents in a more efficient manner. This can have significant implications in various fields like education, research, and administration, where most of the documents are in the local language. Moreover, developing an OCR system for Kannada and Malayalam can contribute to preserving the cultural heritage of India by enabling the digitization and archiving of important historical documents and literature in these languages. Additionally, it can facilitate the translation of these documents into other languages, opening new opportunities for research and dissemination of knowledge. The related study for handwritten character recognition is mentioned in Sect. 2. In Sect. 3, a brief explanation of the proposed method is given. Section 4 provides the results of the experimentation. Sections 5 and 6 provide concluding remarks and ideas for further study.

2 Related Work

Feature extraction and feature categorization is accomplished using the edge hinge method, which is recommended in [1]. Its reliance on single-scale statistical data, in the author's opinion, is the main disadvantage of employing statistical feature extraction. Handwritten Telugu, Kannada, Malayalam, and Tamil texts are used to generate datasets from a variety of authors. Following that, a 300 dpi, grayscale scanner is used to scan the documents, and the results are stored as JPG files. For Tamil handwritten characters in [2], the recommended method has a 94.52% success rate. The Visual Geometry Group technique (VGG 16) is used for feature extraction. The author points out that this approach may be extended to distinguish Tamil words and to access data from ancient temple scripts kept in Bhrami, Vattezhuthu. Future text-to-speech conversion is potentially a possibility thanks to advancements in technology. Each of the 156 Tamil characters in the database has 100 samples. Each character has a folder made up of 100 samples, totalling 156 folders in the collection. The photographs are scaled into 224 $\times$ 224 using the VGG 16 method despite the fact that each image has a different size. Pasha and Padma [3]'s total accuracy for Kannada characters is 91.00%, and it is 97.60% for both. Artificial neural networks (ANN) are employed to read handwritten Kannada characters and numerals. There are a total of 4350 training samples. For identification, a total of 1450 testing samples are needed. There are a total of 4350 training samples. For identification, a total of 1450 testing samples are needed. The [4] character recognition accuracy is 95.59%. For character recognition, the advanced maximally stable extremal regions (AMSER) method is employed. Adding more geometric components, such as edges and blobs, may speed up recognition. The temple walls have 31 inscriptions dating to the reigns of Koperi Varman, Raja Raja I, Rajendra Chola I, and Kulothunga Chola. These inscriptions were captured on camera using a Canon EOS-600D digital SLR. The accuracy rates in [5] were 97.55%, 94%, 92.5%, and 95.2% for Kannada, Telugu, Tamil, and Malayalam numbers, respectively. Extracting zone-based hybrid features. For higher recognition rates, more classifiers must be upgraded. There have been gathered 2000 samples of Kannada numbers from 200 authors, 2000 Telugu numerals from 200 authors, 2000 Tamil numerals from 200 authors, and 2000 Malayalam numerals from 200 authors. The documents are scanned with the HPscan jet 5400c at a resolution of 300 dpi. In [6], it is 98% accurate. For character and pattern recognition, FAST R-CNN is used. By adding more convolutional layers, the accuracy rate can be enhanced. The dataset used is known as Chars74K, and it consists of a collection of images divided into 657+ categories. The dataset for each class, which comprises of 25 handwritten characters, is increased using a variety of augmentation techniques. OCR abandoned accuracy in [7]. The TL-7'T Henzalatha font by CDAC, sold in India, used as the teaching typeface. This recognition system does not include many Telugu literary and cultural works. The system's recognition precision in [8] is 82.04%. The support vector machine (SVM) method is used to distinguish Tamil characters. The system is trained using a total of 35,441 characters, created by 117 different users, and representing 106 different Tamil characters. It is possible to extract more features to increase discriminating power. LeNet and ResNet are the two character recognition models used in [9]. This system uses a comparison of these two models. Pre-processing, dataset creation, dataset augmentation, FAST R-CNN modelling, classification, and testing are the main steps in this method. Reading Malayalam characters that have been handwritten is difficult. In [10], text line segmentation is created from the input images of handwritten papers using the sliding window and skewing approaches. The segmentation technique gave an average DR, RA, and FM of 88%, 87%, and 88%, respectively. A line segmentation method is evaluated using 20 Tamil handwritten text document images. For 125 different classes obtained from the HP dataset, SVM was used in [11] to provide results that were 90.31% accurate, 87.26% accurate for the TDS dataset, and 89.83% accurate for the Thirukkural database. Datasets of handwritten Tamil characters were employed in the analysis from HP Labs India (HP Labs India, 2017; HP Labs India, 2013). A total of 56,250 samples (450 samples × 125 characters) and 169 writer variants were employed. For 125 character classes, there are 300 samples in the most recent dataset (HP Labs India, 2017), which comprises 117 textual variations. The author of [12] obtains a new state-of-the-art accuracy of 94.4% on the IWFHR-10 dataset. FAST R-CNN is used for feature extraction and recognition. The dataset contains roughly 500 samples for each of the 156 Tamil characters written by regional Tamil authors. Even though all of the datasets are supervised, it is interesting that the performance of our network degrades if even one convolutional layer is eliminated. Additionally, they may also experiment with unsupervised data. In [13], Fourier transforms are used in principal component analysis (PCA) for feature extraction and data processing. There are a total of 12,400 characters used across 200 unique authors. About 93.8% of the time was spent correctly recognizing objects. 95.1% accuracy was found for South Indian scripts. The author intends to expand on this work to include other Indian scripts. The proposed approach in [14] provides 94% accuracy. For the purpose of reading handwritten characters, SVM is utilized. The HCL2000 files feature Chinese characters and are derived from datasets from MNIST, CENPARMI, UCOM, and IAM. Using VGG NET 19, a system for classifying handwritten Kannada characters is proposed in [15]. The dataset consists of enormous data samples totalling around 92,000 with 46 classes, and training samples totalling about 31,654 of the handwritten Kannada character set. With validation, the accuracy was roughly 73.51%. The accuracy of the dataset could be improved by increasing its resolution and increasing the frequency of incidents of respect. The suggested procedure in [16] provides accuracy of 98.2%. The classification methodology is carried out in three stages: feature computation, text discrimination coefficient calculation, and decision model. 650 samples of handwritten text and 500 samples of printed text are included in the dataset. The datasets are produced by segmenting Telugu words in application documents that have already been printed. This study is applicable to printed text images with high levels of distortion and very low resolution. Using deep convolution neural networks, or Alex nets, the training of degraded patterns of character picture data is carried out in [17]. There are approximately 497 classes in the datasets. The total datasets include 10,400 data samples. For printed character samples, the accuracy rate is 87%, and for handwritten text, it is 80%. To increase accuracy, feature extraction can be improved. Hu-Moments features, statistical features, and feature classification into different groups are computed using the approach in [18]. For feature extraction, *K*NN and SVM are utilized. The accuracy rate of *K*NN is 76.47%. 212 samples make up the entire dataset. As potential future improvements, the author mentions many classifications. Back-propagation neural networks are used in [19] for this paper. A–Z typed characters in various sizes and types make up the dataset. From 90 to 94% identification rate, the accuracy varies as the number of cycles increases. Major changes in scale, rotation, or translation cannot be accommodated. Even though these variations can be taken into account by using a few pre-processing processes. Convolution neural networks are employed in [20] for the recognition of Arabic characters. Arabic handwritten characters totalling 16,800 are in the database. The recognition percentage for the test photos was 94.9% with accuracy. The author mentions future work that will focus on enhancing the performance of handwritten Arabic character recognition. [21] implements a handwritten character recognition system for Bangla that is based on FAST R-CNN. This study makes use of 20,000 handwritten characters in various shapes and iterations. Out of 2500 test instances, the proposed BHCR-FAST R-CNN misclassified 351 of them, with an accuracy rate of 85.96%. However, the approach incorrectly identified 954 characters out of the 17 500 training characters, resulting in an accuracy percentage of 94.55%. According to the results, FAST R-CNN training could be improved, leading to improved performance. The author of [22] suggests using a convolutional deep model to identify Bengali handwritten characters. Datasets from the BanglaLekha-isolated dataset are used to practise the recognition methods. It is 98.66% accurate for numerals (10 character classes), 94.99% accurate for vowels (11 character classes), 91.60% accurate for compound letters (20 character classes), 91.23% accurate for alphabets (50 character classes), and 89.93% accurate for almost all Bengali characters (80 character classes). By expanding the networks' capacity, even more distinguishing characteristics can be extracted. In [23], a feed-forward, back-propagation neural network for Gujarati handwritten digits is presented. For handwritten Gujarati digits, it has an accuracy rate of 81.66%. There are 2798 total photos used in this experiment, assuming 10 responses per respondent. Of these 278 sets of various digit graphics, 11 sets were produced using a common font. By strengthening feature abstraction methods and pre-processing methods, performance can be increased. Convolutional neural networks (FAST R-CNNs) with the best configuration were used in [24, 25] to conduct recognition. A novel self-adaptive lion algorithm (SALA), which is the conceptual advancement of the traditional lion algorithm (LA), is used to specifically fine-tune the fully linked layer and weights. The data is gathered from HP Labs in India. These examples consist of 169 folders and thousands of photos.

3 Proposed Method

The process for character recognition using OCR models is shown in Fig. 1. It incorporates a number of phases, including as feature extraction, model comparison, input picture pre-processing, and performance assessment. We will use input images from kaggle datasets made just for Malayalam handwritten characters and Kannada characters in this method. Pre-processing the input photographs as the first stage in the workflow is done to increase the uniformity and quality of the images. In order to do this, it is usual practise to use techniques like normalization, noise reduction, and picture enhancement. The workflow's subsequent phase, feature extraction, comes after pre-processing. In order to do this, pertinent traits must be gleaned from the input photographs that have been analysed. Character form, texture, and orientation extraction is the main goal of OCR models. Comparing several OCR models comes next after the characteristics have been gathered. We will assess the effectiveness of several OCR algorithms, such as FAST R-CNN, Efficient B7, VGG 16, and VGG 19, on our input photographs in this step. A section of the dataset will be used to train each model, and a different test set will be used to evaluate it. Performance evaluation is the workflow's last stage. We will use measures like accuracy, precision, recall, and $F1$-score to evaluate each OCR model's performance. These measures will offer a thorough analysis of each model's performance. We will ultimately choose the top-performing OCR model for our particular work based on the performance evaluation.

3.1 Dataset Description

Kaggle provides two separate datasets for character recognition tasks in the Kannada and Malayalam scripts: Kannada characters and Malayalam handwritten characters. The files consist of images of handwritten letters, each of which is labelled with the letter it stands for. The Kannada characters collection contains 1225 photos different Kannada characters. The images were acquired from a number of sources, including books and online Websites, and have the accompanying labels. The images are in grayscale and have a resolution of 1200 by 900. The Malayalam handwritten character dataset has 6090 pictures that represent 20 different Malayalam characters. The material from the sources, which include diaries, posters, and billboards, is annotated on the images. The photographs are colour and 300×300 pixels in size. Both datasets can be used for convolutional neural networks (FAST R-CNNs) and other deep learning techniques including character recognition model training and evaluation. It is important to keep in mind that not all Kannada and Malayalam characters are represented in these databases, making them unsuitable for all character recognition tasks. Overall, these datasets provide a useful starting point for developing and testing character recognition models for the Kannada and Malayalam scripts according to Fig. 2.

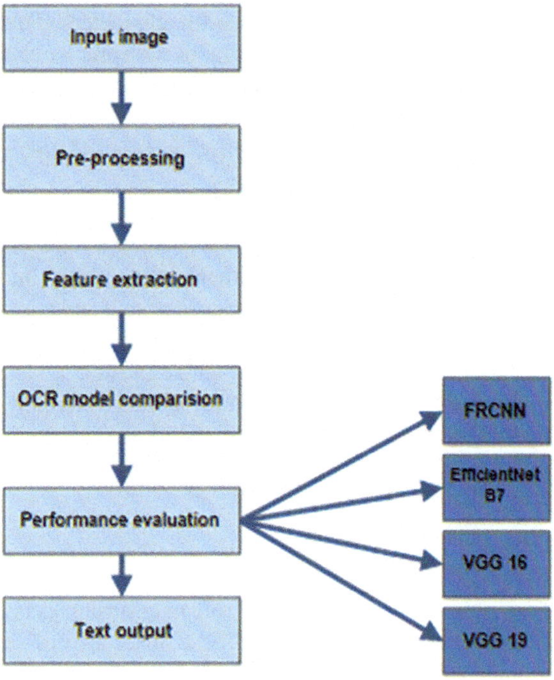

Fig. 1 Proposed architecture

Fig. 2 Sample dataset

3.2 Pre-processing

OCR's pre-processing technique uses segmentation to separate characters and thresholding to turn grayscale images into binary. These methods clean up the image and lessen noise for simpler character recognition. Pre-processing is essential for increasing the OCR models' efficiency and accuracy.

- Grayscaling: As it simplifies the image and improves recognition accuracy, grayscaling is an important step in OCR-based handwritten character recognition systems. Grayscaling reduces a colour image to grayscale, giving each pixel a single intensity value between black and white. This reduction makes the image easier to process and lowers computing complexity. Grayscaling also improves the contrast between the handwritten characters in the foreground and the background, which makes it easier to see the characters. This contrast improvement makes feature extraction easier and helps with precise character recognition. Grayscaling also serves to lessen the influence of noise and artefacts, increasing the accuracy of recognition. Overall, by preparing the image for future processing, grayscaling considerably improves the effectiveness and precision of OCR-based handwritten character recognition systems.

- Thresholding: It is a common method used in pre-processing images for character recognition using OCR models because it reduces noise and simplifies the image. By converting a grayscale image into a binary image based on a threshold value, we can discriminate between foreground and background pixels and highlight the areas of the image that are most crucial for character recognition. This makes it easier for OCR models to recognize and distinguish between individual characters. Because it is a computationally efficient method that can be used to a range of images, thresholding is a well-liked pre-processing technique for OCR.

3.3 Bounding Box

Bounding boxes play a critical role in ensuring accurate results in OCR-based handwritten character recognition. These boxes are utilized to localize and segment individual characters within an input image. Through techniques like contour detection, the system identifies regions of interest corresponding to distinct characters, which are then enclosed within rectangular bounding boxes. By serving as precise boundaries around each character, these boxes enable effective isolation from the background. This isolation enhances the system's ability to focus solely on the characters themselves, reducing the impact of surrounding noise or clutter. Once the characters are localized within their respective bounding boxes, subsequent processing steps such as feature extraction and classification can be performed on each character individually. This approach leads to precise and reliable character recognition. The bounding box information is crucial for preserving the spatial relationship between characters, facilitating accurate interpretation and transcription of handwritten text.

Figure 3 shows the pre-processing methods performed on the Kannada image dataset. This includes thresholding and segmentation for the sake of detecting the characters in that particular image.

Figure 4 shows the pre-processing methods performed on the Malayalam image dataset. This includes thresholding and segmentation for the sake of detecting the characters in that particular image.

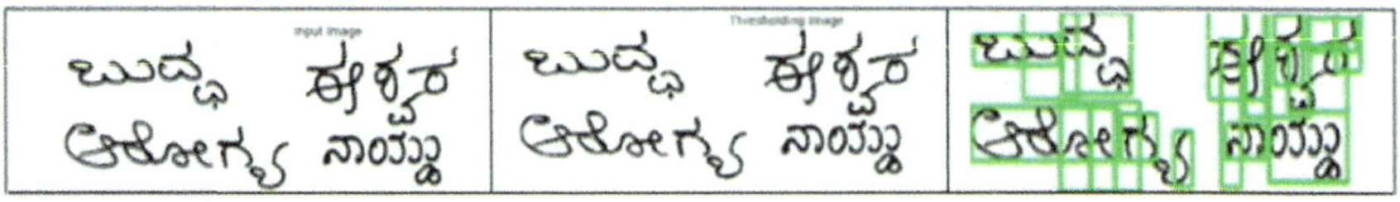

Fig. 3 Pre-processing on Kannada dataset

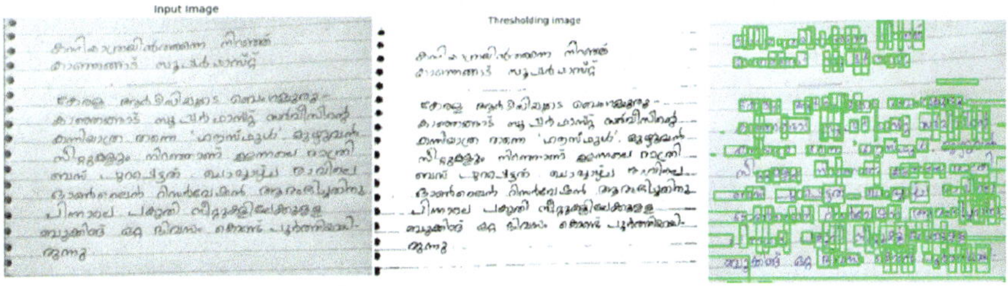

Fig. 4 Pre-processing on Malayalam dataset

3.4 Feature Extraction

Feature extraction for OCR-based handwritten character recognition, several key features are utilized. One important feature is glyph height, which captures the vertical extent of a character by measuring the distance between its highest and lowest points. This feature provides insights into the character's size and proportions. Glyph width is essential as it determines the horizontal extent by measuring the distance between the character's leftmost and rightmost points.

- Fundamental characteristics utilized in OCR-based handwritten character recognition to describe a character's size and dimensions are glyph height and glyph width. By measuring the distance between a character's highest and lowest points, a glyph's height can be calculated. It offers important details regarding the character's overall dimensions. However, glyph width measures the space between a character's leftmost and rightmost points to calculate the character's horizontal extent. The character's breadth and shape can be understood thanks to this function. OCR systems can efficiently capture the spatial characteristics of characters by taking into account glyph height and glyph width. This enables accurate recognition and discrimination between various handwritten characters.
- Skewness is another significant feature used to quantify the asymmetry of the curves or edges within a character. It assesses the distribution of these curves and provides positive or negative skewness values. Positive skewness indicates a longer or heavier tail on the right side, while negative skewness suggests a longer or heavier tail on the left side. Skewness aids in identifying any slant or tilt present in the character, thereby assisting in accurate recognition.
- Kurtosis is employed to measure the peakedness or flatness of the curve or edge distribution within a character. By examining the shape of the distribution, positive or negative kurtosis values are obtained. Positive kurtosis indicates a relatively sharp peak, while negative kurtosis suggests a flatter distribution. This feature helps in characterizing the overall shape and concentration of the curves or edges within a character.
- Mean, median, and standard deviation are statistical features used to analyse the curves and edges of each character. The mean provides the average value of the curves or edges, indicating their central tendency. The median represents the

middle value and is less affected by extreme values. These features offer insights into the typical positioning or orientation of the curves within the character. Additionally, standard deviation measures the variability or dispersion of the curves or edges, reflecting the complexity and intricacy of the character.

Implementing these feature extraction techniques enables an OCR system to effectively capture and represent important characteristics of handwritten characters. These features serve as valuable inputs for subsequent classification algorithms or machine learning models. By considering the size, shape, orientation, and complexity of characters, these features contribute to accurate recognition and discrimination among different characters.

3.5 Model Comparison

A critical phase in the development of character recognition using OCR models is OCR model comparison. To compare the performance of various OCR models on the same set of input images and identify the model that performs the best, this is done. Convolutional neural networks (FAST R-CNN), Efficient Net B7, VGG 16, and VGG 19 are common OCR models for character recognition. Each model is trained on a portion of the dataset before being compared, and its performance is assessed using a different test set. To decide which model is most appropriate for a given character recognition job, model performance is often assessed using measures like accuracy, precision, recall, and $F1$-score.

3.6 Perform Evaluation

- FAST R-CNN: FRCNN-based feature extraction finds and extracts traits for character recognition utilizing OCR models using input photographs that have already been processed. This method uses a set of convolutional filters, referred to as masks, to make use of the edges, corners, and textures of the input image. The masks, which are developed to recognize specific patterns in the image, convert the image into a set of feature maps that may be used as input to the OCR model. It has been proven that implementing this method will boost recognition precision while bringing down the OCR model's computational complexity.
- VGG16: The three hidden layers of the VGG16 model—32-, 62-, and 128-bit layers—have demonstrated success in character recognition challenges. In terms of performance evaluation, VGG16 has been contrasted against several OCR models using a range of evaluation measures. It has demonstrated competitive performance on handwritten character identification tasks using the Kannada and Malayalam scripts. The model was enhanced on smaller datasets to yield even better results when it was applied to transfer learning. In general, the VGG16 OCR

Table 1 Classification report for all models

Model	Precision	Recall	$F1$-score	Accuracy (%)
VGG16	1.00	0.84	0.91	91.19
	0.84	1.00	0.91	
VGG19	1.00	0.85	0.92	92.90
	0.85	1.00	0.92	
Fast RCNN	1.00	0.90	0.94	94.58
	0.90	1.00	0.95	
EfficientNet	1.00	0.93	0.97	96.30
	0.93	1.00	0.97	

model is reliable, effective, and suitable for a wide range of character recognition applications.

- VGG19: In order to evaluate the performance of the VGG19 model for character recognition, we trained its three hidden layers with 32-, 62-, and 128-bit layers. The model's findings were encouraging, showing how useful this architecture is for jobs requiring Kannada and Malayalam character identification. As with any OCR model, additional study and optimization may be needed to improve its performance in specific application cases. Combining the VGG19 model and the recommended architecture provides a strong framework for creating accurate and durable OCR systems for tasks requiring character recognition in the Kannada and Malayalam scripts.
- Efficient Net B7: The architecture of a modern neural network Several computer vision tasks have seen encouraging results using Efficient Net B7. We trained the Efficient Net B7 model using four kernels of varied sizes to extract characteristics from our input photographs for our OCR model comparison. Precision, recall, and $F1$-score are just a few of the parameters we used to evaluate the model's performance. Our results showed that the Efficient Net B7 model with four kernels outperformed other models in terms of its accuracy in Kannada and Malayalam character recognition.

Table 1 presents the performance metrics, including precision, recall, and $F1$-score, for class 1 and class 0, with accuracy achieved by deep learning models.

4 Results and Discussions

In our research, we evaluated the effectiveness of VGG16, VGG19, FAST R-CNN, and Efficient Net B7 on handwritten character datasets in Kannada and Malayalam. The percentage of successfully detected characters is captured by the accuracy statistic we utilized for evaluation. Our results showed that FAST R-CNN was second with accuracy of 94.55%, Efficient Net B7 was first with accuracy of 96.66%, VGG19 was third with accuracy of 92.23%, and VGG16 was last with accuracy of 91.38%. The ability of Efficient Net B7 to efficiently learn and extract relevant information from the input images using its state-of-the-art architecture and various kernel sizes is what allows it to achieve the highest degree of accuracy. The process begins by passing the input image through deep learning models designed specifically for handwritten character recognition. These models are designed to examine the image and determine which characters are shown there. Based on their exposure to training data, the models learn to recognize various handwritten characters using techniques including feature extraction, pattern recognition, and classification. The models use their acquired knowledge and algorithms to create predictions about the characters in the image during the recognition process. The models produce output in the form of recognized texts after the recognition process is finished. The handwritten characters of the input image are represented by these recognized texts. The recognized texts are often saved in a separate text file, frequently with a.txt extension, to aid in further processing and accessibility. Each recognized letter or word is saved in the text file as a series of characters which is depicted in Fig. 5. Easy access, post-processing, analysis, and system integration are all made possible by this storage type. It is possible to do post-processing tasks like natural language processing, data indexing, and information retrieval with ease by saving the recognized texts in a separate text file. Additionally, the accessibility, searchability, and compatibility with different digital systems and apps are improved by converting handwritten writings into a digital version.

Figure 6 illustrates the recognition process of a Malayalam test image, which includes a noisy background. Despite the challenging conditions, the system demonstrates the capability to successfully recognize a majority of the characters from the input image. This achievement highlights the effectiveness of the recognition algorithm in handling noisy backgrounds and accurately identifying the characters present in the image.

- Performance evaluation graph: The graph displayed below illustrates the fluctuations in precision, recall, and $F1$-score values corresponding to different deep learning models. This representation provides insights into the performance variations observed across the models. Precision, recall, and $F1$-score are commonly used metrics to evaluate the effectiveness of classification models, especially in scenarios involving imbalanced data or when the cost of false positives and false negatives differ (Fig. 7).

a b

Fig. 5 **a** Input image, **b** Output image of Kannada script

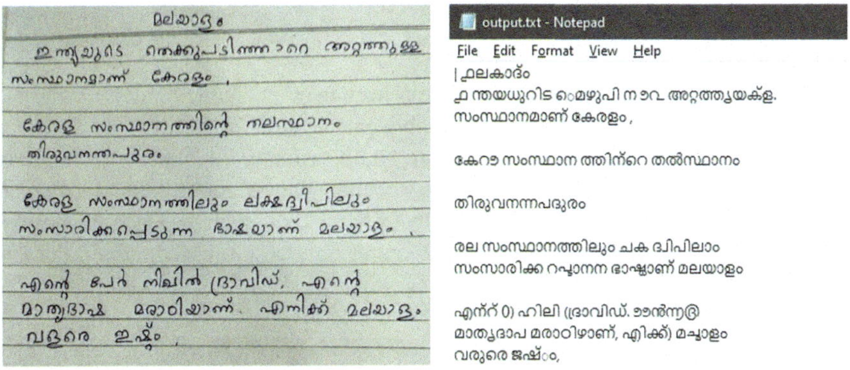

a b

Fig. 6 **a** Input image, **b** Output image of Malayalam script

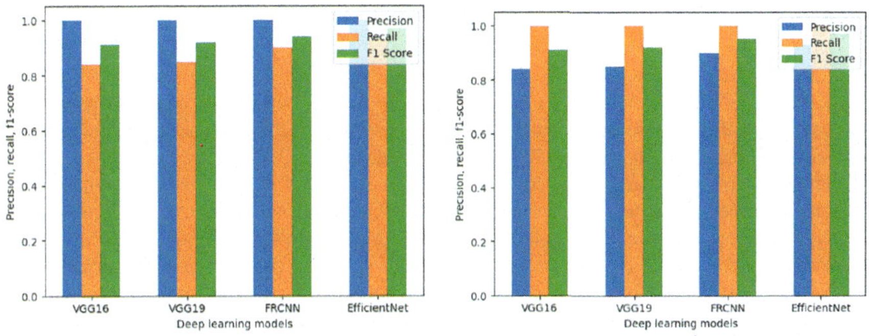

a b

Fig. 7 Precision, recall, $F1$-score values for our proposed deep learning models **a** class 0, **b** class 1

5 Conclusion

We have examined a character recognition method utilising OCR model that involves pre-processing, feature extraction, model comparison, and performance evaluation. We used this process to the Kaggle datasets for Kannada characters and Malayalam handwritten characters using four different OCR models—VGG16, VGG19, FAST R-CNN, and Efficient Net B7. Our results showed that Efficient Net B7 outperformed the other models, with a recognition accuracy of 96.66%. This is probably due to its complex architecture and utilization of several kernels. Although FAST R-CNN had an accuracy of 94.55%, VGG19 had an accuracy of 92.23%, and VGG16 had an accuracy of 91.38%, we also found that all four models performed fairly well.

6 Future Work

There are several ways that OCR models' performance for character identification in Kannada and Malayalam can be considerably enhanced in the future. One approach that might be employed is to look at more complex feature extraction techniques, such as using deep learning models to acquire more nuanced representations of character qualities. The investigation of various OCR models, such as ensemble models that combine the benefits of diverse architectures or attention-based models, should also be a primary goal of future study. Investigating the use of data augmentation techniques may be helpful to increase the size and diversity of the training dataset. Finally, there may be opportunities to apply OCR models to other related tasks, such as handwritten text recognition or document digitization, which could have important uses.

References

1. Aravinda CV, Reddy KUK, Meng L, Prabhu GA (2020) South Indian character recognition using statistical feature extraction and distance classifier. In: 2020 international conference on advanced mechatronic systems (ICAMechS). IEEE, pp 302–307
2. Pragathi MA, Priyadarshini K, Saveetha S, Banu AS, Aarif KM (2019) Handwritten Tamil character recognition using deep learning. In: 2019 international conference on vision towards emerging trends in communication and networking (ViTECoN). IEEE, pp 1–5
3. Pasha S, Padma MC (2015) Handwritten Kannada character recognition using wavelet transform and structural features. In: 2015 international conference on emerging research in electronics, computer science and technology (ICERECT). IEEE, pp 346–351
4. Kumar AN, Geetha G (2019) Character recognition of ancient South Indian language with conversion of modern language and translation. Caribb J Sci 53(20):2019–2031
5. Rajashekararadhya SV, Ranjan PV (2009) Zone based hybrid feature extraction algorithm for handwritten numeral recognition of South Indian scripts. In: International conference on contemporary computing. Springer, Berlin, Heidelberg, pp 138–148
6. Asha K, Krishnappa HK (2018) Kannada handwritten document recognition using convolutional neural network. In: 2018 3rd international conference on computational systems and information technology for sustainable solutions (CSITSS). IEEE, pp 299–301
7. Negi A, Bhagvati C, Krishna B (2001) An OCR system for Telugu. In: Proceedings of sixth international conference on document analysis and recognition. IEEE, pp 1110–1114
8. Shanthi N, Duraiswamy K (2010) A novel SVM-based handwritten Tamil character recognition system. Pattern Anal Appl 13(2):173–180
9. Pius NK, Johny A (2020) Malayalam handwritten character recognition system using convolutional neural network. Int J Appl Eng Res 15(9):918–920
10. Dixit S, Suresh HN (2013) South Indian Tamil language handwritten document text line segmentation technique with aid of sliding window and skewing operations. J Theoret Appl Inf Technol 58(2)
11. Raj MAR, Abirami S, Shyni SM (2023) Tamil handwritten character recognition system using statistical algorithmic approaches. Comput Speech Lang 78:101448
12. Vijayaraghavan P, Sra M (2014) Handwritten Tamil recognition using a convolutional neural network. In: 2018 international conference on information, communication, engineering and technology (ICICET), pp 1–4

13. Aradhya VM, Kumar GH, Noushath S (2008) Multilingual OCR system for South Indian scripts and English documents: an approach based on Fourier transform and principal component analysis. Eng Appl Artif Intell 21(4):658–668
14. Hamdan YB (2021) Construction of statistical SVM based recognition model for handwritten character recognition. J Inf Technol 3(02):92–107
15. Rani NS, Subramani AC, Kumar A, Pushpa BR (2020) Deep learning network architecture based Kannada handwritten character recognition. In: 2020 second international conference on inventive research in computing applications (ICIRCA). IEEE, pp 213–220
16. Rani NS, Vasudev T (2017) An unsupervised classification of printed and handwritten Telugu words in pre-printed documents using text discrimination coefficient. In: Proceedings of the first international conference on computational intelligence and informatics. Springer, Singapore, pp 689–700
17. Rani NS, Chandan N, Jain AS, Kiran HR (2018) Deformed character recognition using convolutional neural networks. Int J Eng Technol 7(3):1599–1604
18. Rani NS, Verma SK, Joseph A (2016) A zone based approach for classification and recognition of Telugu handwritten characters. Int J Electr Comput Eng 6(4):1647
19. Patel CI, Patel R, Patel P (2011) Handwritten character recognition using neural network. Int J Sci Eng Res 2(5):1–6
20. El-Sawy A, Loey M, El-Bakry H (2017) Arabic handwritten characters recognition using convolutional neural network. WSEAS Trans Comput Res 5(1):11–19
21. Rahman MM, Akhand MAH, Islam S, Shill PC, Rahman MH (2015) Bangla handwritten character recognition using convolutional neural network. Int J Image Graph Sig Process (IJIGSP) 7(8):42–49
22. Purkaystha B, Datta T, Islam MS (2017) Bengali handwritten character recognition using deep convolutional neural network. In: 2017 20th international conference of computer and information technology (ICCIT). IEEE, pp 1–5
23. Desai AA (2010) Gujarati handwritten numeral optical character reorganization through neural network. Pattern Recogn 43(7):2582–2589
24. Lincy RB, Gayathri R (2021) Optimally configured convolutional neural network for Tamil handwritten character recognition by improved lion optimization model. Multimed Tools Appl 80(4):5917–5943
25. Benaddy M, El Meslouhi O, Es-saady Y, Kardouchi M (2019) Handwritten tifinagh characters recognition using deep convolutional neural networks. Sens Imaging 20(1):1–17

CavFind: A Novel Algorithm to Detect Cavities on Protein Structure

Swati Adhikari and Parthajit Roy

Abstract Various essential functions in living organisms are performed by binding of proteins with other molecules (ligands). Proper detection and analysis of ligand binding locations (cavities) leads towards the success of the overall drug designing system. The main challenge towards that goal is that the problem is computationally hard. In the present study, a soft computing-based algorithm has been proposed that is capable of detecting cavities on protein's structure. The proposed algorithm is based on Voronoi decomposition and implemented by applying the self-organizing map (SOM) clustering algorithm. The proposed algorithm is evaluated with 48 protein–ligand complexes and compared with two existing algorithms like Fpocket and ConCavity and compared with the results stored in two existing databases like BioLiP and CaviDB. The proposed algorithm is also capable of computing some descriptors of the detected cavities.

Keywords CavFind · Protein–ligand interaction · Ligand binding site · Different forms of binding sites · Cavity/pocket detection methods

1 Introduction

Recognition of different types of cavities in the structures of biomolecules is a very critical task, and such type of activities are very much essential in the context of drug designing. Conventional cavity detection methods are very expensive and time consuming. These are made time and cost effective by the application of computer-aided techniques. Nearly all computer-aided models that detect cavities are categorized as three types—geometry-based, energy-based, and evolutionary-based [1]. Among these methods, the first one is quite time and space consuming, e.g. Fpocket

S. Adhikari · P. Roy (✉)
Department of Computer Science, The University of Burdwan, Golapbag, Purba Bardhaman, West Bengal 713104, India
e-mail: proy@cs.buruniv.ac.in; roy.parthajit@gmail.com

B. R. Devi et al. (eds.), *Proceedings of Fifth International Conference on Computer and Communication Technologies*, Lecture Notes in Networks and Systems 897,
https://doi.org/10.1007/978-981-99-9704-6_6

[2] and CavVis [3]. The second one is dependent on the techniques of filtering procedures, force field parameterizations and scoring functions [1]. Some of them are [4, 5]. On the other hand, the third one depends on multiple sequence alignments to locate the cavities [1]. The standard of sequence alignment tools and the dataset size also affects the performance of these methods, e.g. GarLig [6].

There are some modern approaches towards development of cavity detection methods which integrates different tools of artificial intelligence (AI) with the basic three cavity detection methods. The existing in-silico models are grouped as four types based on their 3D structures, similarity between sequence and structure templates, machine learning and deep learning algorithms that are used [7]. All have their respective merits and demerits. The main drawbacks of 3D structure-based methods are that some of these methods are time and space consuming as well as fails to make any prediction when it is not possible to have a 3D structure. The major drawback of template-based methods is that the success with these methods is solely dependent on the performance of the alignment tool that is being used, and these methods are unable to provide any information regarding area, volume and shape relating to the predicted sites.

To overcome these shortcomings, research is going on to improve the efficiency and accuracy of the existing models by integrating machine learning-based methods like support vector machine, clustering algorithms, etc., and deep learning-based methods like convolutional neural networks with the methods based on 3D structure and sequence and structure templates in order to yield output regarding prediction of binding sites (location of cavities) and analysis of predicted sites with higher accuracy and improved efficiency. Some of the existing algorithms that come under this category are [8–11]. Some of the latest works are found in developing databases that contain probable ligand binding sites such as BioLiP [12], CaviDB [13] and CavitySpace [14]. Here, BioLiP is a database for prediction of biologically relevant ligand–protein interactions for different PDB structures. CaviDB is a database for cavities along with their features for proteins whose structures are known. CavitySpace is also a database for ligand binding sites in the proteome of Human.

In the present study, we propose an algorithm that utilizes the advantages of geometric and clustering techniques. Here, at first Voronoi decomposition is performed on the protein's structure, and next the neural network-based method SOM has been applied to group the Voronoi vertices in an attempt to detect the cavities. Some descriptors of the detected cavities are also calculated. The proposed algorithm has been tested with the dataset that is used by the Fpocket algorithm.

2 Materials and Methods

In this study, the proposed algorithm is designed in order to find all sorts of cavities on protein's structure by amalgamation of geometry-based technique, Voronoi Tessellation, with that of machine learning-based technique, clustering. Our aim is to find some empty locations on protein structure, surrounded by protein atoms. To reach that goal, the proposed algorithm is implemented by the following four phases:

Phase A—*Voronoi decomposition of the protein's space*: The proposed algorithm employs the notion of alpha sphere theory. A sphere that contacts at least four atoms at its circumference and does not contain any atom inside it is termed as an alpha sphere. Voronoi decomposition of a space leads to formation of alpha spheres with Voronoi vertices as their centres and the sphere atoms determine the binding atoms that are involved in formation of the cavity.

Phase B—*Clustering of alpha spheres*: The concept behind the detection of cavity binding atoms is that when the alpha spheres are clustered, we shall get probable groups of cavity atoms, as the interior of the alpha spheres are empty. Although any clustering technique can be used to group alpha spheres/cavity atoms, proper selection is necessary. Otherwise, it may be the case that after clustering, we shall have groups of alpha spheres/cavity atoms with very small sizes, and some other techniques or clustering steps will be required to merge those groups in order to detect cavities with larger sizes as like it is the case for the Fpocket software [2]. In Fpocket, three clustering steps have been used to group alpha spheres.

In this paper, it is proposed to apply the SOM clustering algorithm to find probable groups of cavity atoms. The main advantage of using the SOM algorithm over other clustering algorithms like K-means and hierarchical clustering algorithms is that the SOM is capable of handling large numbers of input data very efficiently. This phenomenon of the SOM helps in processing a huge number of alpha spheres as there will be several thousands of alpha spheres generated after Voronoi decomposition of protein atoms.

For the SOM clustering algorithm, it is required that the number of SOM prototypes be defined well in advance. Like Tian et al. [15], in the proposed algorithm also, the number of SOM prototypes is defined as $5\sqrt{N}$, where N is the size of the dataset.

Phase C—*Determination of active site*: Active site is one in which ligand atoms are bound with the cavity atoms. Some cavities are also detected on the protein structure in which no ligand is bound. Two criteria are used to determine whether a cavity that is detected by CavFind as binding site is an active site or not, as these criteria are used in PocketPicker [16] and Fpocket algorithm [2] for evaluation purpose. As per these criteria, an active site is one in which (1) the distance between the centroid of any cavity and any ligand's atom is within 4 Å and (2) at least 50% of the ligand atoms reside within 3 Å of at least one alpha sphere. All the ligands are checked to have corresponding binding sites. This phase finally identifies the active sites among the list of cavities (obtained as SOM prototypes) at different places of the protein's surface.

Phase D—*Calculation of list of descriptors*: This phase is used to calculate descriptors of the detected cavities. Some of the descriptors, as defined and explained in Fpocket [2], are calculated for each of the detected cavities which are (1) number of alpha spheres, (2) density of the cavity, (3) polarity, volume and charge score of the cavity and (4) mean local hydrophobic density.

The proposed algorithm is named as CavFind.

Flow chart of the CavFind algorithm: This is shown in Fig. 1 and is depicted below.

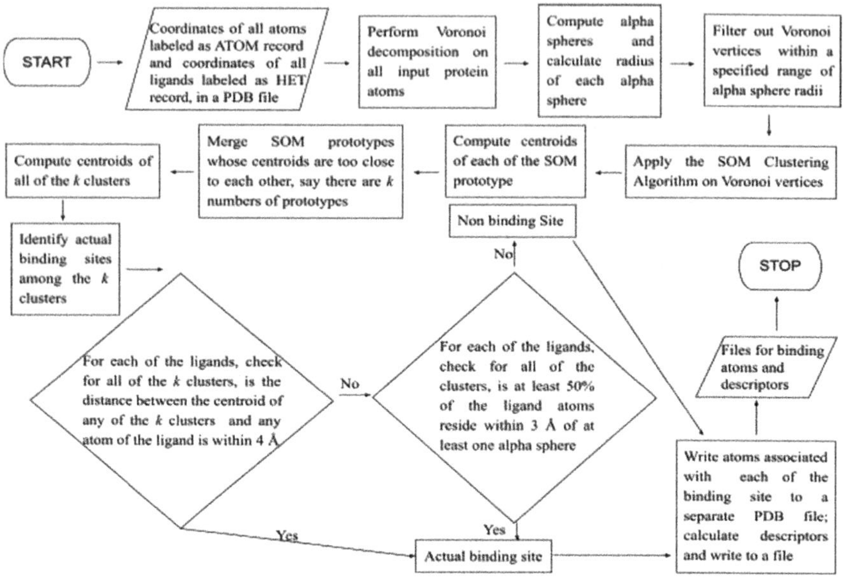

Fig. 1 Flow chart of the CavFind algorithm

Dataset: To test the performance of the CavFind algorithm, we have used the dataset of 48 protein–ligand complexes [16] which have 48 bound and unbound conformations. Here, all structures have been obtained from the database of RCSB PDB [17].

3 Results and Discussion

This section analyses the results obtained after the application of the proposed algorithm on the test dataset. Table 1 shows different statistics regarding successful prediction of cavities on some of the test structures.

From Table 1, it is seen that in one structure, the active sites for not all the ligands are properly identified (highlighted in italics) by the proposed algorithm. In one structure, one actual binding site is shared by two ligands, as these ligands reside very close to each other (highlighted in bold). Figure 2a depicts the same. Here, the red site is shared by two ligands which are too close to each other.

Table 1 Statistics regarding successful prediction of cavities on test structures by using the CavFind algorithm

Struct	No. of ligs	Ligs in active site (Resn_Ch Id_Res Id)	Non detected binding sites of ligs (Resn_Ch Id_Res Id)	No. of AS	No. of sites which are not AS
1snc	2	CA_A_150; THP_A_151		2	9
1stp	1	BTN_A_300		1	8
2ctc	2	ZN_A_308; HFA_A_309		2	24
2h4n	**3**	**ZN_A_262; ZN_A_263; AZM_A_264**		**2**	**16**
2pk4	1	ACA_A_100		1	1
3gch	1	OAC_C_246		1	19
3mth	*5*	*MPB_A_200; CL_B_32; ZN_B_31*	*ZN_D_31; CL_D_32*	*3*	*12*
3ptb	2	CA_A_480; BEN_A_1		2	18

Struct Structure; *Lig* Ligand; *AS* Active site

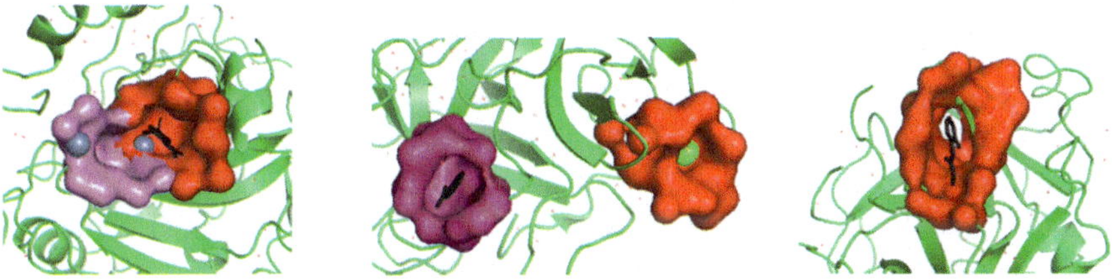

(a) Structure 2h4n (b) Structure 3ptb (c) Structure 3gch

Fig. 2 View of **a** sharing of one binding site by two ligands; **b** clefts and **c** channel

Now let us check whether the proposed CavFind algorithm can detect all types of cavities. Figure 2b [structure 3ptb] shows two binding sites that resemble clefts. The structures 3gch (Fig. 2c) and 1acj (Fig. 3a) are considered for examples of the channel and tunnel, respectively. Here, it is to be noted that no void has been found in the test dataset. In the present study, all figures are created in PyMOL [18].

From Table 1, it is also seen that in every structure some cavities are detected which are not active sites. For example, the structures 2pk4 (Fig. 3b) and 1stp (Fig. 3b) show presence of one and eight such cavities, respectively.

Comparative Study of the Proposed Algorithm: CavFind method is evaluated by computing three extrinsic clustering validity indices, namely precision, recall and accuracy. The capability of the proposed model in predicting the number of cavities

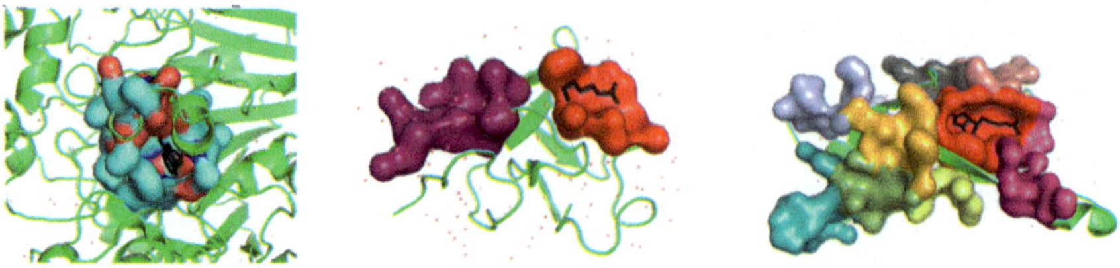

(a) Structure 1acj (b) Structure 2pk4 (c) Structure 1stp

Fig. 3 Views of tunnel and cavities in structures where no ligands associated with them

for the structures of the test dataset is compared with the number of cavities predicted with the Fpocket [2], ConCavity [8], BioLiP database [12] and CaviDB database [13].

The binding sites detected with the MetBP software [19] that detects binding sites for metals and nucleic acids and finds base pairing between them are considered as ground truth information for computation of the precision, recall and accuracy. The number of sites that are detected with the MetBP is considered as positive; otherwise, it is considered as negative. Similarly, the numbers of sites that are predicted by the Fpocket, ConCavity and the BioLiP as binding sites are considered as positives and the numbers of sites that are not predicted by the aforesaid two softwares and the database as binding sites are considered as negatives. Based on these positives and negatives values, these three validity indices of the some of the test structures are computed. Table 2 shows the same.

It is seen from Table 2 that the maximum of metals and nucleic acids binding sites are detected with full precision and full recall values with 100% accuracy by the proposed model, that means, all metals and nucleic acids binding sites of the test structures have been detected where accuracy is 100% which is nearer to the number that is detected with the BioLiP database, Fpocket and ConCavity.

In Table 2, division with zero appears when none of the sites is predicted by the respective models as binding sites. From these two tables, it is clear that the proposed model outperforms these existing tools in predicting cavities on protein's structures.

Table 2 Comparative study of the proposed model, Fpocket, ConCavity and BioLiP

Struct	SOM			BioLiP			ConCavity			Fpocket		
	Pr	Re	Ac	Pr	Re	Ac	Pr	Re	Ac	Pr	Re	Ac
1fbp	1	1	100	1	1	100	1	1	100	1	0.75	75
1gca	1	1	100	1	1	100	Div 0	0	0	1	1	100
1hfc	1	1	100	1	1	100	1	0.33	33	1	0.33	33
2ctc	1	1	100	1	1	100	1	1	100	1	1	100
3ptb	1	1	100	1	1	100	Div 0	0	0	Div 0	0	0
5p2p	1	1	100	0.75	1	75	1	1	100	Div 0	0	0
7cpa	1	0.67	67	1	1	100	1	1	100	1	1	100

Struct Structure; *Pr* Precision; *Re* Recall; *Ac* Accuracy (in %); *Div 0* Division by 0

We have also compared the number of sites detected with the proposed CavFind algorithm with that of the number of sites detected with the online CaviDB database. CaviDB outputs only total number of cavities present in a structure; it does not specify the associated cavity atoms. On the other hand, CavFind outputs associated cavity atoms and the ligand associated with a detected cavity along with total number of cavities.

4 Conclusion

The CavFind algorithm when compared with the two existing algorithms Fpocket and ConCavity shows that this algorithm eliminates the need for integration of multiple clustering algorithms with the result of Voronoi decomposition and no evolutionary information is also required in its implementation. Comparison of the same with the two existing databases BioLiP and CaviDB shows that the output generated with the CavFind algorithm is compatible with the output stored in these two databases. All tables and figures support these facts. So, it can be concluded that the CavFind algorithm can detect all sorts of cavities like cleft, tunnel, channel, etc.

Individual .pdb formatted files are also generated for each cavity that helps in identifying cavity atoms and to visualize the cavities. Different text files also provide some useful information. These files may become helpful for the researchers in this field.

The proposed algorithm may be further improved by applying a genetic algorithm on the output of this algorithm or by application of some other clustering techniques and by incorporating provision for scoring and ranking of the detected cavities.

Acknowledgements Authors are thankful to The University of Burdwan for their support and infrastructure. Authors are also thankful to the RCSB PDB for their online dataset.

References

1. Simões T, Lopes D, Dias S, Fernandes F, Pereira J, Jorge J, Bajaj C, Gomes A (2017) Geometric detection algorithms for cavities on protein surfaces in molecular graphics: a survey. Comput Graph Forum 36(8):643–683
2. Guilloux VL, Schmidtke P, Tuffery P (2009) Fpocket: an open source platform for ligand pocket detection. BMC Bioinform 10(1):168
3. Simões TMC, Gomes AJP (2019) CavVis—a field-of-view geometric algorithm for protein cavity detection. J Chem Inf Model 59(2):786–796
4. Ngan C-H, Hall DR, Zerbe B, Grove LE, Kozakov D, Vajda S (2011) FTSite: high accuracy detection of ligand binding sites on unbound protein structures. Bioinformatics 28(2):286–287
5. Lin Y, Yoo S, Sanchez R (2012) SiteComp: a server for ligand binding site analysis in protein structures. Bioinformatics 28(8):1172–1173

6. Pfeffer P, Fober T, Hüllermeier E, Klebe G (2010) GarLig: a fully automated tool for subset selection of large fragment spaces via a self-adaptive genetic algorithm. J Chem Inf Model 50(9):1644–1659
7. Zhao J, Cao Y, Zhang L (2020) Exploring the computational methods for protein-ligand binding site prediction. Comput Struct Biotechnol J 18:417–426
8. Capra JA, Laskowski RA, Thornton JM, Singh M, Funkhouser TA (2009) Predicting protein ligand binding sites by combining evolutionary sequence conservation and 3D structure. PLoS Comput Biol 5(12):e1000585
9. Jiménez J, Doerr S, Martínez-Rosell G, Rose AS, Fabritiis GD (2017) DeepSite: protein-binding site predictor using 3D-convolutional neural networks. Bioinformatics 33(19):3036–3042
10. Pu L, Govindaraj RG, Lemoine JM, Wu H-C, Brylinski M (2019) DeepDrug3D: classification of ligand-binding pockets in proteins with a convolutional neural network. PLoS Comput Biol 15(2):e1006718
11. Alipanahi B, Delong A, Weirauch MT, Frey BJ (2015) Predicting the sequence specificities of DNA- and RNA-binding proteins by deep learning. Nat Biotechnol 33(8):831–838
12. Yang J, Roy A, Zhang Y (2013) BioLiP: a semi-manually curated database for biologically relevant ligand–protein interactions. Nucleic Acids Res 41(D1):D1096–D1103
13. Velez Rueda AJ, Bulgarelli FL, Palopoli N, Parisi G (2023) CaviDB: a database of cavities and their features in the structural and conformational space of proteins. Database 2023:baad010
14. Wang S, Lin H, Huang Z, He Y, Deng X, Xu Y, Pei J, Lai L (2022) CavitySpace: a database of potential ligand binding sites in the human proteome. Biomolecules 12:967
15. Tian J, Azarian MH, Pecht M (2014) Anomaly detection using self-organizing maps-based K-nearest neighbour algorithm. In: Proceedings of the European conference of the prognostics and health management society 2014, vol 2, no 1. https://doi.org/10.36001/phme.2014.v2i1.1554
16. Weisel M, Proschak E, Schneider G (2007) PocketPicker: analysis of ligand binding-sites with shape descriptors. Chem Cent J 1(1):7
17. Berman HM, Westbrook J, Feng Z, Gilliland G, Bhat TN, Weissig H, Shindyalov IN, Bourne PE (2000) The protein data bank. Nucleic Acids Res 28(1):235–242. URL: rcsb.org
18. Schrodinger L, DeLano W (2020) PyMOL. Available at: http://www.pymol.org/pymol
19. Roy P, Bhattacharyya D (2022) MetBP: a software tool for detection of interaction between metal ion–RNA base pairs. Bioinformatics btac392. https://doi.org/10.1093/bioinformatics/btac392

Tire Imprint Identification and Classification using VGG19

Disha N. Girish and M. Priyanka

Abstract In the fields of forensic research, the automobile industry, and consumer safety, classifying tire imprints is a crucial task. It involves identifying the brand or producer of a tire based on the imprints left on various surfaces. Tire impressions can be valuable evidence in criminal investigations, counterfeit identification, product quality control, and accident reconstruction. Here, we propose a deep learning approach using the VGG19 architecture for tire impression identification and classification. To begin, we collected tire impressions from various two- and four-wheeler MRF tire brands. Our dataset consists of 2693 photos divided into 22 classes, with 15 classes representing 2-wheelers and 7 classes representing 4-wheelers. We processed the images through denoising, normalization, and enhancement steps. We then employed a mask-based segmentation method to extract texture-based information from the tire impressions. This, combined with morphological data, resulted in a comprehensive set of features used in pattern recognition. These features include statistical measurements such as mean, entropy, correlation, contrast, and others. To accurately determine the tire brand, we utilized the Structural Similarity Index Matrix (SSIM) with a minimum similarity index value of 0.7. By comparing the SSIM values, we were able to match the tire impressions to the correct brand. We divided the dataset into 80% training data and 20% test data, which provided the best performance for our proposed technique. The accuracy achieved using this approach was 95.3%.

Keywords Tire imprint classification · Brand identification · VGG19 · Texture feature extraction · Mask-based segmentation

D. N. Girish · M. Priyanka (✉)
Department of Computer Science, School of Computing, Amrita Vishwa Vidyapeetham, Mysore, India
e-mail: priyankam@my.amrita.edu

D. N. Girish
e-mail: dishangirish9@gmail.com

B. R. Devi et al. (eds.), *Proceedings of Fifth International Conference on Computer and Communication Technologies*, Lecture Notes in Networks and Systems 897,
https://doi.org/10.1007/978-981-99-9704-6_7

1 Introduction

It's essential to be able to recognize tire markings in the automotive industry. Tire impressions are a critical source of information for the purposes of monitoring cars, investigating accidents, and examining tires. The industry places a high value on the correct and effective brand-specific classification of tire impressions, since doing so might provide crucial details regarding tire performance, safety, and manufacturing processes. Current methods for tire impression classification and brand recognition are plagued by low accuracy and constrained scalability. This paper proposes a method to overcome these limitations by categorizing tire impressions using the deep convolutional neural network VGG19 model. The VGG19 model has shown better performance, especially in tests of image recognition. This study aims to evaluate the brand-specific tire impression categorization performance of the VGG19 model. Recall, accuracy, and precision metrics are used to evaluate the model on a dataset of tire impressions from various brands. The results of this study have significant implications for the automobile industry, particularly with regard to improving tire quality assurance and vehicle tracking. The results of the study could also help in the development of more accurate and efficient methods for brand recognition and categorization of tire impressions. This work is noteworthy since it is the first to examine the use of deep learning for categorizing tire marks. The results of this work show that tire impression classification using deep learning models is very accurate and efficient. This might revolutionize the automobile industry by improving tire quality control, accident investigation, and vehicle monitoring. The linked study for classifying tire imprints is mentioned in Sect. 2. In Sect. 3, a brief explanation of the suggested strategy is given. Section 4 provides a description of the experimental environment. The list of all the experimental results and supplementary comments are included in Sect. 5. Sections 6 and 7 provide concluding remarks and ideas for further study.

2 Related Work

In the field of forensic research, tire imprint classification is crucial for tasks such as criminal investigations, counterfeit identification, product quality control, and accident reconstruction. Several research studies have been conducted using various methods to achieve accurate tire pattern identification and classification. In one study [1], the Fast Fourier Transform (FFT) method was employed to identify tire patterns. An 8-bit gradation BMP bitmap was used for the match test, and the standard stock image FFT achieved an accuracy of 85%. Another research [2] focused on tire track categorization and feature extraction using the Scale-Invariant Feature Transform (SIFT) algorithm and template matching methods. The dataset consisted of 35 tire track photographs labeled as "clean tracks" and randomly colored with spray brushes in black and white. The study achieved a 91% accuracy rate, with accuracy increasing as the dataset size grew. A different approach was taken in [3], where Gabor wavelets, Principal Component Analysis (PCA), and Support Vector Machines (SVM) were utilized for tire track classification and identification. Various features, such as groove and wave extraction from tire tread, were employed for accurate pattern recognition. The dataset included 15 tire tread patterns from 7 different manufacturers, and data augmentation was performed by creating vertical and horizontal mirror reflections. Tire parameters were used to train an Artificial Neural Network (ANN) in [4], with a focus on predicting unknown curve parameters with measurement errors. Synthetic curves were generated using the MF model and various tire curve combinations. The accuracy rate achieved was 85%. In [5], a binary decision tree was constructed using the fuzzy C-means method for tire pattern classification. The dataset comprised ten samples from each of six pattern types. A study [6] proposed creating an average dynamic tire footprint by averaging computed curves from video clips. Contrast enhancement was employed as a pre-processing step. The dataset consisted of 116 video clips captured in an ATM from 19 different tires. Zhen et al. [7] utilized the wavelet transform technique to enhance tire marks captured at accident sites. A technique described in [8] involved extracting 3D tire patterns or impressions on the ground using a gypsum casting method and extracting tire patterns on clothes using polarization photography and color-separation photography. Morphological characteristics and trace extraction were employed for tire pattern identification. In [9], Hu moments were used for feature analysis of tire tread patterns. DXF analysis was employed to determine pattern information, and the dataset was created using AutoCAD software. Two methods for digitizing imprints were discussed in [10]: 3D optical surface scanning and taking impressions using a scale. To track tire impressions, the ROTOCOM Wheel Tracker (RCWT) was used in [11]. The study collected 280 tire tread impressions from both sides of the RCWT, including seven distinct tire tread patterns, five different tire loads, and four different tire inflation pressures. It was found that the traditional circular contact area used in pavement design significantly overestimated the actual tire-pavement contact area by up to 92% when comparing effective and conventional contact areas. In [12], various techniques were described for tire pattern analysis, including converting photos to JPEG format, grayscale conversion, removing white pixels from image margins, black-and-white conversion, and segmenting the image into 16 parts. The study utilized the Neural Network Toolbox from MATLAB, enabling the creation of different neural network types for efficient processing of large datasets. The accuracy rate achieved was 92.02%. In [13], tire tread sample location was determined using ellipse detection with the Hough transform. The original picture was transformed into a binary image using a Canny edge detector, as the Hough transform disregards color information. After sequential processing, the best ellipses in the image were selected. Fuzzy sets were used to define the parameter values, and the Takagi–Sugeno fuzzy inference system was employed. Data for the study was gathered from online databases. In [14], Py-GCIMS (Pyrolysis-Gas Chromatography-Mass Spectrometry) was combined with linear discriminant analysis to improve sample discrimination. The method allowed for the identification and categorization of trace rubber residues to the specific production batch of a tire, particularly when paired with Principal

Component Analysis (PCA) and Linear Discriminant Analysis (LDA). The dataset consisted of 59 tire samples from accident-related automobiles, with 58 samples from summer tires and one sample from a winter tire. According to [15], an SVM classifier was created using traits from a set of training images. The process of classifying test tires involved feature extraction and tire classification using the trained SVM classifier. Feature extraction was performed based on a low-dimensional approximation obtained through Principal Component Analysis (PCA). To improve resilience to changes in lighting and perspective, features were extracted from the frequency representation of the binary edge map of the tire tread image. The study achieved an accuracy rate of 81.81%. The dataset included 140 winter tires, 70 all-season tires, and 70 summer tires. In [16], machine learning characteristics such as mean, variance, entropy, standard deviation, skewness, and kurtosis, along with properties of efficientNetB7, were employed in a fusion strategy. The features from EfficientNet were reduced using PCA, and the combined characteristics of the test picture were subjected to linear SVM for classification. The dataset contained three classes with a total of 1097 photos: 120 samples for benign cases, 561 samples for malignant cases, and 416 samples for normal cases. In [17], a comparison was made between chest X-ray pictures for Covid-19 classification. The ResNet-50 model, trained on the ImageNet dataset, and a CNN model were evaluated. A dataset of 12,240 photos was used for training and testing, ensuring that the models learned characteristics from the lung region only. Both the CNN and ResNet-50 models achieved a 96% accuracy in categorizing chest X-ray pictures as positive or negative for Covid-19, with the ResNet-50 model performing slightly better in real test samples. In [18], the region of interest in wrist bone pictures was determined through edge-based segmentation and Otsu thresholding. The image was then passed into a Logistic Regression model for age prediction and a CNN classifier for gender categorization. The dataset included 12,612 left wrist X-ray pictures, with 95% accuracy.

3 Proposed Method

3.1 Training and Testing Process

The proposed method entails the compilation of the dataset, feature extraction, and classification of tire brands with a VGG19. For feature extraction, images are preprocessed and subjected to morphological and Canny edge processes. The tire is separated from the background using a mask-based segmentation technique, and the VGG19 is trained using back-propagation with several convolutional and pooling layers. The loss function is optimized using the Adam optimizer, and over-fitting is avoided by evaluating the model on the testing set at the end of each iteration. Overall, the suggested system correctly divides tire brands into categories based on their tread designs as shown in Fig. 1.

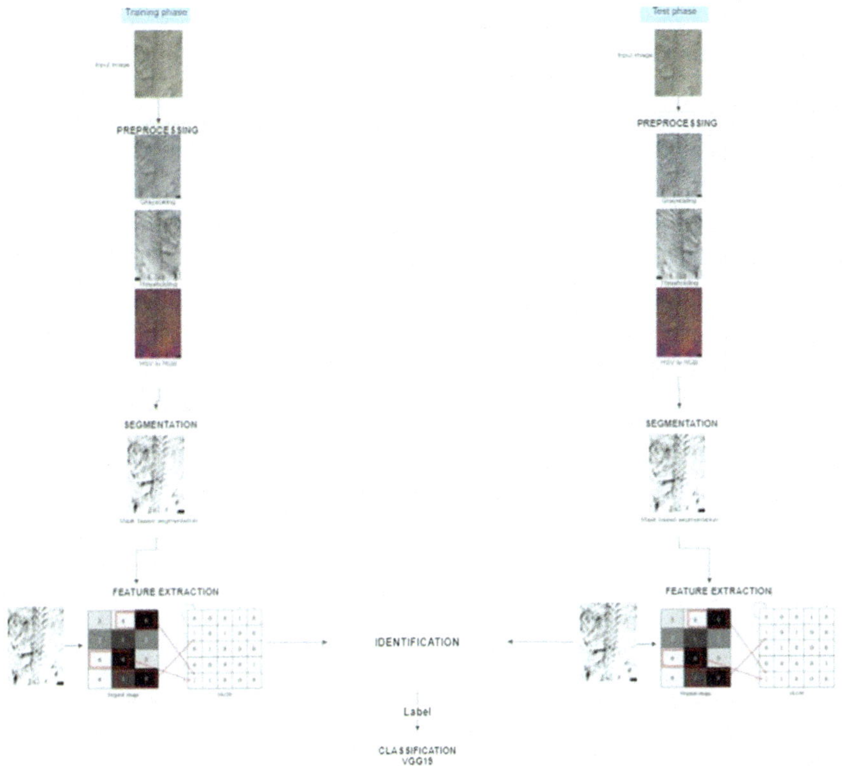

Fig. 1 Proposed architecture

Input Image. The input image is made up of several tire brand imprints. Different lighting, viewpoints, and image quality may be visible in these imprints. The input picture serves as the starting point for all subsequent processing processes, including pre-processing, segmentation, feature extraction, and classification using deep learning models. The suggested system for classifying tire brands accurately identifies and categorizes tire brands based on their tread patterns by combining image processing and deep learning techniques. The procedure starts with gathering and pre-processing the tire photos, followed by feature extraction using morphological operations and edge detection, and then training a VGG19 for classification. The photographs are pre-processed by turning them into grayscale, using thresholding to boost contrast, and changing the color space from HSV to RGB for better visual representation. To identify the edges of the tread, Canny edge detection is utilized after feature extraction using morphological approaches has determined the size and shape of the tire tread patterns. To increase classification accuracy, the tire is separated from the background using a mask-based segmentation technique. Back-propagation is used to train the VGG19 on the retrieved features, and the Adam optimizer is used to optimize it. To avoid over-fitting, the model is assessed on the testing set after each epoch. The training procedure is carried out repeatedly until the model's accuracy on the testing set is adequate. Overall, the suggested system provides an automated and precise approach for identifying tire brands that can be applied to a variety of tasks, including tire inspection and maintenance. By eliminating the need for manual inspection and lowering the possibility of mistakes, it improves the process' efficiency and dependability.

3.2 Dataset Description

The dataset comprises of 2693 photos of MRF tire treads taken with an iPhone 8plus's 12 MP camera at a resolution of 1080 × 1920 pixels. Images of 22 different classes of MRF tire treads, including 7 classes of 4-wheelers and 15 classes of 2-wheelers, all exhibiting different tire patterns, are included in the dataset as shown in Fig. 2. Physically taking images at various sites, such tire manufacturers, and annotating them with pertinent information, like the tire brand, was necessary to collect a sizable dataset of tire impressions. To ensure a broad set of data, the photos were independently obtained under various lighting circumstances and viewing angles. To increase contrast and improve visual representation, pre-processing procedures such as grayscale conversion, thresholding, and conversion from HSV to RGB color space were used. For feature extraction and segmentation, morphological procedures and the Canny edge detector were utilized, then mask-based segmentation for tire isolation. In order to categorize the tire brands according to their tread patterns, a VGG19 was trained on the dataset. To assess the model's performance, the dataset was divided into training and testing sets, each with 80% of the data.

To better comprehend the dataset before using pre-processing techniques, Table 1 shows each tire brand along with its resolution in pixels and the number of dataset

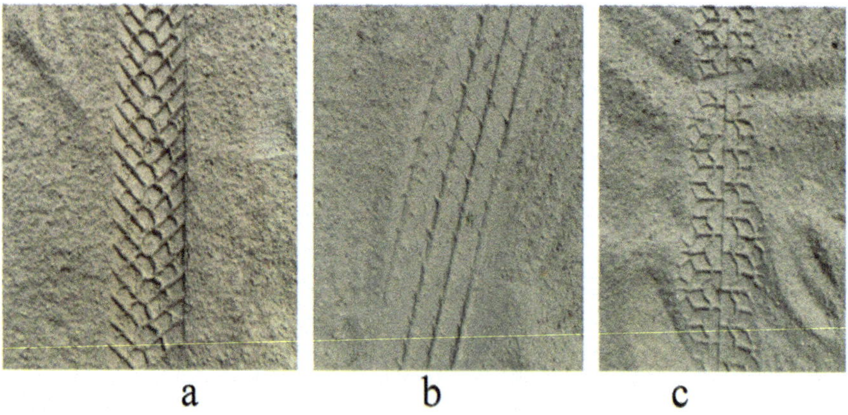

Fig. 2 Sample dataset. **a** MOGRIP meteor M, **b** Nylogrip Plus, **c** perfinza-4-wheeler

Table 1 Dataset description

Brand	Resolution	No. of samples
MRF-2 wheeler		
MRF MOGRIP meteor M	1920 × 1080	132
MRF Nylogrip Plus	1920 × 1080	116
MRF rib	1920 × 1080	104
MRF zapper	1920 × 1080	98
MRF zapper FN	1920 × 1080	102
MRF Nylogrip zapper FG 9090-12 54J tubeless scooter	1920 × 1080	100
MRF MOGRIP meteor-M 90100-10 53J tubeless scooter tire	1920 × 1080	108
MRF Nylogrip Plus2	1920 × 1080	104
MRF Revz—C	1920 × 1080	145
MRF Revz—MG	1920 × 1080	108
MRF Revz—S	1920 × 1080	80
MRF nylo grip EZEERIDE	1920 × 1080	259
MRF ZTX-S	1920 × 1080	134
MRF MOGRIP MOTO-D	1920 × 1080	101
MRF zapper KURVE	1920 × 1080	102
MRF 4-wheeler		
MRF perfinza	1920 × 1080	126
MRF perfinza 1	1920 × 1080	126
MRF muscle master	1920 × 1080	120
MRF ZVTV	1920 × 1080	120
MRF ZVTS	1920 × 1080	112
MRF wanderer	1920 × 1080	134
MRF ZLX	1920 × 1080	162

samples. There are 15 two-wheelers and 7 four-wheelers on the table. There are 22 classes with a 1920 × 1080 pixel width and height.

3.3 *Pre-processing*

The images will be pre-processed to improve their quality and enhance important features. We will convert the images to grayscale and apply thresholding to remove noise. We will also convert the images from HSV to RGB color space to standardize the color channels.

Grayscale images contain only shades of gray, representing the intensity levels of the original image. By converting the colored tire imprint images to grayscale as shown in Fig. 3, the focus is solely on the intensity information, simplifying the subsequent processing steps. Grayscale images also reduce the computational complexity compared to color images, making it more efficient for analysis.

Thresholding is a popular technique in image processing since it cleans up the image and lowers noise. The portions of the image that are most important for character identification can be highlighted by turning a grayscale image into a binary image based on a threshold value as shown in Fig. 3, which allows us to distinguish between foreground and background pixels.

HSV (Hue, Saturation, Value) is an alternative color space representation that separates color information from brightness information. By converting the tire imprint images from HSV to RGB depicted in Fig. 3, the aim is to enhance the color information while maintaining the spatial structure. RGB is a widely used color space, making it compatible with various image processing techniques and deep learning architectures. Pre-processing tire imprint images to enhance relevant information while removing noise and artifacts can be challenging. Extracting meaningful features from the images using techniques like mask-based segmentation

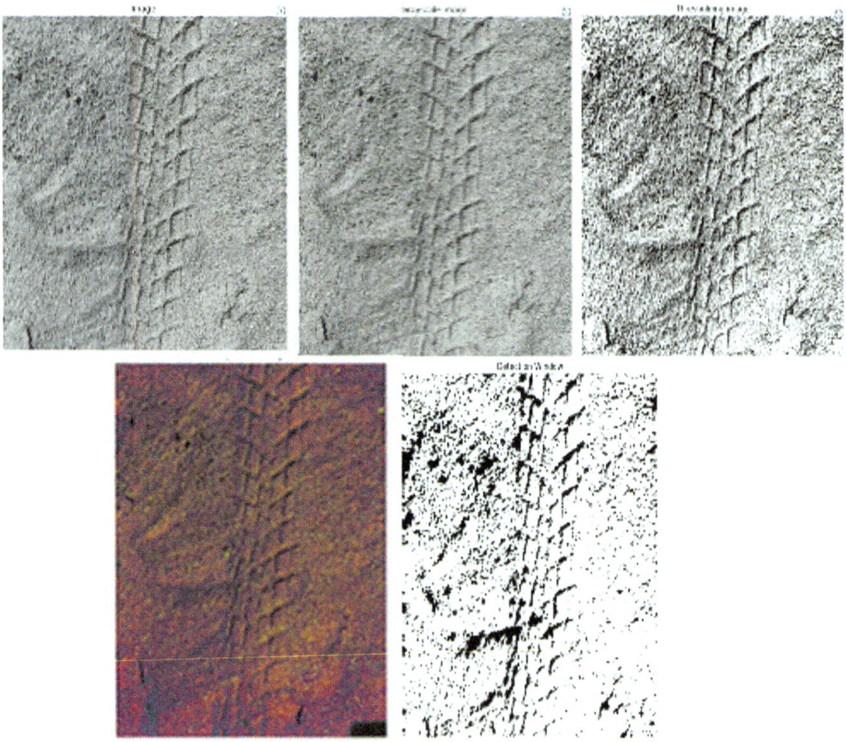

Fig. 3 Pre-processing stages performed on MOGRIP moto D(MRF) type tire imprint

and denoising requires careful consideration of parameter settings and may require domain expertise.

3.4 Segmentation

Mask-based segmentation is a crucial component of the proposed approach for tire imprint classification. Figure 3 depicts that this technique is employed to isolate the specific tire pattern region from the pre-processed images, ensuring that only the relevant information is considered for further analysis and classification. In mask-based segmentation, texture features are utilized to create masks that accurately define the tire pattern region of interest. Texture features capture the spatial arrangement and variations in pixel intensities within the tire imprint images. The texture features are computed, and appropriate thresholds or statistical measures are applied to distinguish the tire pattern region from the surrounding background.

3.5 Feature Extraction

In addition to the morphological feature extraction, the proposed method also incorporates tire template matching. Tire template matching involves comparing the extracted features from an unknown tire imprint to predefined templates representing different tire brands. In this context, the extracted features from the unknown tire imprint, including morphological and texture features, are compared to the features stored in the templates. The template matching process measures the similarity between the features of the unknown imprint and the features of each template, enabling the identification of the most closely matching template. By incorporating tire template matching, the proposed method enhances the classification accuracy by leveraging predefined knowledge about the distinctive features of different tire brands. Utilizing gray-level co-occurrence matrices (GLCM), which are frequently employed in image texture analysis, the texture features are produced. One of the characteristics recovered by the GLCM method is the distribution of the image's gray levels. The suggested method can effectively capture a variety of features that are important for classifying tires by integrating morphological and textural information. This approach allows for more precise and reliable tire classification and brand recognition. The combination of morphological feature extraction, mask-based texture extraction, and tire template matching provides a comprehensive and robust framework for tire imprint classification. It harnesses various aspects of the tire imprints, including shape, texture, and brand-specific characteristics, resulting in accurate and efficient identification of tire imprints in applications such as vehicle tracking, accident investigation, and tire quality control.

3.6 Model Comparison

VGG19. A deep convolutional neural network called VGG19 has demonstrated impressive performance in a number of image categorization tasks. The 19-layer VGG19 model as shown in Fig. 4 was trained using the extensive ImageNet dataset. The proposed approach in this situation employs a modified VGG19 with unique layers for tire brand classification. On the tire dataset, training is conducted using the pre-trained weights of the VGG19 model. The experimental findings demonstrate the modified VGG19 model's usefulness in tire brand classification tasks, as it achieves the highest accuracy among the compared models.

The overall process involves collecting input images of tire imprints, followed by pre-processing to enhance image quality. Segmentation is then performed to extract regions of interest. Feature extraction captures important characteristics, and a model is trained using the extracted features. Finally, the VGG19 model is used for testing and brand classification of tire imprints.

4 Experimental Setup

In the experimental setup, a large dataset of tire imprints is gathered, including a range of brands. Taking into account changes in lighting, viewpoints, and image quality, the dataset is meticulously chosen to ensure a representative sample of each brand. The dataset is separated into training and testing subsets, and a balanced distribution of impressions from each brand is maintained in both. The collected characteristics are linked to the appropriate brand labels during the training phase using the training dataset and the VGG19 model. To improve the model, appropriate loss functions and optimization procedures are applied. During the testing step, the trained VGG19

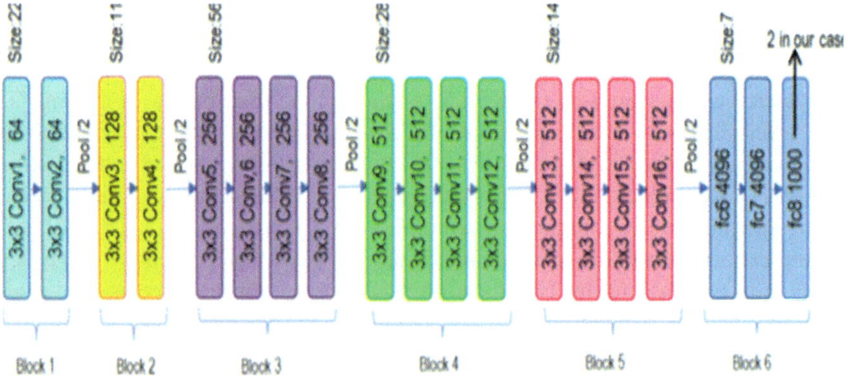

Fig. 4 VGG19 architecture

model is evaluated using a different testing dataset. According to brand, the model categorizes tire impressions, and performance measures including accuracy, precision, and recall are computed to assess the model's effectiveness. By utilizing a variety of datasets, adequate VGG19 model training, and rigorous assessment on unseen data, the experimental setting assures the robustness and generalizability of the presented strategy.

4.1 Dataset

The tire imprint collection dataset is separated between training and testing subsets throughout the dataset split process. Eighty percent of the data are in the training subset, while 20% are in the testing subset. The model may be adequately trained on a huge quantity of data, thanks to this divide in order to grasp patterns and correlations, and the testing subset is used to examine the model's performance on concealed tire impressions in order to precisely gauge its categorization capabilities.

4.2 Pre-processing

Gray Scaling. Gray scaling is a stage in the pre-processing process that turns color photos into grayscale ones. This conversion is carried out to make the photos more streamlined and to highlight the shape of the object rather than its color. The tire photos are gray-scaled in the proposed tire classification system to remove the impact of color on classification accuracy. The color differences between the various tire manufacturers are erased by converting the tire photos to grayscale, and only the tread pattern information is used for categorization. Gray scaling also lessens the complexity of the image data, which facilitates processing using machine learning and image processing methods.

Thresholding. Thresholding plays a vital role in tire imprint classification as it enables the separation of the foreground (imprint) from the background. In this study, we employed Otsu's thresholding method to determine an optimal threshold value based on the intensity distribution of the grayscale tire imprint images. Otsu's thresholding maximizes the between-class variance, resulting in an effective separation of the imprint from the background pixels as shown in (1).

$$\begin{aligned}&\text{If pixel\_value} > \text{threshold\_value:}\\&\text{new\_pixel\_value} = 255\ \#\ \text{set to white}\\&\text{else:}\\&\text{new\_pixel\_value} = 0\ \#\ \text{set to black}\end{aligned} \tag{1}$$

The image processing stage compares the pixel values of each picture pixel using a threshold value. Whether a pixel value is higher or lower than the threshold depends on the threshold value. If the value of a pixel is higher than the cut off, it is given an additional value of 255, which stands for white. In contrast, the value of a pixel is given an updated value of 0, signifying black, if it is equivalent to or below the threshold. The outcome of this operation is a binary picture with black background and white foreground pixels for the characters. To improve the picture for improved character identification, the threshold value may be chosen using a variety of ways.

HSV to RGB. HSV to RGB conversion is a critical pre-processing step in the proposed tire brand classification system. The HSV color space represents color in terms of hue, saturation, and value, which makes it easier to separate color information from intensity information. However, most deep learning frameworks require RGB color images as input. Therefore, the HSV images are converted to RGB images using a conversion algorithm that maps HSV values to RGB values. This step allows the VGG19 to receive RGB images as input, which improves its accuracy in tire brand classification. Overall, the HSV to RGB conversion is a crucial step in the pre-processing pipeline, enabling the VGG19 to better differentiate between tire tread patterns and classify the tire brand accurately.

4.3 Segmentation

The proposed approach adopts mask-based segmentation due to the limitations faced by traditional edge-based techniques like Canny edge detection in accurately extracting the tire pattern region. Mask-based segmentation provides a more robust and accurate method for isolating the regions of interest containing tire imprints from the background and other irrelevant elements. This technique involves creating a binary mask that precisely outlines the tire pattern region, allowing for a more precise and reliable extraction. By utilizing mask-based segmentation, the proposed approach overcomes the challenges posed by Canny edge detection and ensures the accurate isolation of the tire imprints for subsequent analysis and classification tasks. Mask-based segmentation involves creating a binary mask that separates the regions of interest from the background.

4.4 Feature Extraction

Haralick texture features, commonly referred to as Gray-Level Co-occurrence Matrix (GLCM) features, are used to extract texture properties. The Haralick formula involves computing various statistical measures from the GLCM to characterize the texture properties of an image. Features like mean, entropy, correlation, contrast, and other statistical measures, along with morphological features, collectively form a comprehensive set of features used in pattern recognition. These features capture different aspects of the data, providing valuable information for analysis and classification tasks. Morphological feature extraction involves analyzing the shape, structure, and spatial properties of objects within an image using morphological operations. These operations are applied to binary or grayscale images and can provide valuable information for various image analysis tasks. Morphological operations like opening and closing are applied. After applying morphological operations, we obtain morphological features like area, perimeter, compactness, eccentricity, convexity, etc., which are used to capture every bit of patterns that can be extracted from the given image. Assume that the pixel pairings in the GLCM are represented by the pair (i, j). $Pi(i, j)$ stands for the normalized value of the (i, j) element in the GLCM, which measures the likelihood that the pixel pair (i, j) will appear. i and j stand for the values of the pixel pairs' gray levels.

Entropy. Entropy is a metric used in Gray-Level Co-occurrence Matrix (GLCM) analysis to express how much information or randomness is present in the distribution of gray-level values for adjacent pixels in a picture. It conveys the image's textural pattern's ambiguity or complexity. Entropy in GLCM is often expressed in mathematical Eq. (2) as follows:

$$\text{Entropy} = -\sum(i, j)\left[Pi(i, j) * \log 2(Pi(i, j))\right] \tag{2}$$

Correlation. Linear relationship between the gray-level values of adjacent pixels in a picture is characterized by correlation, a metric used in Gray-Level Co-occurrence Matrix (GLCM) analysis. It gauges how interconnected or related the pixel brightness in the image's various orientations is. Equation (3) defines the correlation formula as follows:

$$\text{Correlation} = \sum(i, j)[(i - \mu i)(j - \mu j)Pi(i, j)/(\sigma i \sigma j)] \tag{3}$$

Contrast. In texture analysis, the Gray-Level Co-occurrence Matrix (GLCM) is frequently utilized for calculating the texture properties of an image. The GLCM calculates the statistical correlations between pairs of pixels in a picture based on the gray-level values of the pixels. One of the texture metrics generated by the GLCM is contrast. Equation (4) defines the equation for contrast.

$$\text{Contrast} = \sum(i, j)\left[(i - j)^2 * Pi(i, j)\right] \tag{4}$$

Inverse Difference Moments. The local homogeneity or uniformity of pixel intensities in a picture is described by inverse difference moments (IDM), a textural metric used in Gray-Level Co-occurrence Matrix (GLCM) analysis. It calculates the inverse of the difference between neighboring pixels' gray-level values. In general, the definition of inverse difference moments (IMD) is given in Eq. (5):

$$\text{IDM} = \sum (i, j)\left[Pi(i, j)/\left(1 + (i - j)^2\right)\right] \quad (5)$$

4.5 Template Matching

It is the process of tire template matching involves comparing characteristics from an unidentified tire imprint to predefined templates that represent different tire brands. The Structural Similarity Index Matrix (SSIM) is used for template matching in the proposed method. A popular image quality metric called SSIM compares the brightness, contrast, and structural data of two photos to determine how comparable they are. The SSIM index ranges from − 1 to 1, with 1 representing the equivalence of the two images. In this case, 0.7 is the value chosen as an index to choose the tire brand based on the minimum match index achieved when SSIM is applied.

4.6 VGG19

The widely recognized deep convolutional neural network (CNN) model VGG19 was created especially for applications requiring image recognition. This model has 19 layers, including layers for max-pooling that make it easier to extract complex information from pictures and convolutional layers with narrow receptive fields. Its architecture has excelled in achieving outstanding performance in a variety of picture classification tasks. The depth and effective parameter utilization of the VGG19 model, which allows it to identify and analyze intricate patterns and structures within images, is credited with its effectiveness. In the proposed approach, the VGG19 model is trained using a learning rate set to 0.001. The model's ability to learn complex patterns and extract high-level features from tire imprints makes it suitable for accurate brand classification. Its architecture and training scheme contribute to its effectiveness, making VGG19 a suitable choice for tire imprint classification tasks. The Adam Optimizer, a widely used optimization method, is frequently used to train deep learning models. To improve the training process, this optimizer combines the benefits of two well-known optimization techniques, namely root mean square propagation and adaptive gradient. The Adam optimizer's capacity to dynamically change the learning rate for each parameter, which results in better performance and quicker convergence during training, is its key feature. The VGG19 model is specifically optimized for tire imprint categorization in the suggested method. The model's parameters are optimized during the training phase to increase its precision in classifying tire imprints. The Adam optimizer is used, and its learning rate is set to 0.001.

5 Results and Discussion

On the testing set, the suggested method achieved a remarkable overall accuracy of 95.3%. The model performed well across all classifications, although the MRF brand had the worst accuracy, coming in at 89.7%. Our suggested strategy produced greater accuracy while needing fewer parameters than the VGG19 model. Only 1,650,957 parameters made up our suggested technique, compared to 143,667,240 in the VGG19 model. The accuracy of our method was on par with that of other cutting-edge techniques in the literature. Our approach, however, has the benefit of using a more straightforward design and requiring less processing power. Figure 5 displays a confusion matrix, which is presented as a square matrix with 22 rows and 22 columns representing the true and predicted classes, respectively. Every component of the matrix denotes the frequency or percentage of instances that, given instances' true and expected class combinations, are properly or wrongly categorized. Due to its high accuracy, the suggested method can be used in industrial settings for automated tire categorization and quality monitoring. Future research may focus on enhancing the model's precision and enlarging the dataset to include more tire manufacturers and models. Precision, recall, F1 score, and Intersection over Union (IoU) are some examples of frequently used segmentation evaluation metrics. The following are the formulas for these measurements:

Precision. Precision is determined by the proportion of correctly segmented pixels to all positively (segmented) pixels:

$$\text{Precision} = \text{TP}/(\text{TP} + \text{FP}) \quad (6)$$

In (6), where FP (False Positives) is the number of pixels that were mistakenly categorized as positive and TP (True Positives) is the number of pixels that were correctly segregated.

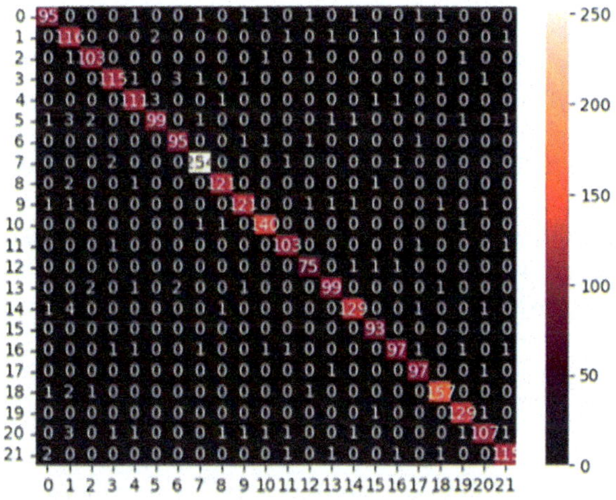

Fig. 5 Confusion matrix

Recall. Recall, also known as responsiveness or true positive rate, measures the percentage of successfully segmented pixels to all positive pixels in the ground truth:

$$\text{Recall} = \text{TP}/(\text{TP} + \text{FN}) \tag{7}$$

In (7), where FN (False Negatives) denotes the quantity of successfully segmented positive pixels.

F1 Score. The F1 score provides a single metric that balances both metrics, as it is the harmonic mean of precision and recall:

$$\text{F1 Score} = 2 * (\text{Precision} * \text{Recall})/(\text{Precision} + \text{Recall}) \tag{8}$$

In (8), it considers both the precision and recall values and is useful when there is an imbalance between the positive and negative classes.

For various partitions of the training and test datasets, the given classification report in Table 2 shows various performance metrics including accuracy, recall, and F1-score. Impressive results are obtained when the VGG19 model is applied, especially when the dataset is divided into training data (80%) and test data (20%). 95.3% accuracy is a remarkable accuracy rating for this split. This difference highlights how sensitive the model's effectiveness is correlated with its allocation of data from both tests and training. When utilizing the VGG19 model for tire impression identification and classification, it is imperative to take into account the effect that dataset splits have on the model's performance.

5.1 Performance Evaluation Graph

The graph plot showing the training loss and validation loss as a function of the training process's epoch count is shown in Fig. 6. The training dataset's iterations or passes are represented by the epochs on the x-axis. The loss, which is a metric indicating the effectiveness or inaccuracy of the model, is shown on the y-axis. The training loss is shown by the blue line, which displays how the loss value varies during the course of the model's training over several epochs. As the model begins to learn and modify its parameters, the training loss is initially rather large. The loss steadily drops as training goes on, showing that the model is becoming better at correctly predicting the training data. The model's performance on a distinct validation dataset that it was not exposed to during training is represented by the orange line, which is known as the validation loss. With more epochs, the validation loss gets smaller.

Table 2 Classification report for different training and test dataset split

Dataset split	Classification report				Accuracy (%)
	No. of classes	Precision	Recall	F1-score	
50–50	0	1.00	0.93	0.97	91.26
	1	0.71	1.00	0.83	
	2	0.70	0.70	0.70	
	3	0.70	0.70	0.70	
	4	0.35	0.70	0.47	
	5	0.00	0.00	0.00	
	6	0.57	0.40	0.47	
	7	0.70	0.70	0.70	
	8	0.70	0.70	0.70	
	9	0.30	0.50	0.37	
	10	0.10	0.12	0.11	
	11	0.10	0.10	0.10	
	12	0.00	0.00	0.00	
	13	1.00	0.39	0.56	
	14	0.71	0.71	0.71	
	15	0.73	1.00	0.84	
	16	1.00	0.80	0.89	
	17	0.72	1.00	0.84	
	18	1.00	0.61	0.76	
	19	0.68	1.00	0.81	
	20	1.00	0.60	0.75	
	21	0.95	1.00	0.97	
	Accuracy			0.69	
	Macro avg.	0.62	0.62	0.60	
	Weighted avg.	0.70	0.69	0.67	
60–40	0	1.00	0.93	0.97	92.97
	1	0.90	0.90	0.90	
	2	0.90	0.90	0.90	
	3	0.91	1.00	0.95	
	4	0.62	1.00	0.77	
	5	0.80	0.40	0.53	
	6	1.00	0.70	0.82	
	7	0.80	0.80	0.80	
	8	0.80	0.80	0.80	
	9	0.60	1.00	0.75	
	10	0.60	0.75	0.67	

(continued)

Table 2 (continued)

Dataset split	Classification report				Accuracy (%)
	No. of classes	Precision	Recall	F1-score	
	11	0.44	0.60	0.51	
	12	0.64	0.19	0.30	
	13	1.00	0.45	0.62	
	14	0.60	0.71	0.65	
	15	0.73	1.00	0.84	
	16	1.00	0.91	0.95	
	17	0.77	1.00	0.87	
	18	1.00	0.68	0.81	
	19	0.75	1.00	0.86	
	20	1.00	0.60	0.75	
	21	0.93	1.00	0.96	
	Accuracy			0.81	
	Macro avg.	0.81	0.79	0.77	
	Weighted avg.	0.84	0.81	0.80	
70–30	0	1.00	0.93	0.97	94.68
	1	0.91	1.00	0.95	
	2	1.00	0.80	0.89	
	3	0.82	0.90	0.86	
	4	0.62	1.00	0.77	
	5	0.83	0.50	0.62	
	6	0.90	0.90	0.90	
	7	0.90	0.90	0.90	
	8	0.82	0.90	0.86	
	9	0.80	0.67	0.73	
	10	0.70	0.88	0.78	
	11	0.42	0.70	0.53	
	12	0.00	0.00	0.00	
	13	1.00	0.82	0.90	
	14	0.90	0.90	0.90	
	15	0.85	1.00	0.92	
	16	1.00	0.80	0.89	
	17	0.84	1.00	0.91	
	18	1.00	0.87	0.93	
	19	0.88	1.00	0.94	
	20	1.00	0.91	0.95	

(continued)

Table 2 (continued)

Dataset split	Classification report				Accuracy (%)
	No. of classes	Precision	Recall	F1-score	
	21	0.99	1.00	1.00	
	Accuracy			0.87	
	Macro avg.	0.83	0.84	0.82	
	Weighted avg.	0.86	0.87	0.86	
80–20	0	1.00	0.87	0.93	95.28
	1	0.83	1.00	0.91	
	2	0.91	1.00	0.95	
	3	1.00	0.90	0.95	
	4	0.71	1.00	0.83	
	5	1.00	0.60	0.75	
	6	0.77	1.00	0.87	
	7	1.00	0.70	0.82	
	8	0.91	1.00	0.95	
	9	1.00	0.83	0.91	
	10	0.89	1.00	0.94	
	11	0.51	0.90	0.65	
	12	1.00	0.09	0.16	
	13	1.00	1.00	1.00	
	14	1.00	1.00	1.00	
	15	1.00	1.00	1.00	
	16	1.00	1.00	1.00	
	17	1.00	1.00	1.00	
	18	1.00	1.00	1.00	
	19	1.00	1.00	1.00	
	20	0.91	1.00	0.95	
	21	1.00	0.97	0.98	
	Accuracy			0.92	
	Macro avg.	0.93	0.90	0.89	
	Weighted avg.	0.94	0.92	0.91	

6 Conclusion

In conclusion, the suggested method combines image processing and deep learning strategies to accurately categorize tire brands based on their tread patterns. The accuracy of the classification is increased through the application of morphological feature extraction, texture feature extraction, Canny edge detection, and mask-based segmentation. The VGG19 model performs better with a greater bit depth of 128

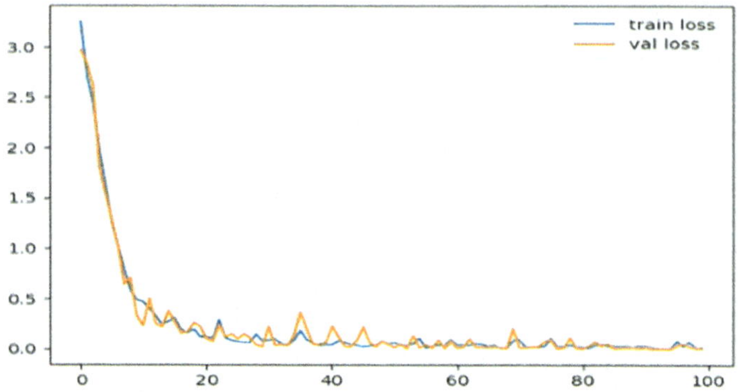

Fig. 6 Graph plot for training loss and validation loss for every increasing epoch

and obtains the maximum accuracy. The suggested technique offers encouraging tire brand categorization outcomes and has the potential to be applied in automated tire inspection industrial applications. However, there are still many limitations and difficulties that must be resolved, including the dataset's accessibility and quality as well as the model's complexity. Future research can concentrate on increasing the dataset to include more tire brands and different tire tread patterns, as well as investigating other deep learning architectures and approaches to enhance the classification's accuracy and effectiveness. Overall, the proposed method offers a starting point for additional investigation and advancement in the area of automated tire examination. Training the deep learning model, like VGG19, is computationally demanding and time-consuming. Hyper-parameter optimization, selecting suitable loss functions, and addressing over-fitting or under-fitting are crucial for good classification performance. Generalizing the model to unseen tire imprints and ensuring robustness to lighting conditions, angles, and scales is vital. Tire tread patterns can vary significantly, posing challenges in accurately classifying similar patterns. Deploying the method in real-world applications requires considering performance, computational efficiency, and reliability. These challenges involve optimizing the model, improving generalization, and addressing real-time processing requirements, ensuring the successful integration of the solution into existing systems.

7 Future Work

The suggested approach could be improved in subsequent research to handle more intricate tire tread textures and patterns. By fine-tuning pre-trained CNN models like ResNet or EfficientNet, transfer learning could be incorporated as one potential enhancement. This might shorten the amount of time needed to train the model to a high level of accuracy. To further increase the model's resilience, it is also possible to look into the usage of generative adversarial networks (GANs) to generate synthetic tire images and augment existing data.

References

1. Da Wang S, Xiao JY, Liu ZQ, Niu JL (2008) Study on tire imprint image recognition based on FFT image differencing algorithm. In: 2008 fifth international conference on fuzzy systems and knowledge discovery, Oct 2008, vol 2. IEEE, pp 504–508
2. Yilmaz AG, Nabiyev N. Recognition of tire track patterns using SIFT and template matching. El-Cezeri 9(2):634–643
3. Huang DY, Hu WC, Wang YW, Chen CI, Cheng CH (2010) Recognition of tire tread patterns based on Gabor wavelets and support vector machine. In: Computational collective intelligence. Technologies and applications: second international conference, ICCCI 2010, Kaohsiung, Taiwan, 10–12 Nov 2010. Proceedings, Part III 2. Springer, Berlin Heidelberg, pp 92–101
4. Olazagoitia JL, Perez JA, Badea F (2020) Identification of tire model parameters with artificial neural networks. Appl Sci 10(24):9110
5. Jung SW, Bae SW, Park GT (1994) A design scheme for a hierarchical fuzzy pattern matching classifier and its application to the tire tread pattern recognition. Fuzzy Sets Syst 65(2–3):311–322
6. Nava R, Fehr D, Petry F, Tamisier T (2019) Objective tire footprint segmentation assessment from high-speed videos. Tire Sci Technol 48(4):315–328
7. Zhen W, Shiwu L, Yunpeng W (2010) Tire-imprints image enhancement algorithm based on wavelet transformation. In: 2010 international conference on optoelectronics and image processing, Nov 2010, vol 2. IEEE, pp 666–670
8. Jin Y (2013) Characteristics of vehicle tire tread pattern and its application in forensic science. Adv Mater Res 706:1241–1245
9. Zhang F, Li P, Li Y, Liu Y, Yang Z (2017) Tire tread pattern recognition and feature analyzing based on Hu moment. In: 2017 global conference on mechanics and civil engineering (GCMCE 2017), June 2017. Atlantis Press, pp 122–125
10. Buck U, Albertini N, Naether S, Thali MJ (2007) 3D documentation of footwear impressions and tire tracks in snow with high resolution optical surface scanning. Forensic Sci Int 171(2–3):157–164
11. Muniandy R, Moazami D, Hamid H, Hassim S (2014) Characterization of effective tire contact area for various tread patterns. Instrum Sci Technol 42(1):15–26
12. Michalíková A, Pažický B (2019) Classification of tire tread images by using neural networks. In: 2019 IEEE 15th international scientific conference on informatics, Nov 2019. IEEE, pp 000107–000112
13. Michalíková A, Vagač M (2015) A tire tread pattern detection based on fuzzy logic. In: Flexible query answering systems 2015: proceedings of the 11th international conference FQAS 2015, Cracow, Poland, 26–28 Oct 2015, vol 400. Springer, p 381
14. Sarkissian G, Keegan J, Du Pasquier E, Depriester JP, Rousselot P (2004) The analysis of; tires and tire traces using FTIR and Py-GC/MS. Can Soc Forensic Sci J 37(1):19–37
15. Bulan O, Bernal EA, Loce RP, Wu W (2012) Tire classification from still images and video. In: 2012 15th international IEEE conference on intelligent transportation systems, Sept 2012. IEEE, pp 485–490
16. Suresha R, Devika KM, Prabhu A (2022) Support vector machine classifier based lung cancer recognition: a fusion approach. In: 2022 international conference on edge computing and applications (ICECAA), Oct 2022. IEEE, pp 1–8
17. Priyanka M, Sreekumar S, Arsh S (2022) Detection of COVID-19 from the chest X-ray images: a comparison study between CNN and ResNet-50. In: 2022 IEEE 2nd Mysore sub section international conference (MysuruCon), Oct 2022. IEEE, pp 1–7

18. Priyanka M, Divyashree M, Madhu V (2022) Computer vision-based approach for estimating age and gender using wrist X-ray images. In: 2022 IEEE 2nd Mysore sub section international conference (MysuruCon), Oct 2022. IEEE, pp 1–7

Uncovering Insights in Agroforestry: A Text Mining Analysis Using Web Scraping and Topic Modeling

Parisa Monika, Desu Devi Sri, and M. Suneetha

Abstract Big data is so prevalent, text mining has become an essential tool to provide an overview and easy understanding of vast amounts of data. In this article, we present a study on agroforestry that collects information from Google Scholar using web scraping techniques. We describe the evolution of a web scraper using Python modules and packages to extract the obligatory data from their source HTML page. The collected data is organized into a CSV file and subjected to various topic modeling strategies, such as Latent Dirichlet Allocation (LDA), Bidirectional Encoder Representations from Transformers (BERT), and Latent Semantic Analysis (LSA) to analyze and present the data as topics. The article highlights the significance of web scraping in retrieving data and offers insights into the domain of agroforestry facilitating further analysis. This research demonstrates the power of text mining and topic modeling in exploring large datasets and deriving valuable insights.

Keywords Web scraping · Google Scholar · Text mining · Topic modeling · Agroforestry

1 Introduction

Agroforestry is a form of land management that combines shrubs and trees with livestock or agricultural products. It is a sustainable system of land use that combines the advantages of crops and trees to produce a system that is more diverse and fruitful. Agroforestry may boost biodiversity, water quality, and soil health while also generating many streams of income that are beneficial to the economy. Text mining methods are beneficial in analyzing enormous volumes of data available like Google Scholar for in-depth analysis of available data to learn something new that might be better farming practice and with the increased evolution of data, it has become essential to use advanced tools like web scraping and topic modeling to

P. Monika (✉) · D. Devi Sri · M. Suneetha
Department of Information Technology, VR Siddhartha Engineering College, Vijayawada, India
e-mail: parisamonika7@gmail.com

B. R. Devi et al. (eds.), *Proceedings of Fifth International Conference on Computer and Communication Technologies*, Lecture Notes in Networks and Systems 897,
https://doi.org/10.1007/978-981-99-9704-6_8

extract and analyze data efficiently [1]. The objective of this study is to show the efficacy of text mining techniques in exploring large datasets and deriving valuable insights [2]. Ultimately, the knowledge collected from this research can help people better grasp the term agroforestry.

2 Related Work

In all accident reports, the latent Dirichlet allocation (LDA) technique is utilized to obtain various topics from the documents and the keywords associated [3]. Sectoral motifs are categorized in order to comprehend the nature of the accident process. LDA and text mining are employed to extract feature process of an accident. Examining management concerns pertaining to the industry, source, activity, and accident outcome while taking into consideration the particulars of industrial sites is something that safety managers and policy makers do. Five patterns in sectors are identified based on feature of accident process. Between 1993 and 2020, the publication of 175 peer-reviewed online articles was taken place. This paper outlines the existing studies and undertakes a bibliometric survey of the literature on poverty and environmental deterioration [4]. Based on some text mining data, clustering is utilized to determine how environmental deterioration increases poverty. Finally, we suggest that one of the directions for future research should be to include several databases to support the conclusions of environmental deterioration and poverty. They found that Asia is the first continent in sustainable forest management [5]. They used text mining approach on the Scopus dataset. They used LDA algorithm to determine the quantity of publications. They discovered the modifications made in the SFM field by employing this method year by year using text mining to locate research narratives on sustainable forest management. After thorough analysis of the data from the articles on the role of agroforestry in climate change mitigation and adaptation in Sub-Saharan Africa. The authors identify the most common agroforestry practices and tree species used in the region, as well as the potential benefits and problems of putting agroforestry into practice to lessen climate change and the agroforestry adoption by small-scale farmers in Africa [6]. The authors identify the factors that influence farmers' decisions to adopt practices of agroforestry.

3 Methodology

3.1 Design Strategy

In the present research, the keyword used is agroforestry and as a contribution to the Google Scholar search which displays various articles related to it. As the Google Scholar is usually the combination of articles from various websites. Based on the various html formats of the web page, a code is designed to web scrape. The information web scraped will be pushed into a CSV file and then perform the topic modeling analysis on the data [7]. The data web scraped must undergo text preprocessing through which the textual data can be made ready for analysis. Usually, the text preprocessing is performed through a lots of steps included in it and after the preprocessing is being completed, we get the corpus of the data related to the agroforestry. The topic kind of analysis can be made using some of the topic modeling techniques.

3.2 System Architecture Diagram

The architecture that is followed in the data collection and topic modeling on agroforestry is shown in Fig. 1.

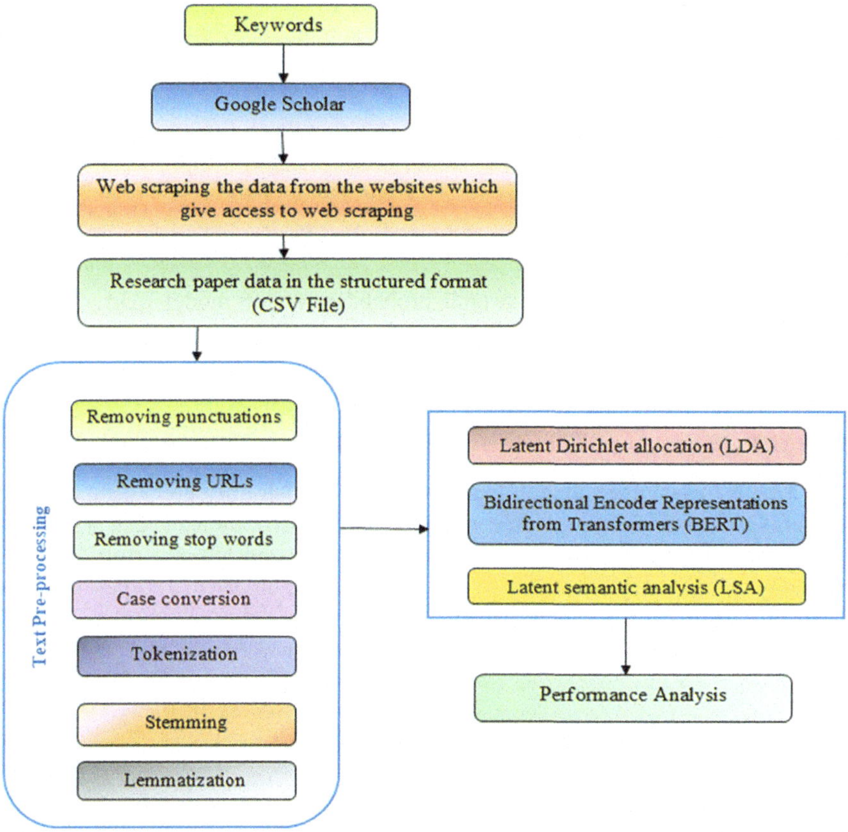

Fig. 1 System architecture

Data Extraction

A total of 3300 articles have been collected from the Google Scholar and the articles we collected are published between 2017 and 2022 i.e., last five years of data. These articles are the data from websites that have been acquired through web scraping like science direct and osti.gov.in.

Text Preprocessing

Preparing raw text data for analysis or natural language processing involves cleaning and altering it. This process is known as text preprocessing.

3.3 Methods Used

1. ***LDA (Latent Dirichlet Allocation)***

Latent Dirichlet Allocation (LDA) is a well-liked topic modeling method to extract topics from a given corpus [8]. Text in a document can be categorized using LDA to a certain topic.

LDA model can be mathematically represented by the following equation.

$$P(W, Z, \varphi, \alpha, \beta) = \pi_{i=1}^{k} P(\varphi_i; \beta) \pi_{j=1}^{M} P\left(\theta_j; \alpha\right) \pi_{t=1}^{N} P\left(Z_{j,t} | \theta_j\right) P\left(W_{j,t} | \varphi Z_{j,t}\right) \quad (1)$$

$P(W, Z, \varphi, \alpha, \beta)$ represents overall probability of the model LDA, $P(\varphi_i; \beta)$ represents Dirichlet distribution of topics over terms, $P\left(\theta_j; \alpha\right)$ represents Dirichlet dispersion of documents across each topic, $P(Z_{j,t}|\theta_j)$ represents topic probability present in a document, $P(W_{j,t}|\varphi Z_{j,t})$ represents word probability present in a topic.

2. ***BERT (Bidirectional Encoder Representations from Transformers)***

It is a built-in language model depends on the Transformer architecture, which is useful in natural language processing for the classification of text, answering questions, and named entity recognition. BERT is also capable of performing topic modeling, which involves identifying the topics that are hidden in a corpus of textual data.

The TF-IDF scores can be evaluated using the below formula:

$$C - \text{TF} - \text{IDF} = \frac{t_i}{w_i} \times \log\left(m / \sum_{j}^{n} t_j\right) \quad (2)$$

t represents the frequency of each word for each class *i*, *w* represents the total number of words. Average number of words per class *m* divided by the total frequency of the word across all the *n* classes.

3. ***LSA (Latent Semantic Analysis)***

It is a statistical technique that identifies motifs in the connections between individual texts and documents in a corpus of text. A lengthy text is reduced using the mathematical technique of singular value decomposition (SVD) in the form of a matrix with count of the word for every document (Rows represent distinct words, whereas columns represent each document). By using this method, the number of rows is decreased while the similarity between the columns is preserved. The cosine similarity between any two columns is then used to compare the documents. Values near 1 indicate relatively similar.

4 Observations and Results

4.1 Metadata from Google Scholar

The metadata on agroforestry from the Google Scholar search engine consist of 3300 papers, complete with title, abstract, published year, and publication, have been extracted here and pushed the data into CSV file in the structured format.

In the collected corpus, words with their respective counts are placed in descending order in Table 1, the word agroforestry is having the highest count among the corpus.

The words within the corpus can be visualized in the word cloud in which the word with high count will be visible with large size and the words with low count will be evident in smaller size.

Figure 2 shows the words with a word having weight greater than 10.

The inter topic distance map is displayed in the left portion of the output of the topic modeling analysis using LDA, which has two components [1]. When utilizing PCA to construct the map, the right side of the output displays a bar graph with the various analytics for each topic, showing that each topic has its own special set of terms and frequency distribution.

The output for LDA which has two partitions is shown in Fig. 3.

Based on the calculated TF-IDF scores of the words, a word with highest score will be on the bar graph at the top. TF-IDF scores of words in Topic 5 are wildfire_0.0621535, fire_0.0328584, vegetation_0.0247968, nonstate_0.0244353, and

Table 1 Words with maximum frequency from the corpus collected

S. No.	Index	Word	Count
1	240	Agroforestry	13,495
2	6888	System	9375
3	7190	Tree	6550
4	6476	Soil	5562
5	7454	Crop	4045

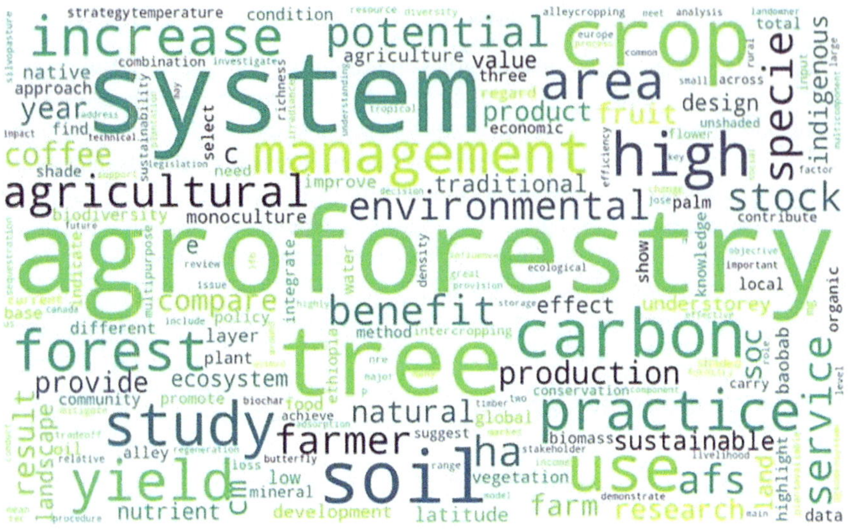

Fig. 2 Words having their weights in a document greater than 10

the word provincial_0.0223714. Figure 4 displays the bar graph for each topic that can be analyzed based on the words in each topic.

The similarity matrix that is developed will show the connections between the various problems. The ratings for similarity will fall between 0.5 and 1. The commonality of the topics is depicted in the BERT model as shown in Fig. 5.

Topics are identified in Latent Semantic Analysis (LSA) as a collection of terms. By examining the connections among words in a document-term matrix and using singular value decomposition (SVD) to make the matrix less dimensional, LSA is able to pinpoint these subjects. The topics from the LSA are displayed in Fig. 6.

4.2 Observations

High frequency words can be helpful in creating a variety of ideas. We must compute the coherence score in order to examine the effectiveness of our LDA. The LDA's coherence score is 0.35. Since a suitable perfect LDA model is having a coherence score that is closest to 0.38, model LDA is the suitable model for the current topic modeling investigation. The main benefit of topic modeling is the creation of many new subjects from the data we have [9]. The coherence value provided by LSA is 0.4035. BERT is a good model, since it has a coherence score 1 for topic modeling in text mining on Agroforestry.

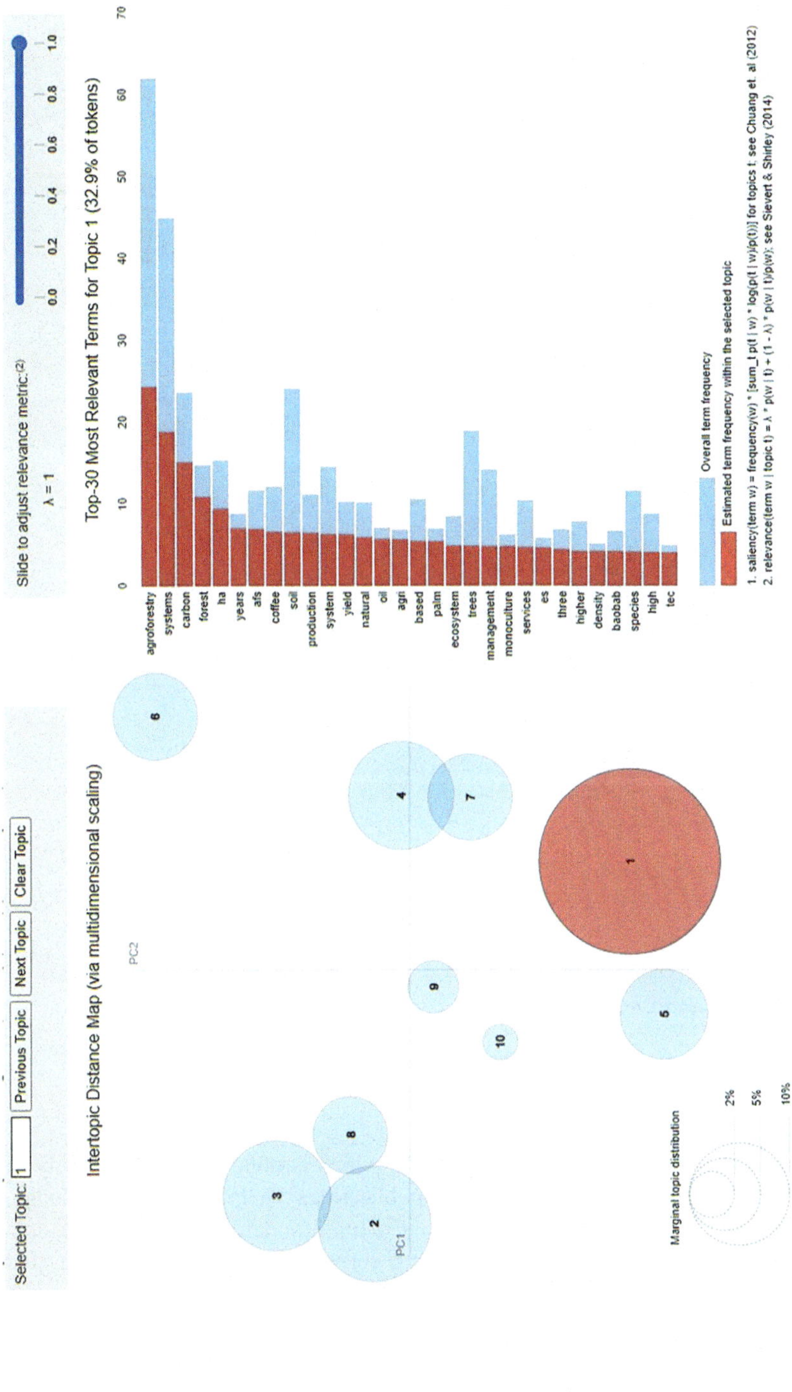

Fig. 3 Graphical visualization for Topic 1 using LDAvis

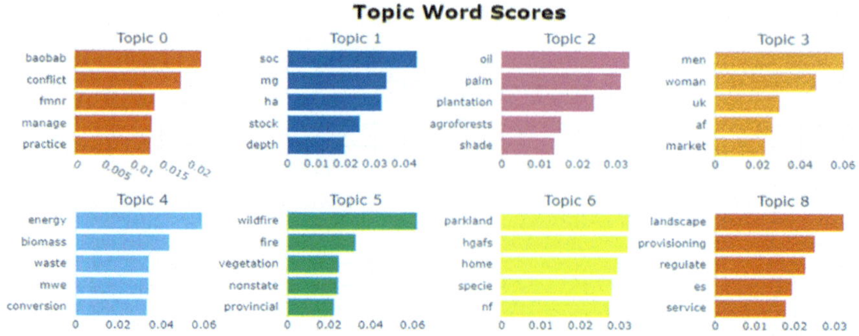

Fig. 4 A bar graph based on BERT for each topic

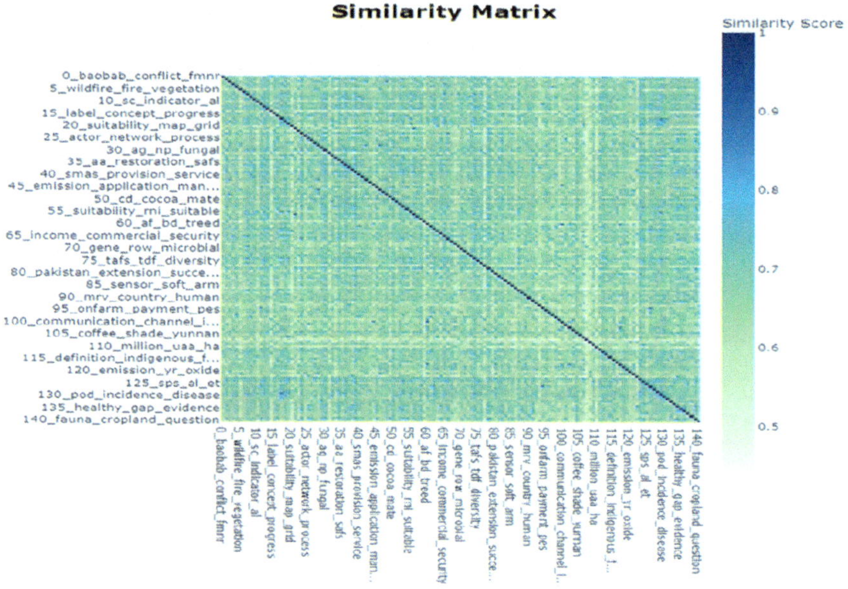

Fig. 5 Similarity matrix of BERT

5 Conclusion and Future Study

The textual analysis of the agroforestry data, which is obtained from a variety of sources, including Google Scholar research publications is helpful to gain knowledge. Different subjects can be identified with the aid of topic modeling tools, and each topic has a unique word frequency derived from the corpus. Finding the terms connected to the relevant keyword is made easier with the help of the corpus that was taken from the analytics. LDA is the best model in comparison for the present work.

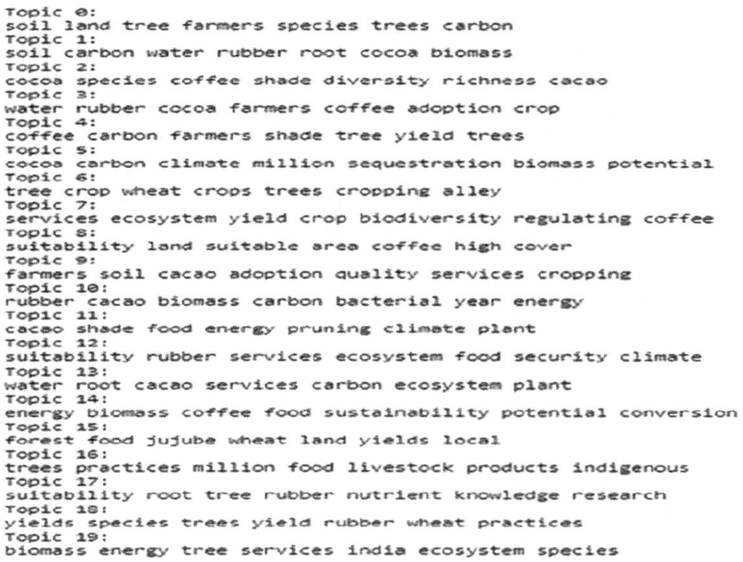

Fig. 6 Topics obtained from LSA model

As of right now, this project has done agroforestry analysis. Gaining fresh perspectives requires research of other agroforestry-related terms. Other topic modeling methods like PLSA and PLDA may be included in future research. The dashboard should be easier to grasp and more visually appealing. A dashboard can be created to present the study findings in an approachable and understandable manner, extending its reach to a larger audience by addition of sophisticated topic modeling methods and approachable data.

References

1. Hong S, Kim J, Yang E (2022) Automated text classification of data of higher education buildings using text mining and machine learning techniques
2. Zuliani A, Contiero B, Schneider MK, Arsenos G, Bernués A, Dovc P, Gauly M, Holand Ø, Martin B (2020) Topics and trends in Mountain Livestock Farming research: a text mining approach
3. Suh Y (2021) Sectoral patterns of accident process for occupational safety using narrative texts of OSHA database
4. Burki MAK, Burki U, Najam U (2021) Environmental degradation and poverty: a bibliometric review
5. Schober A, Šimunović N, Darabant A, Stern T (2021) Identifying sustainable forest management research narratives: a text mining approach
6. Ndang'ang'a KM, Finlayson JF, Muthuri CC, Atieno JRO (2020) An exploratory analysis of agroforestry research in Africa: a bibliometric study
/newpage

7. Lee SH, Cho YW, Im ET, Gim G-Y (2019) A study on customer satisfaction analysis of public institutions using social text mining
8. Gupta RK, Agarwalla R (2022) Prediction of research trends using LDA based topic modeling
9. Zhang Y, Jiang X, Mentzer AJ, McVean G, Lunter G (2022) Topic modeling identifies novel genetic loci associated with multimorbidities in UK Biobank

Generative Adversarial Networks: A Comprehensive Review

Pavani Kotha, V. Janardhan Babu, and Sreejyothsna Ankam

Abstract Numerous models in the deep learning field have been created as a result of the rise in processing capacity. A generative model called Generative Adversarial Networks (GAN) first appeared in 2014. Many architectures of GAN have been proposed in the process of research conducted on GAN. Any GAN architecture is the result of the competition between two networks, the Generator and Discriminator, to determine the distribution of the sampled data. This process helps in many applications like text to image conversion, style transfer, generating new images, attribute transfer, photo enhancement. This paper helps in knowing about the main working principle of any GAN architecture, recent advances, and applications of GAN.

Keywords Generative Adversarial Network (GAN) · Generator · Discriminator · Vanilla GAN · Style GAN · BiGAN

1 Introduction

1.1 Motivation Behind GAN

Generative Adversarial Networks (GANs) are designed to generate data that is realistic and resembles a specific distribution. GANs are a kind of deep learning model that Ian Goodfellow and his associates first developed in 2014. The core concept of GANs is to throw a generator and a discriminator, two neural networks, against one another in a game-like environment. Since GANs don't need labeled training

P. Kotha (✉)
Department of CSE, SVPCET, Puttur, India
e-mail: kothapavani1990@gmail.com

V. Janardhan Babu
Department of AI, SVPCET, Puttur, India

S. Ankam
Department of AIML, GRIET, Bachupally, Hyderabad, India

B. R. Devi et al. (eds.), *Proceedings of Fifth International Conference on Computer and Communication Technologies*, Lecture Notes in Networks and Systems 897,
https://doi.org/10.1007/978-981-99-9704-6_9

data, they fall under unsupervised learning. They make an effort to come close to the training data's actual distribution as they gain knowledge from unlabeled examples. Realistic and imaginative data samples are produced by GANs, which are frequently difficult for human viewers to identify from actual data. This skill has inspired interest in a variety of industries, including art, design, and entertainment, where realistic content development is crucial.

1.2 Structure of GAN

Generative models in deep learning are the base for implementing Generative Adversarial Networks. GAN generates realistic images from latent space. Generator and Discriminator are the two networks that make up every GAN architecture. The networks are trained in tandem with one another. Starting with random noise, the generator creates an output that is fed into the discriminator along with samples of actual data. The discriminator makes an effort to differentiate between the generator's output and the original sample. The categorization task is carried out by the discriminator. The generator aims to produce samples that are similar to the starting sample. The discriminator's objective is to appropriately classify the samples. Both networks organize themselves through back propagation as a result of the learning process. When the generator is creating samples that are as similar to the original sample as possible, the training process finds a Nash equilibrium and the discriminator is unable to increase its accuracy. They engage in combat with one another similar to a game of min–max while learning (Fig. 1).

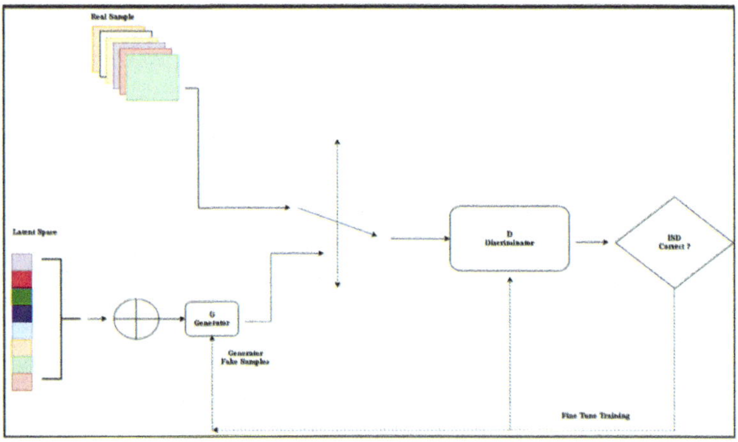

Fig. 1 Architecture of GAN

1.3 Loss Functions and Training Challenges

Here, we lay down the mathematical foundation and GAN training strategy. Considering that the genuine data distribution's labels are 1, while the data's labels in the created data distribution are 0, respectively. Initially, the discriminator should be optimized by fixing the generator. During discriminator training, cross entropy is minimized. The loss function looks like this:

$$L_{\mathrm{d}} = \log(D(x)) + \log(1 - D(G(z)))$$

We must maximize the loss function to make the discriminator optimal. The generator must be optimized by minimizing the following equation.

$$L_{\mathrm{g}} = \log(1 - D(G(z)))$$

The generator and discriminator equations can be combined to form the following expression for the overall loss function $L(D, G) = \min_{\mathrm{G}} \max_{\mathrm{D}}\left\{E_{x\sim P_{\mathrm{data}(x)}} \log(D(x)) + E_{z\sim P_{z(z)}} \log(1 - D(G(z)))\right\}$.

Maximizing the loss function serves as the training criteria for the discriminator by fixing the generator.

$$D_{\mathrm{G}}^{*} = \max_{\mathrm{D}}\{L(G, D)\}$$

The loss function can be reformulated as

$$L(G, D) = \int x P_{\mathrm{data}(x)} \log(D(x))\mathrm{d}x + \int z P_{z(z)} \log(1 - D(G(z)))\mathrm{d}z$$

$$\int x P_{\mathrm{data}(x)} \log(D(x))\mathrm{d}x + \int x P_{g(x)} \log(1 - D(x))\mathrm{d}x$$

Taking the above equation's derivative, the discriminator's optimal value

$$D_{\mathrm{G}}^{*} = \frac{P_{\mathrm{data}(x)}}{P_{\mathrm{data}(x)} + P_{g(x)}}$$

For the discriminator *D*, this is the best option. The equation above can be used to calculate the ratio of the two probability densities by the GAN discriminator. *D*(*x*) shows how likely it is that *x* was derived from actual data as opposed to made-up information. If the input data are genuine data, the discriminator seeks to close the gap between *D*(*x*) and 1. If the input data comes from the created data *G*(*z*), the generator *G* wants to get *D*(*G*(*z*)) as close to 0 as possible, while the discriminator wants it as close to 1. The generator's goal is to reverse *D* and reduce the loss function, giving us the best possible G^{*}.

$$G^* = \min_{G} L\left(D_G^*, G\right)$$

The minimum of the aforementioned requirements is met if and only if $P_{\text{data}(x)} = P_{g(x)}$. At that point $D_G^* = 1/2$. In order to learn the GAN's parameters, we must train the discriminator D to maximize the input data's accuracy using either the generated data $G(z)$ or the real data x. The reduction of $\log(1 - D(G(z)))$ should likewise be trained into Generator G.

2 Variants of GAN

2.1 *Vanilla GAN*

Vanilla GAN is the simplest model of GAN. Generator and discriminator are implemented using multilayer perceptrons. Weights are learned using Stochastic Gradient Descent (Fig. 2).

The objective function of Vanilla GAN is evaluated as $L(D, G) = \min_G \max_D \left\{E_{x\sim P_{\text{data}(x)}} \log(D(x)) + E_{z\sim P_{z(z)}} \log(1 - D(G(z)))\right\}$. This GAN architecture suffers from various problems like mode collapse, vanishing gradients, and convergence failure [1, 2]. When all of the generator's outputs are the same, mode collapse happens. The discriminator detects the generated samples as fraudulent with ease. Vanishing Gradient occurs when the earliest layers of the network fail to update their weights as a result of a little gradient change. A failure to converge occurs when the generator and discriminator are unable to reach equilibrium (Fig. 3).

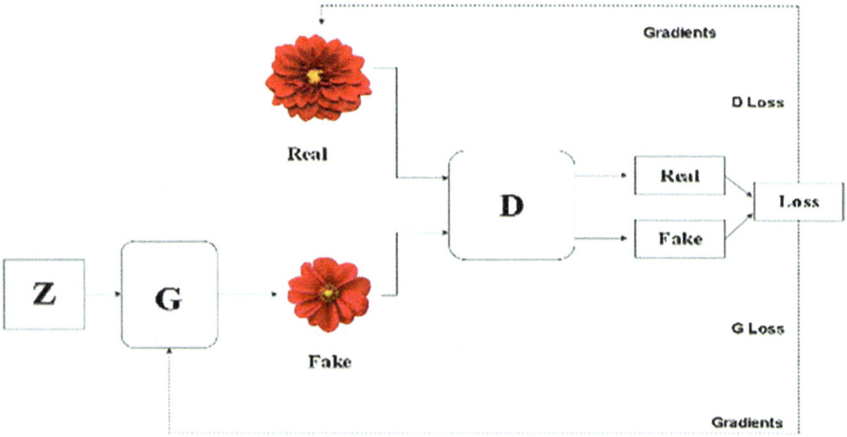

Fig. 2 Architecture of GAN

Algorithm Pseudo code for Vanilla GAN

1: **Inputs:**
Batch size n
Number of epochs itr
Step size s
Learning rate η
Generator input $x \sim P_{data}(x)$
Target Data $z \sim P_g(z)$

2: **Initialize:**
$\theta_G, \theta_D, s, \eta$, itr

3: **for** p= 1 to itr **do**
for k=1 to s **do**
Sample n noise vector $\{z_1, z_2, z_3 \ldots\ldots z_n\}$ from $P_g(z)$
Sample n data vector $\{x_1, x_2, x_3 \ldots\ldots x_n\}$ from $P_{data}(x)$
Update θ_D by gradient descent

$$L(\theta_D) = [\frac{1}{n}\sum_{i=1}^{n} log(D(x_i\theta_D)) + \frac{1}{n}\sum_{i=1}^{n} log(1 - D(z_i, \theta_G))]$$

$$\mu_D = \Delta_\theta L(\theta_D)$$

$$\theta_D^{K+1} = \theta_D + \eta\mu_D$$

end for
Sample n noise vector $\{z_1, z_2, z_3 \ldots\ldots z_n\}$ from $P_g(z)$
Update θ_G by gradient descent

$$L(\theta_G) = \frac{1}{n}\sum_{i=1}^{n} log(1 - D(G(z_i, \theta_G))$$

$$\mu_G = \Delta_\theta L(\theta_G)$$

$$\theta_G^{K+1} = \theta_G - \eta\mu_G$$

end for

Fig. 3 Pseudo code for Vanilla GAN

2.2 *DCGAN*

A brand-new class of CNNs dubbed Deep Convolutional Generative Adversarial Networks (DCGANs) was proposed by Radford et al. [3]. DCCGAN was created using CNN, but with the following modifications. Nointerconnected hidden layers in DCGAN. Fractional-strided convolutions were used in the generator and discriminator in place of layer pooling, integrating the generator and discriminator with the batch normalization concept. ReLU activation is applied to all generative model layers except the top layer, using Leaky ReLU in the layer of the discriminative model (Fig. 4).

Layer (type)	Output Shape	Param #
dense_7 (Dense)	(None, 12544)	1254400
batch_normalization_21 (BatchNormalization)	(None, 12544)	50176
re_lu_14 (ReLU)	(None, 12544)	0
reshape_7 (Reshape)	(None, 7, 7, 256)	0
conv2d_transpose_21 (Conv2DTranspose)	(None, 7, 7, 128)	819200
batch_normalization_22 (BatchNormalization)	(None, 7, 7, 128)	512
re_lu_15 (ReLU)	(None, 7, 7, 128)	0
conv2d_transpose_22 (Conv2DTranspose)	(None, 14, 14, 64)	204800
batch_normalization_23 (BatchNormalization)	(None, 14, 14, 64)	256
re_lu_16 (ReLU)	(None, 14, 14, 64)	0
conv2d_transpose_23 (Conv2DTranspose)	(None, 28, 28, 1)	1600

Fig. 4 Architecture of generator

2.3 BiGAN

Bidirectional generative adversarial networks (BIGAN), a suggested algorithm by Donahue et al., are mentioned. To create a kind of GAN that can learn rich representations for us in applications like unsupervised learning, this idea is being put out. Along with generator and discriminator, the design also includes an encoder that converts data points to the latent feature space of the generative model. The data generated by the generator and the latent space, as well as the data and the latent space compared to the latent space and generated input, are used as the input to the discriminator. Hence the training objective is defined as (Figs. 5 and 6)

$$\min_{\text{G,Enc}} \max_{\text{D}} \left\{ E_{x \sim P_{\text{data}(x)}} \log(D(x, \text{Enc}(x))) + E_{z \sim P_{z(z)}} \log(1 - D(G(z))) \right\}$$

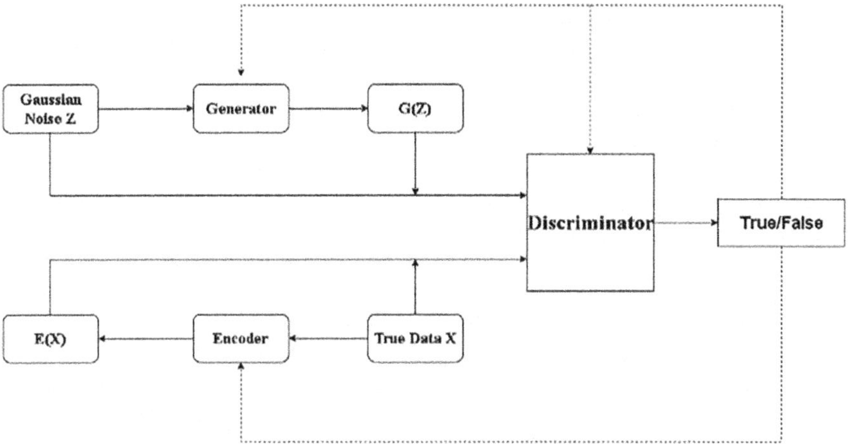

Fig. 5 Architecture of BiGAN

3 Evaluation Metrics

3.1 Quantitative Metrics

Quantitative metric is the process of producing exact numerical ratings that are used to assess the effectiveness of created images.

Maximum Mean Discrepancy

Drawing samples from both created images and real data independently determines the dissimilarity between the distributions. This is accomplished with the aid of a kernel, a function that may be used to approximately represent the dataset's distribution. The distance decreases when the distributions get closer.

Wasserstein Critic

To determine the distance between the true data distribution and the produced data distribution, the Wasserstein distance [4] was proposed in WGAN. The Wasserstein distance between the genuine information and the created data is calculated using the critic.

Frechet Inception Distance

Heusel et al. [5] proposed FID. It is a quantitative statistic that draws features from produced and actual images that were previously trained using an Inception v2 model. It incorporates the collected features into a continuous multivariate Gaussian distribution. After that, the Frechet distance is used to determine the sample quality $\text{FID} = ||\mu_x - \mu_g||^2 - \text{Tr}(\sigma_x - \sigma_g - 2\sqrt{\sigma_x \sigma_g})$. In order to obtain better-quality GAN, we must reduce the FID score. FID works good in terms of computational

Algorithm Pseudo code for BIGAN

1: **Inputs:**
Batch size n
Number of epochs itr
Step size s
Learning rate η
Generator input $x \sim P_{data}(x)$
Target Data $z \sim P_g(z)$
2: **Initialize:**
$\theta_G, \theta_D, s, \eta$, itr
3: **for** p= 1 to itr **do**
for k=1 to s **do**
Sample n noise vector $\{z_1, z_2, z_3 --- z_n\}$ from $P_g(z)$
Sample n data vector $\{x_1, x_2, x_3 .--- x_n\}$ from $P_{data}(x)$
Update θ_D by gradient descent
f(z)=G(z)
f1(x)=Enc(x)
Concatenate [f(z),z]
Concatenate[x,f1(x)]
$$L(\theta_D) = [\frac{1}{n}\sum_{i=1}^{n} log(D(x_i\theta_D), Enc(x)) + \frac{1}{n}\sum_{i=1}^{n} log(1 - D(z_i, \theta_G), z)]$$
$\mu_D = \Delta_\theta L(\theta_D)$
$\theta_D^{K+1} = \theta_D + \eta\mu_D$
end for
Sample n noise vector $\{z_1, z_2, z_3 --- z_n\}$ from $P_g(z)$
Update θ_G by gradient descent
f(z)=G(z)
f1(x)=Enc(x)
Concatenate [f(z),z]
Concatenate[x,f1(x)]
$$L(\theta_G) = \frac{1}{n}\sum_{i=1}^{n} log(1 - D(G(z_i, \theta_G), z)$$
$\mu_G = \Delta_\theta L(\theta_G)$
$\theta_G^{K+1} = \theta_G - \eta\mu_G$
end for

Fig. 6 Architecture of BiGAN

efficiency and resilience. Nevertheless, it only considers the distributions of the first two order moments. However, it makes the unreliable assumption that features have a Gaussian distribution. A sufficient sample size must be used to calculate FID (often above 50 K). Smaller sample sizes may cause the FID to be overestimated.

R-Precision

To evaluate GAN models like text to image synthesis, Xu et al. [6] suggested the R-precision. It assesses the text's coherence in artificially created images. We select *R* text descriptions that are consistent with a single image and *N-R* text descriptions that are inconsistent. Compute the cosine similarity between the *N* sentence level features and image features obtained by the respective text and image encoders. The final step is to rank the potential text descriptions according to degree of similarity.

R-precision is determined as r/R if it is discovered that r text descriptions agree with the highest rated R results.

4 Conclusion and Future Directions

The GANs fundamental theory, structure, variants, evaluation metrics, applications are all examined and summarized in this study. It provides a comprehensive grasp of the literature on GAN. In this work, a wide range of GAN applications are presented. In order to improve GAN efficiency, new and improved solutions to both new and old problems in GANs must be investigated. Although the field of GAN is a promising one, there are still issues with it, such as unstable training, non-convergence, a lack of agreement on an assessment metric, the need for additional processing power, and the complexity of the model. GAN is a fascinating and practical study area with many applications. Avoiding the issue of mode collapse in GANs is still a study area that has to be explored. The existence of Nash equilibrium and the theory of a GAN model's convergence are two further research fields. We can combine GAN with variational autoencoders. There will be a significant influence on many applications when assessment measures for comparing non-image data are proposed.

References

1. Raj VB, Hareesh K (2020) Review on generative adversarial networks. In: 2020 international conference on communication and signal processing (ICCSP), Chennai, pp 0479–0482. https://doi.org/10.1109/ICCSP48568.2020.9182058
2. Salimans T, Goodfellow I, Zaremba W, Cheung V, Radford A, Chen X (2016) Improved techniques for training GANs. In: Proceedings of the 30th international conference on neural information processing systems, NIPS'16, Barcelona, Spain, Dec 2016. Curran Associates Inc., pp 2234–2242. ISBN 978-1-5108-3881-9
3. Radford A, Metz L, Chintala S (2015) Unsupervised representation learning with deep convolutional generative adversarial networks. arXiv preprint arXiv:1511.06434
4. Gulrajani I et al (2017) Improved training of Wasserstein GANs. Adv Neural Inf Process Syst 30
5. Heusel M et al (2017) GANs trained by a two time-scale update rule converge to a local Nash equilibrium. In: NIPS
6. Xu T, Zhang P, Huang Q, Zhang H, Gan Z, Huang X, He X (2018) AttnGAN: fine-grained text to image generation with attentional generative adversarial networks. In: Proceedings of the IEEE/CVF conference on computer vision and pattern recognition, June 2018, pp 1316–1324
7. Mirza M, Osindero S (2014) Conditional generative adversarial nets. arXiv preprint arXiv: 1411.1784
8. Salimans T et al (2016) Improved techniques for training GANs. Adv Neural Inf Process Syst 29
9. Zhang R, Isola P, Efros AA, Shechtman E, Wang O (2018) The unreasonable effectiveness of deep features as a perceptual metric. In: Proceedings of the IEEE/CVF conference on computer vision and pattern recognition, June 2018, pp 586–595

10. Zhang Z, Xie Y, Yang L (2018) Photographic text-to-image synthesis with a hierarchically-nested adversarial network. In: Proceedings of the IEEE conference on computer vision and pattern recognition, pp 6199–6208
11. Lin H, Fu X, Jing C, Ding X, Huang Y (2018) A2Net: adjacent aggregation networks for image raindrop removal. arXiv:1811.09780. [Online]. Available: https://arxiv.org/abs/1811.09780
12. Donahue J, Krähenbühl P, Darrell T (2017) Adversarial feature learning
13. Isola P et al (2017) Image-to-image translation with conditional adversarial networks. In: Proceedings of the IEEE conference on computer vision and pattern recognition
14. Zhu J-Y et al (2017) Unpaired image-to-image translation using cycle-consistent adversarial networks. In: Proceedings of the IEEE international conference on computer vision
15. Zhang H et al (2017) StackGAN: text to photo-realistic image synthesis with stacked generative adversarial networks. In: Proceedings of the IEEE international conference on computer vision
16. Wang T-C et al (2018) High-resolution image synthesis and semantic manipulation with conditional GANs. In: Proceedings of the IEEE conference on computer vision and pattern recognition
17. Ledig C et al (2017) Photo-realistic single image super-resolution using a generative adversarial network. In: Proceedings of the IEEE conference on computer vision and pattern recognition
18. Karras T, Laine S, Aila T (2019) A style-based generator architecture for generative adversarial networks. In: Proceedings of the IEEE/CVF conference on computer vision and pattern recognition

System Modification Process to Reduce Calculation Complexity in Backward Forward Sweep Algorithm

T. S. N. G. Sarada Devi and T. V. L. N. Pavan Phani Kumar

Abstract In backward forward sweep algorithm, even though forward sweep (FW) to compute bus voltages, backward sweep (BW) is to compute branch currents, junction bus voltage calculation is required as mandatory additional step in backward sweep process. This calculation complexity increases with increase in number of junction buses particularly when loads are also connected to junction buses and consequently, execution time also increases. This paper concentrates on this problem and contributes with System Modification Process (SMP) to reduce Backward Forward (BW-FW) sweep algorithm calculations complexity. In SMP, transfer buses are connected to a given electrical distribution system to avoid calculating junction bus voltage, which is a mandatory step in backward sweep process, thus decoupling the voltage and current calculations. Hence in this present work, conventional backward forward sweep algorithm is modified as Decoupled Backward Forward Sweep (DBFS) algorithm. The system modification process is demonstrated with a 3-node lateral as numerical example and implemented on practical 55 bus three-phase unbalanced distribution system. In Matlab simulation, the efficiency of the DBFS algorithm has been tested with unbalanced radial distribution systems (URDS). The simulation outcomes in terms of number of repetition, execution time obtained, bus voltages per phase, and total computed power per phase with DBFS are compared to conventional BW-FW sweep algorithm results to demonstrate the proposed method efficiency. From the results, the proposed DBFS algorithm not only reduces conventional BW-FW sweep algorithm calculations complexity but also requires less amount of execution time (including SMP) to achieve same results without degrading the efficiency.

T. S. N. G. Sarada Devi (✉)
Department of Electrical Engineering, Malla Reddy Engineering College for Women (Autonomous), Hyderabad, India
e-mail: sarada216@gmail.com

T. V. L. N. Pavan Phani Kumar
Tata Advanced Systems Limited, Bangalore, India
e-mail: pavanphani417@gmail.com

B. R. Devi et al. (eds.), *Proceedings of Fifth International Conference on Computer and Communication Technologies*, Lecture Notes in Networks and Systems 897,
https://doi.org/10.1007/978-981-99-9704-6_10

Keywords Backward forward sweep algorithm · System modification process · Decoupled backward forward sweep algorithm

1 Introduction

The loads connected to the feeders in distribution system are generally multi phase, unbalanced, and distributed loads with non-uniform load curves. Apart from this, it has feeders with "high *R*/*X* ratios, huge number of branches that are radial in structure". This makes the distribution system into complex and the direct methods used for power flow solution of transmission system are not suitable for distribution system (DS) without some assumptions. For finding out the power flow solution of the radial distribution network, several researchers focused on some techniques and those are available in the literature [1–3]. In [4], direct approach method for unbalanced three-phase distribution and in [5], new approach to load flow solutions are developed. For real time analysis of primary distribution system, a three-phase power flow method has been described in [6]. Different methods with simple changes are in [7–14]. However, most of the researchers prefer BackWard-ForWard (BW-FW) Sweep algorithm to find the power flow solution for (multi phase) unbalanced distribution system as this method is fast and gives accurate results. Detailed literature on this algorithm is given in [15]. Further different algorithms have been developed based on BW-FW sweep method. Reference [16] expanded the BW-FW sweep algorithm to handle the lack of slack bus. In [17], the forward sweep calculated bus voltages are multiplied with the ratio of calculated and specified voltages. The matrix-based BW-FW sweep algorithm is highlighted in [18]. Linear three-phase power flow is described in [19]. The methodology in [20] does not require any assumptions related to the *R*/*X* ratio. Breadth-first search method for creation of modified incidence matrix, Nodal admittance matrix usage in BW-FW sweep algorithm [21] have been developed to solve power flows in the presence of micro grids and DGs [22–24]. Different concepts like fixed-point and Contraction Mapping Principle [25], network tree depth concept [26] are implemented in BW-FW method. A complete study on this method has been explained by comparing different methods in [27]. However, the backward sweep process is same in all the reference papers based on BW-FW sweep algorithm.

Problem Identification

In particular, when the loads are also connected (lumped) at junction buses, **calculating the junction bus voltages is an additional step during backward sweep process** in conventional BW-FW sweep algorithm. To overcome this, system modification process is introduced in this paper. The aim of this paper is to reduce the calculations complexity in conventional backward forward sweep algorithm by connecting transfer/pseudo buses to junction buses through the branches of negligible impedance. In next section, backward forward sweep algorithm is explained with an example system.

2 Backward Forward Sweep Algorithm

In given system, all end nodes are said to be "terminal nodes" i.e. only single branch is connected to it. On the other hand, if two or more branches are connected to a node, it is said to be junction node. The backward forward sweep algorithm/ method is again divided into backward sweep process and forward sweep process. The backward sweep starts with the assumption that "terminal node voltages are equal to feeder voltage at substation". The load currents are calculated using available complex power and terminal node voltages. The branch currents are calculated using Kirchhoff's current law.

2.1 Backward Forward Sweep Algorithm for 3 Bus Lateral System

Assume a single-phase lateral with three nodes as shown in Fig. 1. Network data is tabulated in Table 1. Let node 1 voltage be 7200 V and 0°. In given Fig. 1, node 2 is a junction node and node 3 is a terminal node.

Backward Sweep: It begins by assuming that terminal node voltages are equal to feeder voltage at substation.

The load current at Node "k" is calculated from Eq. (1)

$$I_k^m = \mathrm{conj}\left(\frac{S_k}{V_k^{(m-1)}}\right) \tag{1}$$

The branch current through $k-1$ to k line is calculated using

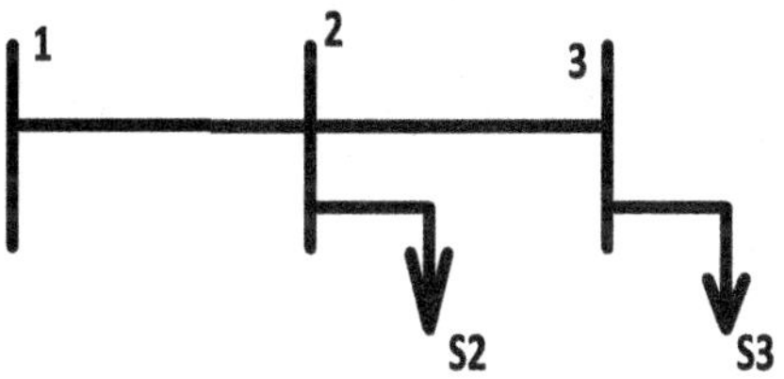

Fig. 1 Single phase lateral with three nodes

Table 1 Line and load data of single-phase lateral

From node	To node	Impedance in Ω	Node	Complex power (kW + jkVAr)
1	2	0.1705 + j0.3409	2	1500 + j750
2	3	0.2273 + j0.4545	3	900 + j500

$$I_{k,k-1}^{m} = \sum I_{k}^{m} \tag{2}$$

Error in voltage (Ver) is calculated by Eq. (3) and it must be less than tolerance (say "er") for convergence.

$$\text{Ver} = \left(V_{k}^{m-1} - V_{k}^{m}\right) < \text{er} \tag{3}$$

where "*m*" is the iteration indicator.

In 3-phase system, Error in voltage per phase is calculated by following equations. The BW-FW sweep process is also implemented per phase and the BW-FW calculation is tabulated in Table 2 along with DBFS.

$$\Delta V_{ia}^{m} = V_{1a}^{m-1} - V_{1a}^{m};\ \Delta V_{ib}^{m} = V_{1b}^{m-1} - V_{1b}^{m};\ V_{ic}^{m} = V_{1c}^{m-1} - V_{1c}^{m}. \tag{4}$$

3 Decoupled Backward Forward Sweep (DBFS) Method

In Fig. 2, load is lumped at junction node also (node 2). With this type of systems, calculating junction bus voltage is a mandatory step in backward sweep process before calculating the load current at junction bus. This calculation complexity further increases with increase in junction buses with load. To reduce this computational complexity, "System Modification Processes" are introduced and incorporated in BW-FW sweep method. Flow chart of modified backward forward sweep algorithm is given in Fig. 3.

Steps in System Modification Process

i. *Identify junction nodes.*
ii. *Connect transfer nodes/pseudo nodes with branch Impedances of negligible value (say 0.0001 Ω) to junction nodes.*
iii. *Transfer junction node loads to its corresponding transfer node.*
iv. *Modify the system in such a way that Loads are placed (lumped) only at terminal nodes.*

3.1 DBFS Method for a 3 Bus Lateral System

Consider same 3 bus single phase lateral as shown in Fig. 1 with same data. Here, bus 2 is junction bus. As a part of system modification process, connect a line segments 2–4 with impedance of 0.0001 + j0.0001 Ω as shown in Fig. 2. The junction bus 2 load is transferred to transfer bus 4 through line segments 2–4. However, in this case, the system size is increased from 3 to 4 nodes, but this additional node (node 4) is handled in the program internally in calculations part only.

Table 2 Comparing two iterations of BFS and DBFS algorithms

BW-FW	MBFS
Backward sweep (1st iteration) $I_3 = \text{conj}\{(900 + j500)/7200/1000\}$ $I_3 = 143\angle - 29$ A $V_2 = V_3 + I_{2-3} * Z_{2-3}$ $7260.1\angle 0.32$ V (***This step is avoided in MBFS method***) $I_2 = \text{conj}\{(1500 + j750)/7260.1\angle 0.32/1000\}$ $I_2 = 231\angle - 26.3$ A $I_{1-2} = I_2 + I_{2-3}$ $I_{1-2} = 332.1 - j171.6$ A $V_1 = V_2 + I_{1-2} * Z_{1-2} = 7200.11 + j0.0841$ V ***Forward sweep*** $V_1 = 7200 + j0.0$ V $V_2 = V_1 - I_{1-2} * Z_{1-2}$ $V_2 = 7084.8 - j83.95$ V $V_3 = V_2 - I_{2-3} * Z_{2-3}$ $V_3 = 7024.87 - j125.03$ V ***Backward sweep 2nd iteration*** $I_3 = \text{conj}\{(900 + j500)/(7024.87 - j125.03)/1000\}$ $I_3 = 126.80 - j73.43$ A $I_{2-3} = I_3$ $V_2 = V_3 + I_{2-3} * Z_{2-3} = 7087.06 - j84.1$ V $I_2 = \text{conj}\{(1500 + j750)/(7087.06 - j84.1)/1000\}$ $I_2 = 210.36 - j108.32$ A $I_{1-2} = I_2 + I_{2-3}$ $I_{1-2} = 337.16 - j181.75$ A $V_1 = V_2 + I_{1-2} * Z_{1-2} = 7206.50 - j0.1505$ V ***Forward sweep*** $V_1 = 7200 + j0.0$ $V_2 = V_1 - I_{1-2} * Z_{1-2}$ $V_2 = 7080.5 - j83.9$ $V_3 = V_2 - I_{2-3} * Z_{2-3}$ $V_3 = 7018.3 - j124.83$ V ***Voltage error checking*** $(V_2)_{\text{ite2}} - (V_2)_{\text{ite1}} = (7084.8 - j83.95) - (7080.5 - j83.9) >$ tolerance value $(V_3)_{\text{ite2}} - (V_3)_{\text{ite1}} = (7024.8 - j125.03) - (7018.3 - j124.83) >$ tolerance value So go to next iteration	***Backward sweep (1st iteration)*** $I_3 = \text{conj}\{(900 + j500)/7200/1000\}$ $I_3 = 143\angle - 29$ A $I_4 = \text{conj}\{(1500 + j750)/7200/1000\}$ $I_4 = 232.9\angle - 26.5$ A $I_{1-2} = I_4 + I_3$ $I_{1-2} = 333.50 - j173.2$ A ***Forward sweep*** $V_1 = 7200 + j0.0$ V $V_2 = V_1 - I_{1-2} * Z_{1-2}$ $V_2 = 7084.09 - j84.15$ V $V_3 = V_2 - I_{2-3} * Z_{2-3}$ $V_3 = 7024.05 - j125.23$ V $V_4 = V_2$ ***Backward sweep 2nd iteration*** $I_3 = \text{conj}\{(900 + j500)/(7024.87 - j125.03)/1000\}$ $I_3 = 126.80 - j73.43$ A $I_{2-3} = I_3 = 126.80 - j73.43$ A $I_4 = \text{conj}\{(1500 + j750)/(7084.05 - j84.16)/1000\}$ $I_4 = 210.45 - j108.37$ A $I_{1-2} = I_4 + I_{2-3}$ $I_{1-2} = 337.25 - j181.8$ A ***Forward sweep*** $V_1 = 7200 + j0.0$ $V_2 = V_1 - I_{1-2} * Z_{1-2}$ $V_2 = 7080.5 - j83.9$ $V_3 = V_2 - I_{2-3} * Z_{2-3}$ $V_3 = 7018.3 - j124.83$ V $V_4 = V_2$ ***Voltage error checking*** $(V_2)_{\text{ite2}} - (V_2)_{\text{ite1}} = (7084.09 - j84.15) - (7080.5 - j83.9) >$ tolerance value $(V_3)_{\text{ite2}} - (V_3)_{\text{ite1}} = (7024.8 - j125.03) - (7018.3 - j124.83) >$ tolerance value So go to next iteration

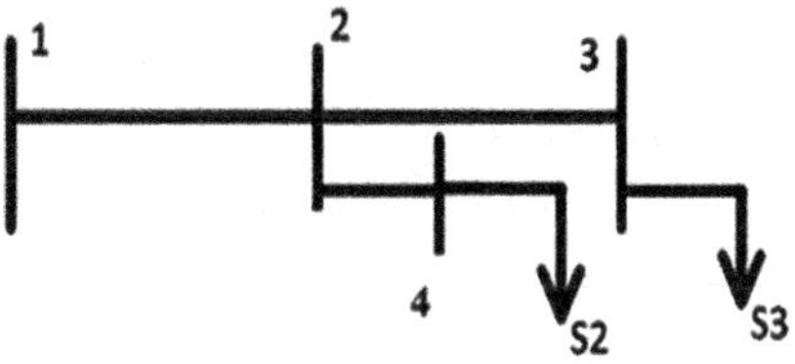

Fig. 2 Modified 1-phase 3 bus lateral with pseudo node (node 4)

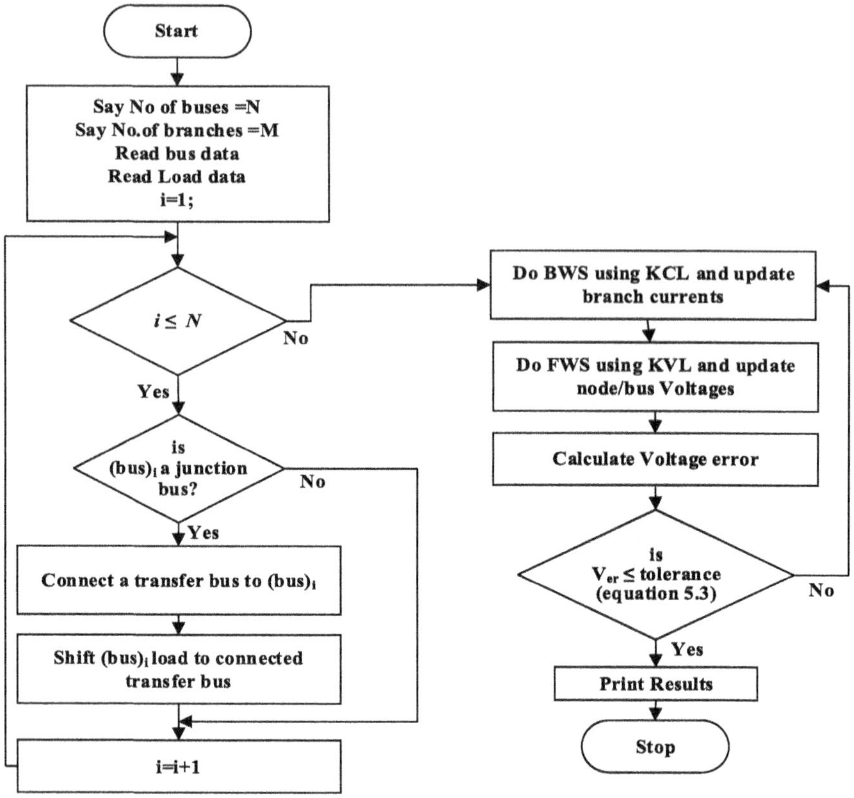

Fig. 3 Flowchart of DBFS method

Backward Sweep: Same as in conventional BW-FW sweep method, the backward sweep in DBFS begins by assuming the voltages at Nodes 3 and 4 (terminal nodes) as 7200 V at an angle 0° (feeder voltage at substation).

The load $I_3 = (S_3/V_3)^*$

$$I_3 = \text{conj}\{((900 + \text{j}500)/7200)1000\}$$
$$I_3 = 143\angle - 29\,\text{A}$$

The current $I_{2-3} = I_3 = 143\angle - 29 = 0.125 - \text{j}0.069\,\text{A}$.

The load current at Node 4 $= I_4 = (S_4/V_4)^* = \text{conj}\left(\frac{1500+\text{j}750}{7200+\text{j}0.00}\right)1000$

$$I_4 = 232.9\angle - 26.5 = 208.33 - \text{j}104.16\,\text{A}$$

The current $I_{1-2} = I_{2-3} + I_4 = 333.5 - \text{j}173.2$ A.

Forward Sweep: Now the forward sweep begins to calculate bus voltages.

The voltage at Node 2 = $V_2 = V_1 - Z_{1-2} * I_{1-2}$, $V_2 = 7084.09 - j84.15$ V.

The voltage at Node 3 = $V_3 = V_2 - Z_{2-3} * I_{2-3}$, $V_3 = 7024.16 - j125.23$ V.

The voltage at Node 4 = $V_4 = V_2 - Z_{2-4} * I_{2-4}$, $V_4 = 7084.05 - j84.16$ V.

This is end of iteration. In the next iteration, BW starts with calculated voltages in place of the initial assumed feeder voltage value. This course of action iteratively repeats until the solution is converged.

With the help of system modification process, node voltage calculations in backward sweep are avoided successfully.

The DBFS iterative procedure stops once the voltage mismatch is less than or equal to tolerance (say 0.0001). First two iterations of BW-FW and DBFS methods are compared and tabulated in Table 2. From Table 2, it is evident that in conventional BW-FW sweep method, voltages of all junction buses need to be calculated in backward sweep process in every iteration, whereas in the proposed DBFS method, this calculation is not required. Hence the amount of complexity of the calculation part is reduced.

4 DBFS Method-Case Study-55 Bus Unbalanced Radial Distribution System (URDS)

The practical 55 bus three-phase unbalanced radial distribution feeder with mutual impedances is taken for consideration in MATLAB environment for testing the effectiveness of proposed DBFS method.

Base values are 1000 kVA and 11 kV. System data is taken from [28]. The proposed DBFS method is implemented on practical 55 bus three-phase URDS. While calculating voltages in forward sweep, voltage drops due to mutual impedances are also added to voltage drops due to self-impedances. Results in terms of bus voltage and total computed power per phase are obtained.

Number of iterations and execution time in two methods are obtained and results are tabulated in Table 3 and illustrated in Fig. 4a. Results in terms of junction bus voltage calculations required per iteration are tabulated in Table 4. Bus voltages per phase are compared to conventional backward forward sweep method results and represented as bar graphs in Fig. 4b. Furthermore, total computed power per phase is obtained with two methods and results are tabulated in Table 5. It is observed that bus voltages are almost same in both the methods with less amount of computational effort in the proposed DBFS method. Total execution time of DBFS method (including system modification process) is very much less with respect to the conventional BW-FW sweep method.

Table 3 Comparing BW-FW, DBFS methods

S. No.	Number of iterations		Execution time in s	
	Existing method	Proposed method	Existing method	Proposed method
1	4	4	0.0623	0.008
2	4	4	0.108	0.0059
3	4	4	0.0105	0.0088
4	4	4	0.0125	0.0072
5	4	4	0.0192	0.0054
Average	4	4	**0.0386**	**0.0071**

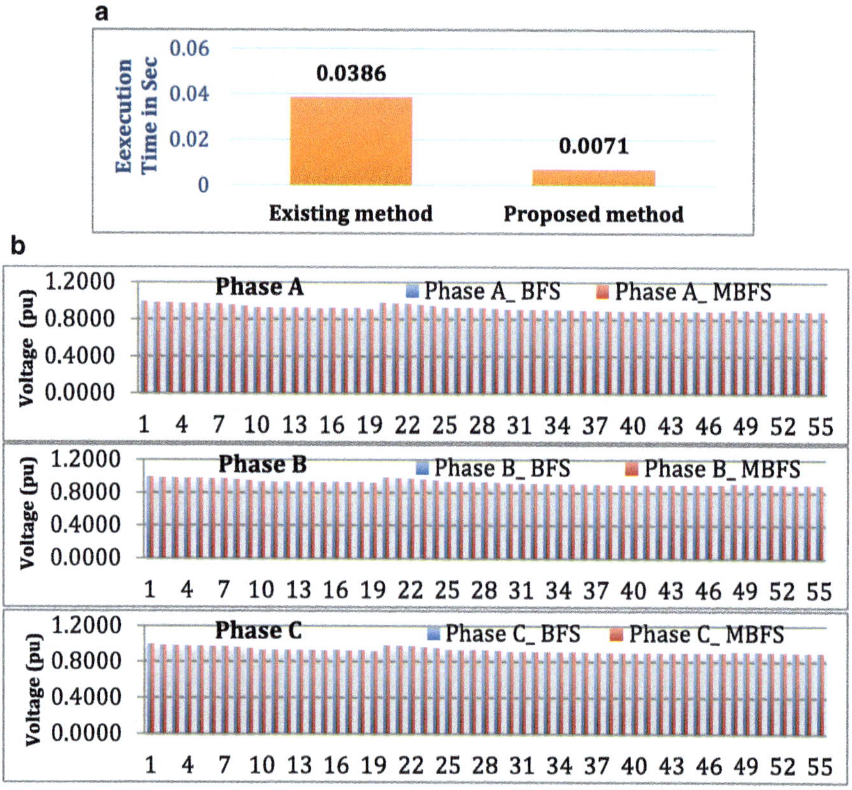

Fig. 4 **a** Comparing execution time of BW-FW and DBFS methods. **b** Comparing bus voltages of BW-FW and DBFS methods

Table 4 Comparing nodal voltage calculations required in BW-FW and DBFS

System	Nodal voltage calculations required per iteration in BW sweep process	
	In BW-FW method	In DBFS method
3 bus	1 time	0
55 bus	21 times	0

Table 5 Computed total power per phase for 55 bus URDS with BW-FW and DBFS

Computed total power in kVA	Phase A	Phase B	Phase C
BW-FW	459	425.29	447.01
DBFS	459.07	425.32	447.05

5 Conclusion

This paper contributes with system modification process (SMP) to reduce Backward Forward (BW-FW) sweep algorithm calculations complexity. In SMP, transfer buses are connected to a given electrical distribution system to avoid calculating junction bus voltage, which is a mandatory step in backward sweep process, thus decoupling the voltage and current calculations.

The system modification process is demonstrated with a 3-node lateral as numerical example and implemented on practical 55 bus three-phase unbalanced distribution system in Matlab simulation. From the results, it is clear that bus voltages are almost same in both the methods with less amount of computational effort in the proposed DBFS method. Total execution time of DBFS method (including system modification process) is very much less with respect to the conventional BW-FW sweep method. In conventional BW-FW sweep method, the voltages of all junction buses need to be calculated in backward sweep process in every iteration, whereas in the proposed DBFS method, this calculation is not required. Hence, the amount of complexity of the calculation part is reduced. Unlike in the conventional BWFW sweep method, backward sweep process focuses only on current calculations and forward sweep process concentrates only on voltage calculations.

References

1. Deb S, Rajan DV, Bose SP, Bhowmik PS (2011) Power flow analysis of power system using power perturbation method. In: IEEE PEAOC
2. Lagace PJ (2012) Power flow methods for improving convergence. In: IEEE IECON 2012
3. Luo GX, Hong HW, Shirmohammadi D, Semlyen A (1988) A compensation based power flow method for weakly meshed distribution and transmission networks. IEEE Trans Power Syst 3(2):753–762
4. Teng J-H (2003) A direct approach for distribution system load flow solutions. IEEE Trans Power Delivery

5. Dhiman J, Thakur T. A new approach to load flow solutions for radial distribution system. In: IEEE/PES T&D
6. Shirmohammadi D, Cheng CS (1995) A three-phase power flow method for real-time distribution system analysis. IEEE Trans Power Syst 671–679
7. Vasquez W, Quilumba F (2016) A load flow method for radial distributions systems with distributed generation. In: IEEE ETCM
8. Casolino GM, Losi A (2015) Load Area application to radial distribution systems. In: IEEE 1st international forum on research and technologies for society and industry leveraging a better tomorrow (RTSI), Turin, Italy, pp 269–273. https://doi.org/10.1109/RTSI.2015.7325109.
9. Martinez JA (2011) Load flow calculations in distribution systems with distributed resources. A review. In: IEEE PES general meeting
10. Mandava S, Ramesh V, Karthikeyan P (2014) A simple load flow method for radial distribution system. In: International conference on advances in electrical engineering (ICAEE), Vellore, India, pp 1–6. https://doi.org/10.1109/ICAEE.2014.6838529
11. Das D, Ghosh S (1999) Method for load-flow solution of radial distribution network. IEE Proc Gener Transm Distrib 146(6)
12. Wu S, Jiang F, Piao Z (2012) Power flow calculation for radial distribution systems with distributed generation. In: IEEE ICMA 2012
13. Bhujel D, Adhikary B (2012) A load flow algorithm for radial distribution system with distributed generation. In: IEEE ICSET, 2012
14. Ravi Teja BR, Murty VVSN (2014) A contribution to load flow in radial distribution system and comparison of different load flow methods. In: IEEE conference EPSCICON
15. William HK (2002) Distribution system modelling and analysis. Text book
16. Díaz G, Gómez-Aleixandre J, Coto J (2016) Direct backward/forward sweep algorithm for solving load power flows in AC droop-regulated microgrids. In: IEEE TSM, 2016
17. Wang HL, Chang GW, Chu SY (2007) An improved backward/forward sweep load flow algorithm for radial distribution systems. IEEE Trans Power Syst 22(2):882–884
18. González-Morán C, Arboleya P, Mohamed B (2017) Matrix backward forward sweep for unbalanced power flow in $\alpha\beta$ 0 frame. Electr Power Syst Res 148:273–281
19. Garces A (2016) A linear three-phase load flow for power distribution systems. IEEE Trans Power Syst 31(1):827–828
20. Stojkov M, Pavić I, Alinjak T (2017) Improvement of backward/forward sweep power flow method by using modified breadth-first search strategy. IET Gener Transm Distrib 11(1):102–109
21. Li J, Chen F, Wang Z (2004) Implementing transformer nodal admittance matrices into backward/forward sweep-based power flow analysis for unbalanced radial distribution systems. IEEE Trans Power Syst 19(4):1831–1836
22. Kumar A, Jha BK, Dheer DK, Singh D, Misra RK (2019) Nested backward/forward sweep algorithm for power flow analysis of droop regulated islanded microgrids. IET Gener Transm Distrib 13(14):3086–3095
23. Ferraz RSF, Ferraz RSF, Medina ACR, Batista OE, Gomes AC, Cortes LRC (2019) Backward-forward sweep load flow algorithm for a feeder with high distributed generation integration. CEEL-Universidade Federal de Uberlândia
24. Issicaba D, Coelho J (2016) Evaluation of the forward-backward sweep load flow method using the contraction mapping principle. IJECE 6(6):3229–3237
25. Ouali S, Cherkaoui A (2020) An improved backward/forward sweep power flow method based on network tree depth for radial distribution systems. IJECE 2020, 11 pages. Article ID 5643410
26. Hocaoglu H, Eminoglu U (2009) Distribution systems forward/backward sweep-based power flow algorithms: a review and comparison study. Electr Power Compon Syst 37:91–110
27. Behera S, Mudgal N, Naik B (2014) Load flow analysis of unbalanced radial distribution systems. Thesis, NIT Rourkela
28. Sarada Devi MSNG, Yesuratnam G (2019) Demand re-allocation in conjunction with feeder input power using iterative power flows method. Int J Recent Technol Eng 8(4). ISSN: 2277-3878

Exploit Reversible Gates to Implement Fast Quantum Sorting Algorithm

Subhasree Bhattacharjee, Kunal Das, Arindam Sadhu, Soumyadip Sarkar, and Bikramjit Sarkar

Abstract This paper proposes an algorithm of divide-compare-swap-based (DCS) sorting that reduces time complexity by utilizing quantum parallelism. A new cost optimized comparator circuit has been designed and simulated using qiskit for comparing two quantum numbers. The sorting process consists of divide, compare-swap and merge. The set of n quantum numbers are divided equally into two sub arrays followed by comparing-swapping of two middle elements. This process continues in two sub arrays till array with single element is found. The sorting circuit performs well in qiskit with time complexity n, which has a significant improvement over existing quantum sorting technique.

Keywords Bitonic sort · Divide-compare-swap · Quantum parallelism · Reversible gate · Swap gate

S. Bhattacharjee (✉) · S. Sarkar
Department of Computer Application, Narula Institute of Technology, Kolkata, West Bengal, India
e-mail: bhattacharjeesubhasree@gmail.com

K. Das
Department of Computer Science, Acharya Prafulla Chandra College, Kolkata, West Bengal, India
e-mail: kunal@apccollege.ac.in

A. Sadhu
Department of Electronics and Communication Engineering, Greater Kolkata College of Engineering and Management, Kolkata, West Bengal, India

B. Sarkar
Department of Computer Science and Engineering, JIS College of Engineering, Kalyani, West Bengal, India
e-mail: bikramjit.sarkar@jiscollege.ac.in

B. R. Devi et al. (eds.), *Proceedings of Fifth International Conference on Computer and Communication Technologies*, Lecture Notes in Networks and Systems 897,
https://doi.org/10.1007/978-981-99-9704-6_11

1 Introduction

Quantum computation follows quantum mechanics that is concerned with two states of a particle that coexist together due to quantum mechanical properties like superposition and entanglement [1]. The main cause of speedup of quantum system compared to classical system is owing to emergence of many outstanding algorithms [2–6]. In spite of the advent of the new evolution, quantum computer is less available to the researchers because of its high cost [7, 8]. Sorting is the operation that is very much required for large applications [9]. Classical bitonic merge sort is generally modeled as a sorting network [10] and this sort is a com- parison-based sorting with parallelism. Bitonic sort [11] has highly parallelizable design. For recursive merge networks [11, 12], bitonic sort is suitable. The spatio temporal complexity of bitonic sort is $O(N \log 4N)$ [11]. In last few years, many novel contributions were reported in the field of quantum sorting [13–15]. Unlike classical system, quantum system exhibits much more parallelism due to its quantum mechanical properties [1]. Cheng and Wang [13] proposed a quantum merge sorting using divide and conquer technique.

In this paper, a division-compare-swap (DCS)-based sorting algorithm for n qubits is presented. For four qubits qA, qB, qC, qD, two stages are involved. In the first stage, qB is compared to qC. If qB is greater than qC, swap occurs. Then (qA, qB') and (qC', qD) are compared and swapped in parallel. So in the first stage, only two comparisons are there. Similar process repeats for the second stage. A total of $n/2$ stages are required and for each stage, two comparison-swaps are done for n qubits. So, total number of comparisons is $n/2*2$, which is equal to n. The DCS sorting algorithm is realized with the help of a newly designed reversible gate called RG gate. We have designed n bit comparator circuit with this gate. The circuit takes two numbers with three qubits each. In general, the comparator circuit takes two n qubit quantum numbers as input, compares them and makes the control bit high when first number is greater than the second one. This control bit is connected with the control input of swap gate, which swaps the two numbers when control bit is 1. The paper is represented as follows. Related information on quantum computing is described in Sect. 2. Section 3 discusses proposed methodology. Section 4 analyzes the sorting complexity, whereas Sect. 5 analyzes simulation and finally, Sect. 6 concludes the paper.

2 Background

In quantum computing, the smallest unit for representing information is called qubit or quantum bit. Due to superposition principle, a qubit can be in either state 1 or 0 or both 10 [16].

In the classical sorting algorithm, there is an input sequence of numbers $q \leq q1$, $q2$, qn > and for that input sequence, the output sequence will be $q' \leq q1'$, $q2'$, … qn' > such that q' is in non-decreasing order [14, 17].

Figure 1 describes the circuit diagram of 4 qubit sorting. In first stage of DCS Sorting, four qubits are equally divided into two parts and middle two elements are compared and swapped if required. Stage 2 repeats the same process. This strategy is depicted in Fig. 1 for four qubits. We have considered 4 qubits as inputs. The inputs are qA, qB, qC and qD. qB and qC are passed to the comparator block where they are compared. If qB is greater than qC, then the comparator block makes the control bit of swap block high and then swap happens. After swapping, qA and qB' are sent to comparator block and parallely qC' and qD are sent to another comparator block. In comparator block, both the pair are compared and swap if $qA > qB'$ and $qC' > qD$. After swapping, stage 2 starts. Same procedure will be repeated in stage 2. For four qubits, two stages are required. Generalized block diagram for n qubits is shown in Fig. 2.

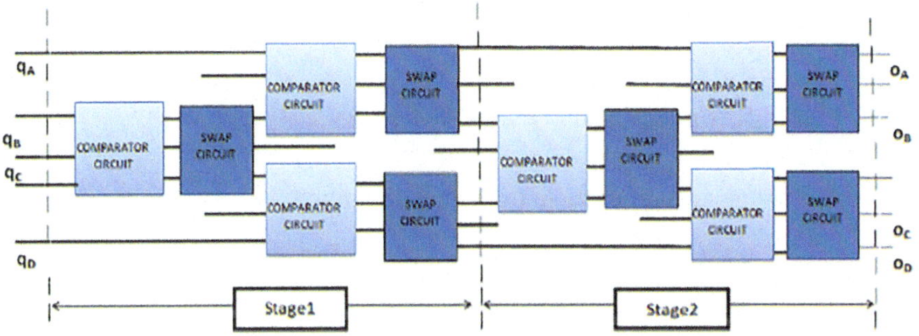

Fig. 1 Block diagram of 4 qubit sorting

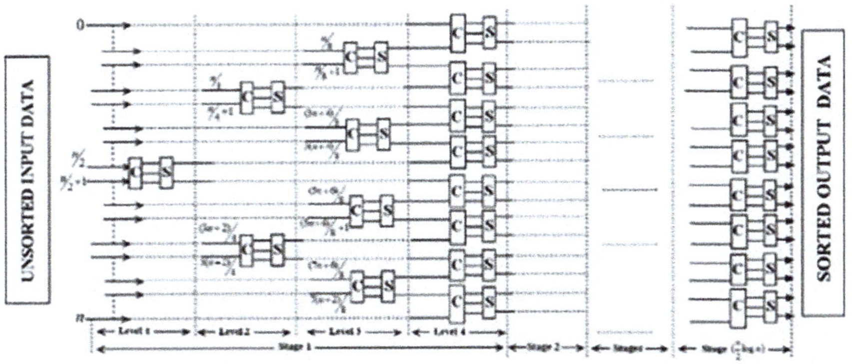

Fig. 2 Circuit structure of 4 qubit sorting

3 Proposed Methodology

The DCS (n) quantum sorting algorithm is discussed in this section. We consider n qubits as input. This algorithm follows divide and compare-swap approach in parallel. The DCS sort algorithm divides n qubits into two equal partitions. Then compares the two qubits that are adjacent to the partition. ($n/2 - i$)th element is compared to $n/2$th element and if ($n/2 - i$)th element is lesser than $n/2$th element, swap occurs. The first partition is then equally divided into two parts and the second partition is also divided equally. Following the previous rule, the adjacent elements are compared and swapped in parallel. Same procedure will continue till single element is obtained. The algorithm is as follows:

Algorithm: DCS Sort

Include Psi as quantum data

STEP 1: Seven quantum register each of 1 qubit is initialized to |0 >

Three pairs of two single qubit registers *qrAi*, *qrBi* initialized to input |Ψ1> and |Ψ2>

Repeat steps 2–4 for $n/2$ times

Step 2:

$$\mathrm{DCS}(n) = \mathrm{DCS}\left(\frac{n}{2}\right) \times \mathrm{DCS}\left(\frac{n}{2}\right) \times \mathrm{Oracle}\left(\frac{n}{2}, \frac{n}{2} - 1\right)$$

STEP 3: Apply quantum oracle for each pair of qubit

$$\left|X_{A2} > |Y_{B2} > \mathrm{to}\right|\overline{X_{A2} \oplus Y_{B2}} >$$

And

$$|X_{A2} > |Y_{B2} > \mathrm{to}\left|X_{A2}\overline{Y_{B2} >}\right.$$

$$q_{rA2} \leftarrow \left|\overline{X_{A2} \oplus Y_{B2}} >\right.$$

$$q_{rB2} \leftarrow \left|X_{A2}\overline{Y_{B2} >}\right.$$

STEP 4: Measure the register

$$q_r \leftarrow q_{rB2} \oplus (q_{rA2} * q_{rB1}) \oplus (q_{rA2} * q_{rA1} * q_{rB0})$$

If *qr* is 1

Then |*XA* > is greater than |*YB* >

STEP 5: End

The Fig. 3 describes Internal structure of our newly proposed reversible RG gate. RG is a four input ($|a\rangle$, $|b\rangle$, $|c\rangle$, $|d\rangle$) gate. First and second output lines give the first two inputs ($|a\rangle$, $|b\rangle$). The third output line results from the XNOR of the first three

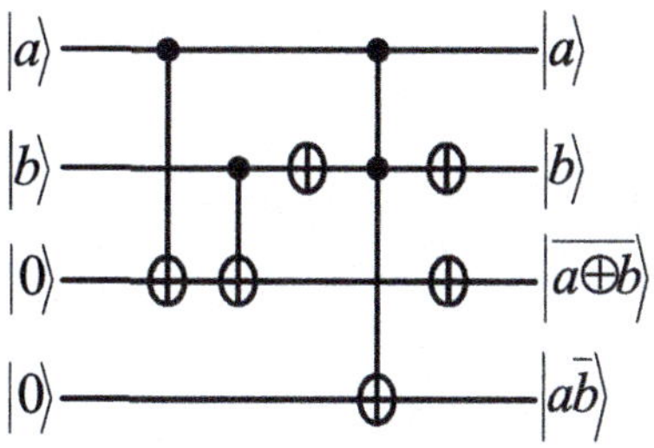

Fig. 3 Internal structure of RG gate

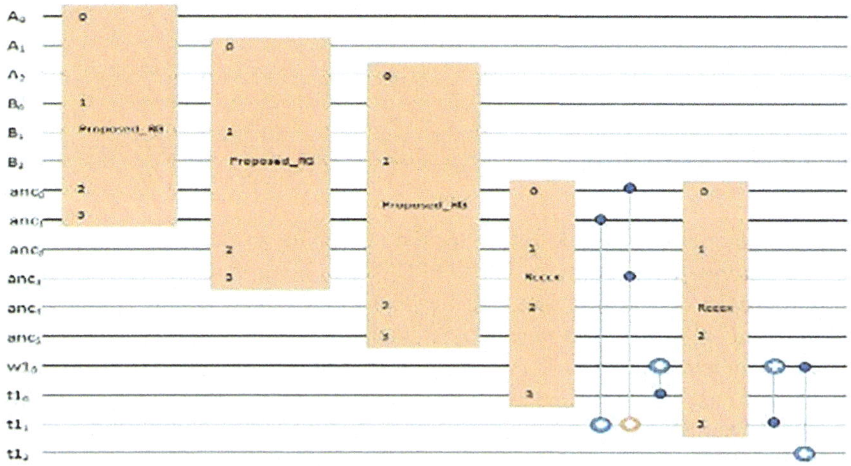

Fig. 4 Block diagram of comparator circuit in Qiskit

inputs. The fourth output line results from $|a\bar{b} \oplus d\rangle$. $|c\rangle$ and $|d\rangle$ bits are initialized to 0 and considered as ancillary bits. Using RG gate, a new comparator circuit is described in Fig. 4.

4 Complexity Analysis

A total of $n/2$ stages are required for sorting n numbers and for each stage, two comparison-swaps are done for n qubits. Thus, total number of comparisons is $n/2*2$, which is equal to n. We have used the following recurrence relation for calculating the complexity

$$T(n) = 2 * T\left(\frac{n}{2}\right)$$

Putting $n/2$ and $n/4$, we can generalize the following recurrence relation as

$$T(n) = 2^i * T(\frac{n}{2^i})$$

Now, if

$$\frac{n}{2^i} = 1, \text{i.e.} n = 2^i$$

So, $i = \log n$. Now, since $T(1) = 1$, then $T(n) = n * T(1) = n$.
Hence, running time of DSC algorithm is

$$T(n) = O(n)$$

5 Qiskit Simulation

There are various software platforms that have the ability to connect to real quantum devices. In this paper, we have simulated DCS sort algorithm over IBM Quantum Experience (IBMQ) QISKit SDK [18] in python 3.6. We have used the version Qiskit v0.29.0 and OpenQASM as quantum assembly/instruction languages. Among the various simulation methods of Aer Simulator, we have used matrix product state simulation method. After importing Aer Simulator from qiskit, we have set the method of Aer Simulator as 'matrix_product_state'. We have simulated Reversible RG gate in qiskit. Figure 5 depicts the simuulated RG gate. Simulation of comparator circuit is described in Fig. 6.

The comparator circuit proposed using RG gate is shown in Fig. 6. Two numbers of 3 bit each have been fed to test the comparator circuit described in the above figure.

The following algorithm is proposed to optimize the circuit of the comparator.

Algorithm 1: Cost Optimized C^nNOT(n)

Include Psi as quantum data

STEP 1: Initialize n registers of 1 qubit each to |0> and initialize number_CCNOT = 1

STEP 2: For control bit c 3 to n

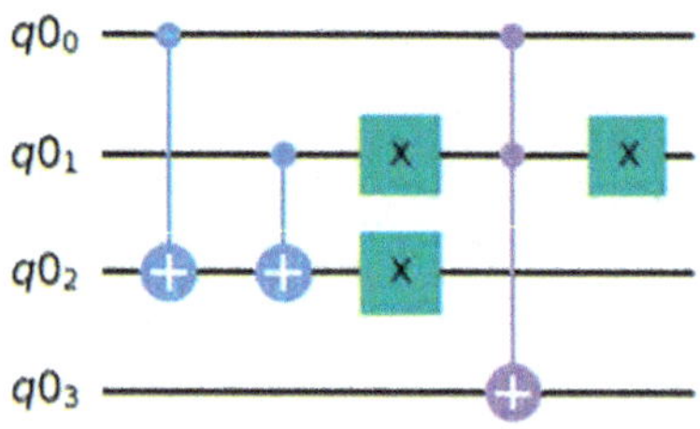

Fig. 5 Realization of RG gate in qiskit

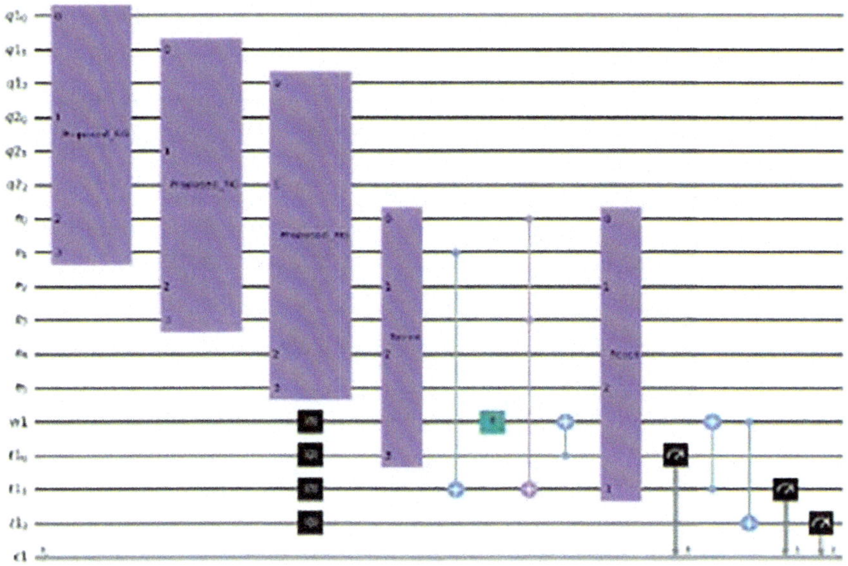

Fig. 6 Comparator circuit in qiskit

number_CCNOT = number_CCNOT + 1

Repeat steps 2

STEP3: Cost_C^{n}NOT ← 5*number_CCNOT

STEP 4: Return Cost_C^{n}NOT

Step 5: End

Algorithm 2: Cost Optimized Comparator(n)

STEP 1: Initialize Cost_RG = 7and Cost_Comparator = 0

STEP 2: Cost_Comparator = Cost_RG*n + Cost Optimized C^{n}NOT(n)

STEP 3: Return Cost_Comparator

Step 4: End

Using algorithms 1 and 2 over the comparator circuit shown in Fig. 6, the cost reduces to 42 from 58. The cost optimized comparator is shown in Fig. 7. A C^{3}NOT gate can be implemented by two CCNOT gates and thus cost reduced from 14 to 10 [19]. Thus, a C^{15}NOT gate can be realized with 14 CCNOT gates and cost reduced up to 53.94%.

The corresponding histogram is demonstrated in Fig. 8. Four numbers of 3 bits each have been taken as inputs. The numbers are A: 101, B: 110, C: 010 and D: 111. Figure 9 demonstrates the corresponding histogram of the sorting circuit. The output of the circuit is in sorted order.

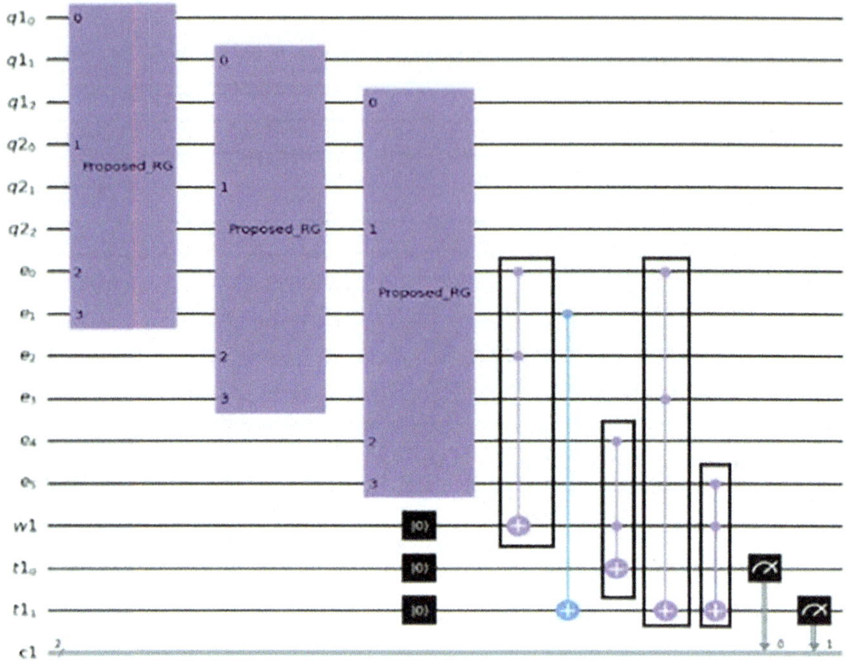

Fig. 7 Optimized comparator circuit in qiskit

Fig. 8 Comparator circuit output in qiskit

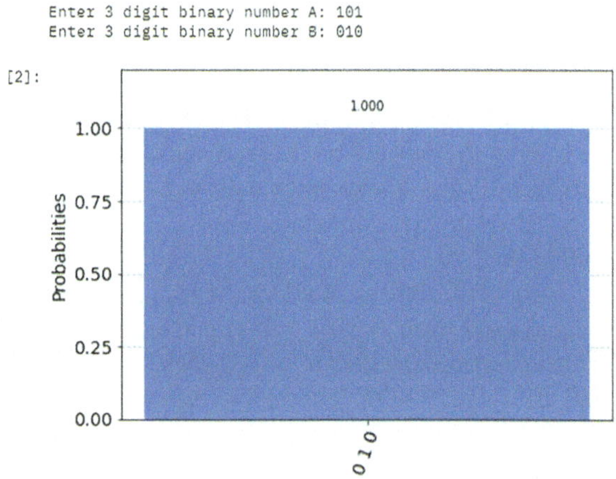

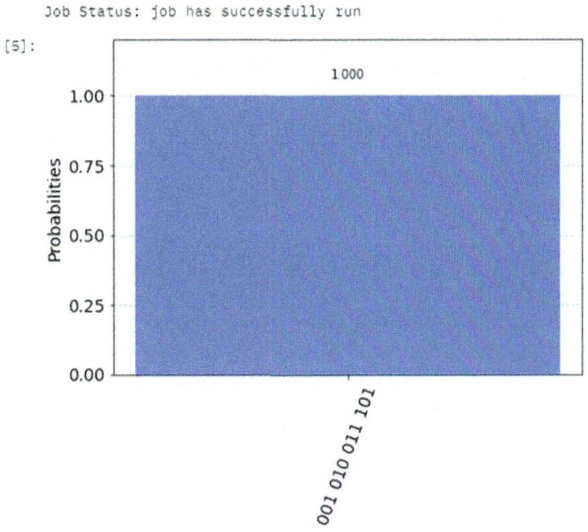

Fig. 9 Output of sorting circuit in qiskit

6 Conclusion

In this paper, a new divide, comparison and swap-based sorting algorithm for sorting n bit quantum numbers has been proposed. By utilizing the feature of quantum parallelism, the time complexity of this sorting reduces to linear time, which is a significant improvement over other quantum sorting technique. We have also designed a new reversible RG gate to implement the comparator circuit for comparing two quantum numbers. The cost of the comparator is optimized by replacing C^nNOT gate by $(n - 1)$ number of CCNOT gate. The time complexity of this sorting is n, which has a significant improvement over other existing quantum sorting techniques.

References

1. Bennett CH, DiVincenzo DP (2000) Quantum information and computation. Nature 404(6775):247–255. https://doi.org/10.1038/35005001
2. Shor PW (1999) Polynomial-time algorithms for prime factorization and discrete logarithms on a quantum computer. SIAM Rev 41(2):303–332. https://doi.org/10.1137/S0036144598347011
3. Grover LK (1996) A fast quantum mechanical algorithm for database search. In: Proceedings of the twenty-eighth annual ACM symposium on theory of computing, Philadelphia, vol 623, pp 212–219. https://doi.org/10.1145/237814.237866
4. Deutsch D (1985) Quantum theory, the Church-Turing principle and the universal quantum computer. Proceed R Soc Lond A Math Phys Sci 400(1818):97–117. https://doi.org/10.1098/rspa.1985.0070
5. Bernstein DJ (2009) Introduction to post-quantum cryptography. Springer, Berlin, Heidelberg
6. Simon DR (1997) On the power of quantum computation. SIAM J Comput 26(5):1474–1483. https://doi.org/10.1137/S0097539796298637

7. Dyakonov M (2019) When will useful quantum computers be constructed? Not in the foreseeable future, this physicist argues. Here's why the case against QUANTUM COMPUTING. IEEE Spect 56(3):24–29. https://doi.org/10.1109/MSPEC.2019.8651931
8. Hsu J (2015) How much power will quantum computing need. IEEE Spect 5:e001028
9. Sedgewick R, Wayne K (2011) Algorithms, 4th edn. Addison-Wesley, Upper Saddle River
10. Batcher KE (1968) Sorting networks and their applications. In: Proceedings of the April 30–May 2, 1968, spring joint computer conference, pp 307–314. https://doi.org/10.1145/1468075.1468121
11. Akl SG (2014) Parallel sorting algorithms. Academic press, New York
12. Johnson J, Douze M, Jegou H (2019) Billion-scale similarity search with GPUs. IEEE Trans Big Data 7(3):535–547. https://doi.org/10.1109/TBDATA.2019.2921572
13. Cheng ST, Wang CY (2006) Quantum switching and quantum merge sorting. IEEE Trans Circ Syst I Reg Papers 53(2):316–325. https://doi.org/10.1109/TCSI.2005.856669
14. Hoyer P, Neerbek J, Shi Y (2002) Quantum complexities of ordered searching, sorting, and element distinctness. Algorithmica 34:429–448
15. Klauck H (2003) Quantum time-space tradeoffs for sorting. In: Proceedings of the thirty-fifth annual ACM symposium on theory of computing, pp 69–76. https://doi.org/10.1145/780542.780553
16. Yanofsky NS, Mannucci MA (2008) Quantum computing for computer scientists. Cambridge University Press, Cambridge
17. Vudadha C, Phaneendra PS, Sreehari V, Ahmed SE, Muthukrishnan NM et al (2012) Design of prefix-based optimal reversible comparator. In: Proceedings of the 2012 IEEE computer society annual symposium on VLSI, pp 201–206. 10.1109%2FISVLSI.2012.49
18. Qiskit, originally authored by Luciano Bello, Jim Challenger, Andrew Cross, Ismael Faro, Jay Gambetta, Juan Gomez, Ali Javadi-Abhari, Paco Martin, Diego Moreda, Jesus Perez, Erick Winston, and Chris Wood. https://github.com/Qiskit/qiskit-terra
19. Qi XM, Chen FL, Wang HT, Sun YX, Guo LM (2014) Novel parity-preserving designs of reversible 4-bit comparator. Int J Theor Phys 53(4):1092–1102

Analyzing the Performance Variations of Naive Bayes, Linear SVM, and Random Forest for Spam Detection: A Comprehensive Study on the &Quot; Spam or Ham" Dataset

Bhawna Ojha, Pradeep Yadav, Rakhi Arora, Nitin Dixit, Gaurav Dubey, and Khemchand Shakyawar

Abstract Spam detection is the process of identifying and filtering unsolicited or unwanted messages, often referred to as spam, from legitimate or desired messages. It is an important task in email systems, messaging platforms, and other communication channels to protect users from unwanted or potentially harmful content. The performance of Naive Bayes, Linear SVM, and Random Forest classification machine learning algorithms on the "Spam or Ham" dataset was evaluated using various metrics, including accuracy, precision, recall, and F1 score. The results indicate that all three classifiers are suitable for this dataset, but Naive Bayes outperforms Random Forest and Linear SVM in terms of precision, recall, and F1 score. This suggests that Naive Bayes can accurately classify spam messages while minimizing the number of false positives. The optimal classifier will depend on features such as dataset size, the specific business problem, and available computing resources. However, based on this specific experiment's results, Naive Bayes appears to be a good choice for classifying spam messages. The accuracy of 0.98 is achieved by Naive Bayes algorithm.

B. Ojha (✉) · P. Yadav · R. Arora · N. Dixit · G. Dubey · K. Shakyawar
ITM-Institute of Technology and Management, Gwalior, India
e-mail: bhawnaojha.tech@itmgoi.in

P. Yadav
e-mail: pradeep.yadav@itmgoi.in

R. Arora
e-mail: rakhi.arora@itmgoi.in

N. Dixit
e-mail: nitin.dixit@itmgoi.in

G. Dubey
e-mail: gaurav.dubey@itmgoi.in

K. Shakyawar
e-mail: khemchand.shakya@itmgoi.in

B. R. Devi et al. (eds.), *Proceedings of Fifth International Conference on Computer and Communication Technologies*, Lecture Notes in Networks and Systems 897,
https://doi.org/10.1007/978-981-99-9704-6_12

Keywords Combined · Naive Bayes · Linear SVM · Precision · Recall · And F1 score

1 Introduction

A branch of computational intelligence called "machine learning" focuses on developing mathematical and statistical frameworks that enable computers to learn from data and make predictions or judgments based on that learning. Giving the algorithm a lot of data and letting it find patterns and connections among that data is the machine learning process [1]. Powerful technology known as machine learning has skyrocketed in prominence in recent years. It is a subfield of artificial intelligence (AI) that concentrates on developing statistical models and algorithms that let machines learn from data and forecast or decide based on that learning. One of machine learning's key benefits is its capacity to spot intricate patterns and connections in data that may be challenging or unattainable for humans to notice. This makes it a very potent instrument for a variety of uses, such as predicting consumer behavior and uncovering fraud [2].

Machine learning comes in a collection of forms, such as reinforcement learning, autonomous learning, and supervised learning. An algorithm is trained using labeled data in supervised learning, where the intended outcome is already known. On the other hand, unsupervised learning uses unlabeled data to train an algorithm where the intended outcome is unknown. Through contact with the surroundings and environmental feedback that takes the shape of incentives or consequences, a living thing develops how to arrive at decisions. This process is known as reinforcement learning [3]. Machine learning is used in numerous applications, including image identification, natural language processing, recommendation systems, and predictive analytics. Decision trees, random forests, support vector machines, neural networks, and deep learning are a few of the more well-known machine learning methods [4].

The three main categories of machine learning are supervised learning, unsupervised learning, and reinforcement learning. Each type of machine learning employs a unique method for learning from data and is best suited to particular problems.

In supervised learning [5], the algorithm is trained using labeled data, which means that each piece of input data has a matching piece of output data. The algorithm's objective is to discover a mapping between incoming data and output value. A supervised learning algorithm, for instance, could be taught to forecast housing costs based on characteristics like the number of bedrooms, the size, and the location.

Regression [6] and classification [5] are the two subcategories of supervised learning. While the output number in classification is a discrete variable, it is a continuous variable in regression. Predicting stock values, home prices, or weather predictions are a few examples of regression issues. Classification issues include determining whether an email is spam or not, categorizing pictures, or spotting fraudulent transactions, to name a few.

In unsupervised learning [7], the algorithm is taught on unlabeled data, which means that each input data point does not have an associated output value. Without any previous knowledge of the output value, the algorithm's aim is to find patterns and relationships in the data. For instance, classifying users on the basis of their buying patterns could be done using an unsupervised learning system. Unsupervised learning can be applied to a number of tasks, such as dimensionality reduction, anomaly spotting, and clustering. Clustering brings data points that are comparable together, while anomaly detection looks up data points that stand above the rest of the data. Reducing dimension is a method that decreases the number of characteristics in information while retaining the most important ones.

Reinforcement instruction is the process during which an algorithm learns a new ability by engaging with people around it and getting knowledge in the way of incentives or consequences. The goal of the algorithm is to identify a strategy that maximizes the total reward over time. One application of a reinforcement learning method is teaching a robot how to find its way through a maze.

Numerous fields, such as automation, control systems, and game playing, employ reinforcement learning. The algorithm must combine exploration and exploitation to find the best course of action while also taking advantage of the available information to increase reward.

2 Literature Review

Peter and Klaus [8] in their research focus on robust regression methods to assign appropriate weight to the variables. In order to adequately utilize the model-consistent data and expand the efficiency of the parameter estimates, the author wants to focus on down-weighting each of the matrix cells rather than complete observations.

Qingzhi [9] suggested an hourly rainfall algorithm utilizing supervised learning for accurate and timely rainfall predictions. The autocorrelation of the rain served as another focus for the author. The immunological components that the algorithm was based on will be shown and described in detail by Andrew and Jon [10]. The topic that of basic algorithm updates that eliminate some of the original version's unnecessary challenges proceeds to be addressed. An interdisciplinary area of research [11–15] that has seen a lot of growth recently is machine learning. It entails creating algorithms that let machines gain knowledge from data and develop over time without explicit programming. There is a lot of research on machine learning, which spans a variety of applications, such as audio and image recognition as well as natural language processing.

In this literature, the creation of learning algorithms like supervised, unsupervised, and reinforcement learning takes up a sizable percentage of the time. Numerous methods for feature engineering, model choice, and regularization have also been studied by machine learning researchers.

The body of research on supervised learning is substantial and covers a wide range of topics, including algorithms, optimization strategies, model choice, and assessment measures. Some well-liked supervised machine learning methods consist of support vector machine learning, decision forests, and neural networks. Numerous applications, such as speech recognition, picture recognition, and natural language processing, have made use of these approaches. These models are typically trained using optimization techniques like gradient descent and stochastic gradient descent, with a variety of changes to increase convergence speed and precision. Model selection, where several model hyperparameters are tweaked to achieve the greatest performance on the validation set, is another crucial component of supervised learning.

Accuracy, precision, recall, and F1-score [16] are among the metrics used to assess these models' performance. Newer assessment methods, like adversarial attacks and robustness testing, are becoming more prominent in addition to standard metrics to evaluate how well these models stand up to adversarial cases.

Overall, as new methods and uses are created, the literature on supervised learning expands and changes. Although there are still many issues to be resolved, supervised learning has shown considerable promise in improving artificial intelligence and resolving complicated issues.

Numerous scientific investigations on algorithms for learning, including support vector machines, decision trees, regression, and deep neural networks, have been included in the literature review on supervised learning. Comparisons have also been made between different algorithms to ascertain their benefits and drawbacks. There have been investigations into how feature selection, data preparation, and hyperparameters adjustment affect the accuracy of model predictions. New algorithms have even been developed in some research that perform better than the traditional ones. The full overview of the most advanced machine learning algorithms is provided by the wide literature study on supervised learning.

The thorough analysis of a research which applied machine learning techniques to the subject of cyber security. With new threats emerging every day, the area of cybersecurity [17] is constantly evolving. Researchers are now using machine learning algorithms to detect and stop assaults in order to keep up with these threats. The identification of malware has been one of the most promising fields of study. Even if a program has never been seen before, machine learning algorithms can analyze its behavior and determine if it is harmful or not. Development of real-time virus detection methods without degrading system performance has been the focus of recent research in this field. The detection of phishing assaults has been a subject of research in another area. Based on their content and other variables, machine learning algorithms can analyze emails and determine if they include phishing content or not. Recent studies have concentrated on creating algorithms that can recognize highly targeted phishing emails that are challenging to spot using conventional techniques. Machine learning techniques have been utilized to enhance access control systems in addition to detection. Login attempts from unknown IP addresses or devices can be recognized as suspicious by algorithms that analyze the data. Overall, machine learning algorithms have shown considerable potential in cybersecurity research, and academics are still looking into new ways to employ these algorithms to protect computer systems and networks.

Machine learning algorithms have been used for a variety of cybersecurity activities, from recognizing suspicious network behavior to detecting and preventing malware infestations. Artificial neural networks (ANNs), [18] decision trees, random forests, support vector machines (SVMs) [19], deep learning, and Bayesian networks are a few of the machine learning techniques that are most frequently employed in the cybersecurity industry.

ANNs are employed for a number of cybersecurity operations, such as spam filtering, malware detection, and intrusion detection. Decision trees are used for classification tasks like malware categorization, phishing email detection, and identifying harmful websites. Random forests, which have been employed for tasks including recognizing harmful network traffic and categorizing malware, combine numerous decision trees to improve accuracy.

SVMs are employed for classification tasks, such as malware and intrusion detection. Artificial neural network-based deep learning has been utilized for applications like virus detection, intrusion detection, and phishing detection. A probabilistic graphical model called a Bayesian network is used for classification and prediction tasks including identifying network threats and detecting intrusions. Overall, machine learning algorithms have shown to be useful in the field of cybersecurity, aiding practitioners and researchers in real-time detection and prevention of cyber threats.

In the field of cybersecurity, unsupervised learning algorithms are frequently employed, notably for tasks like anomaly detection and grouping. Unsupervised learning, in contrast to supervised learning, looks for patterns and anomalies in the data rather than relying on labeled data. Clustering is one of the unsupervised learning techniques that is most frequently applied in cybersecurity. By putting comparable data points together based on their commonalities, clustering algorithms can be used to spot trends in network traffic or classify related virus strains. Anomaly detection is another unsupervised learning technique frequently employed in cybersecurity. Anomaly detection algorithms find strange or unexpected events in data, including unexpected login attempts or abrupt increases in network traffic. These algorithms can be utilized to find dangers that haven't been seen before and may have eluded other detection techniques.

Cybersecurity also makes use of dimensionality reduction methods like Principal Component Analysis (PCA) and t-distributed Stochastic Neighbor Embedding (t-SNE) [20]. Large datasets' dimensionality is decreased by these methods, which facilitates the visualization and analysis of complex data. The ability to identify patterns and anomalies in data that may not be immediately obvious through manual analysis or with conventional detection methods makes unsupervised learning algorithms crucial tools for cybersecurity researchers and practitioners.

3 Methodology

In this research, classification algorithms are discussed. Classification is learning. There are N no. of classification algorithms. Before applying classification, we will discuss the architecture shown in Fig. 1. Here, the data set of "Spam or Ham" is taken for the classification problem. More detailed description of the "Spam or Ham" dataset is given below.

The number of Instances: 5572 SMS messages.

Attributes: Each instance in the dataset is a bag-of-words vector that depicts an SMS message. The 8497 distinct words (or tokens) that appear in the texts make up the feature space. The frequency of each characteristic (word) within the message serves as its representation. The dataset also contains each message's original content.

Class Distribution: With roughly 13% of the communications classified as spam and 87% as ham, the dataset is unbalanced.

Fig. 1 Proposed architecture for model building

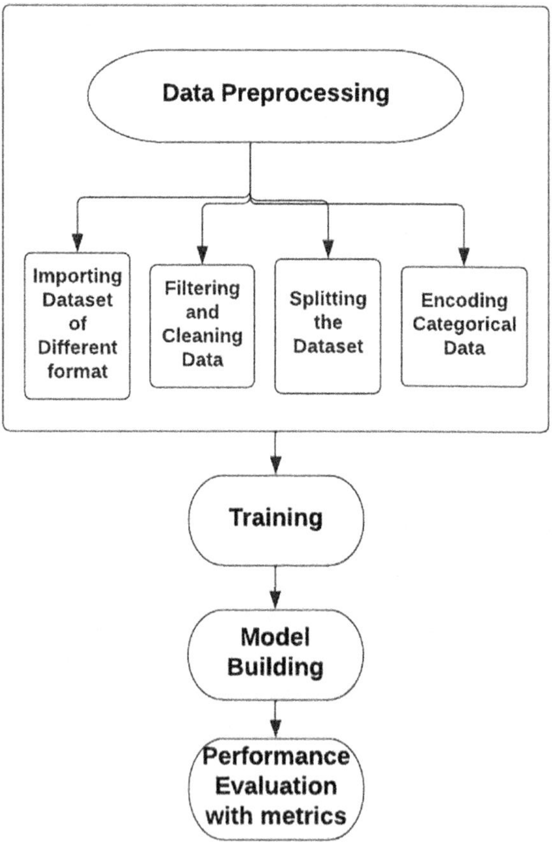

Here are some common pre-processing techniques that can be applied to the "Spam or Ham" dataset:

Text Cleaning: Eliminate any extraneous characters, including punctuation, specific characters, and words with stops (ordinary words like "the," "a," "an," etc.) that have little meaning.

Tokenization: Split the text into individual words or tokens.

Stemming or Lemmatization: Normalize the words by reducing them to their root form (stemming) or dictionary form (lemmatization) to reduce the vocabulary size and avoid redundancy.

Lowercasing: Convert all the text to lowercase to avoid the same word being treated as different tokens due to capitalization.

Removing Rare or Common Words: Remove words that occur too frequently (such as stop words) or too rarely (rare words that are unlikely to provide any useful information for the classification task).

Feature Selection: To lower the overall dimension of the set of features and increase the effectiveness of the model for machine learning, choose a portion of the most meaningful characteristics that are most pertinent to the classification objective.

4 Result

After pre-processing for model building, we have applied classifiers like Naive Bayes, Linear SVM, and Random Forest.

To evaluate the model's performance, performance measures including accuracy, precision, recall, and F1 score are utilized.

$$\text{Accuracy} = (\text{TP} + \text{TN})/(\text{TP} + \text{TN} + \text{FP} + \text{FN}). \tag{1}$$

$$\text{Precision} = \text{TP}/(\text{TP} + \text{FP}). \tag{2}$$

$$\text{Recall} = \text{TP}/(\text{TP} + \text{FN}). \tag{3}$$

$$\text{Precision is equal to } (\text{TP} + \text{TN})/(\text{TP} + \text{TN} + \text{FP} + \text{FN}).$$

$$\text{Precision is equal to } \text{TP}/(\text{TP} + \text{FP}) \cdot \text{Recall is TP}/(\text{TP plus FN}).$$

where TP represents True positives, TN represents True negative, FP shows false positive, and FN shows false negative.

$$\text{F1 Score} = 2 * \text{Pre.} * \text{Rec.}/\text{Pre.} + \text{Rec}\ldots$$

After applying the classifiers on the dataset, the following result is shown in Table 1.

Accuracy: Accuracy is a straightforward metric that measures the overall correctness of the model.

Formula: Accuracy = Number of Correct Predictions Total Number of Predictions Accuracy = Total Number of Predictions Number of Correct Predictions

It is important to note that accuracy may not be the best metric for imbalanced datasets, where one class significantly outnumbers the others.

Precision: Precision is the ratio of correctly predicted positive observations to the total predicted positives.

Formula: Precision = True Positives True Positives + False Positives Precision = True Positives + False Positives True Positives

Precision is a useful metric when the cost of false positives is high.

Recall (Sensitivity or True Positive Rate): Recall measures the ability of the model to capture all the relevant instances of a class.

Formula: Recall = True Positives True Positives + False Negatives Recall = True Positives + False Negatives True Positives

Recall is important when the cost of false negatives is high.

F1 Score: The F1 score is the harmonic mean of precision and recall. It provides a balance between precision and recall.

Formula: F1 Score = 2 × Precision × Recall Precision + Recall F1 Score = Precision + Recall 2 × Precision × Recall

The F1 score ranges from 0 to 1, where 1 indicates perfect precision and recall, and 0 indicates poor performance.

Figure 2 shows the results on various performance parameters.

Table 1 Results

S. No	Classifier	Accuracy	Precision	Recall	F1 score
0	Naive Bayes	**0.98**	0.97	0.91	0.94
1	Linear SVM	0.97	0.96	0.89	0.90
2	Random forest	0.96	0.95	0.84	0.89

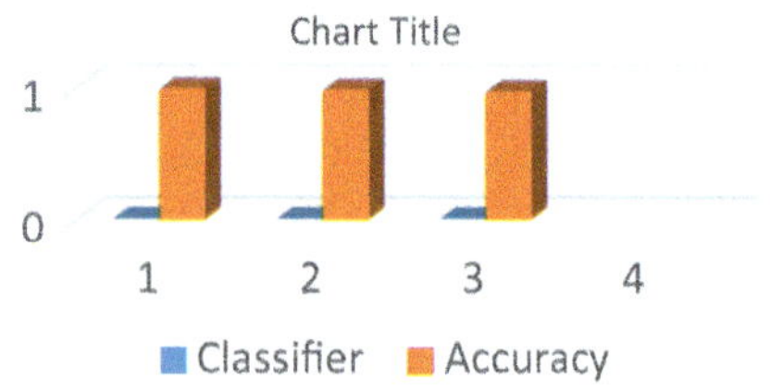

Fig. 2 Classifier versus accuracy

5 Conclusion

Based on Table 1, generated from applying different classification machine learning algorithms on the "Spam or Ham" dataset, we can conclude that all three classifiers, Naive Bayes, Linear SVM, and Random Forest, perform well on this dataset with high accuracy, precision, recall, and F1 score.

In terms of the specific evaluation metrics, we can see that Naive Bayes and Linear SVM perform better in terms of precision, recall, and F1 score compared to Random Forest. This suggests that Naive Bayes and Linear SVM are better at correctly classifying spam messages while minimizing the number of false positives. The accuracy of 0.98 is achieved by Naive Bayes algorithm. Overall, the selection of a classifier will be influenced by a number of variables, including the dimension of the data dataset, the particular business issue, and the computational resources available. However, based on the results from this specific experiment, Naive Bayes appears to be a good choice for classifying spam messages.

References

1. Nadeem MW, Goh HG, Ponnusamy V, Andonovic I, Khan MA (2021) Hussain M (2021) A fusion-based machine learning approach for the prediction of the onset of diabetes. Healthcare 9(10):1393
2. Bell J (2022) What is machine learning? Machine learning and the city: applications in architecture and urban design, pp 207–216
3. Uddin S, Khan A, Hossain ME, Moni MA (2019) Comparing different supervised machine learning algorithms for disease prediction. BMC Med Inform Decis Mak 19(1):1–16
4. Islam S, Amin SH (2020) Prediction of probable backorder scenarios in the supply chain using distributed random forest and gradient boosting machine learning techniques. J Big Data 7:1–22
5. Osisanwo FY, Akinsola JET, Awodele O, Hinmikaiye JO, Olakanmi O, Akinjobi J (2017) Supervised machine learning algorithms: classification and comparison. Int J Comput Trends Technol 48(3):128–138
6. Kavitha S, Varuna S, Ramya R (2016) A comparative analysis on linear regression and support vector regression. In: Proceedings of the 2016 online international conference on green engineering and technologies (IC-GET). IEEE, pp 1–5
7. Schutter A, Shamir L (2015) Galaxy morphology: an unsupervised machine learning approach. Astron Comput 12:60–66
8. Filzmoser P, Nordhausen K (2021) Robust linear regression for high-dimensional data: an overview. Wiley Interdiscipl Rev Computat Stat 13(4):e1524
9. Zhao Q, Liu Y, Yao W, Yao Y (2021) Hourly rainfall forecast model using supervised learning algorithm. IEEE Trans Geosci Remote Sens 60:1–9
10. Watkins A, Timmis J, Boggess L (2004) Artificial immune recognition system (AIRS): an immune-inspired supervised learning algorithm. Genet Program Evol Mach 5:291–317
11. Yadav P, Gupta R, Kumar S (2019) Video image retrieval method using dither-based block truncation code with hybrid features of color and shape. In: Engineering vibration, communication and information processing: ICoEVCI 2018, India. Springer, Singapore, pp 339–348
12. Yadav P, Gupta R, Kumar S (2017) Content based image retrieval using Dither Block Truncation coding with similarity comparison algorithm. In: Proceedings of the 2017 international conference on computer, communications and electronics (comptelix). IEEE, pp 489–493

13. Singhal S, Yadav P (2022) Machine learning techniques for ıntrusion detection system: a survey. In: Data, engineering and applications: select proceedings of IDEA 2021. Springer, Singapore, pp 29–40
14. Yadav P, Mishra M, Bhargava CP (2019) Probability prediction using ımproved method in delay-tolerant network. In: Data communication and networks: proceedings of GUCON 2019. Springer, Singapore, pp 13–19
15. Gupta R, Yadav P, Kumar S (2017) Race identification from facial images using statistical techniques. J Stat Manag Syst 20(4):723–730
16. Huang H, Xu H, Wang X, Silamu W (2015) Maximum F1-score discriminative training criterion for automatic mispronunciation detection. IEEE/ACM Trans Audio Speech Lang Process 23(4):787–797
17. Shaukat K, Luo S, Varadharajan V, Hameed IA, Xu M (2020) A survey on machine learning techniques for cyber security in the last decade. IEEE Access 8:222310–222354
18. Khorashadi-Zadeh H, Sanaye-Pasand M (2005) Correction of saturated current transformers secondary current using ANNs. IEEE Trans Power Deliv 21(1):73–79
19. Tang Y, Zhang YQ, Chawla NV, Krasser S (2008) SVMs modeling for highly imbalanced classification. IEEE Trans Syst Man Cybern B 39(1):281–288
20. Chatzimparmpas A, Martins RM, Kerren A (2020) T-visne: Interactive assessment and interpretation of t-sne projections. IEEE Trans Vis Comput Graph 26(8):2696–2714

A Novel Encryption Framework to Improve the Security of Medical Images

M. Senthilkumar, K. Suthendran, S. V. Suji Aparna, Mahesh Kotha, S. Kirubakaran, Srinivasarao Dharmireddi, and Voruganti Naresh Kumar

Abstract In medical imaging, image security is crucial yet difficult to implement. To protect medical images, numerous studies have been carried out. Image confidentiality may be achieved using encryption without the risk of data loss. Traditional encryption approaches cannot be used to protect e-health data directly because of restrictions on data size, redundancy, and capability, particularly when medical data are transported via unencrypted networks. Because images differ from text, they have a greater potential for loss of information and privacy; patients may no longer be able to access their data. Researchers have recognized these security risks and have suggested several image encryption approaches to address the issue. However, the researchers discovered that the proposed solutions still have several application-specific security issues. To provide an effective image encryption method for the healthcare sector, this study offers a novel cuttlefish particle swarm-optimized ciphertext-policy weighted attribute (CPSOCPWA) algorithm. The security and computation time of the suggested method is examined, assessed, and then contrasted with those of typically encrypted methods. It has been determined how well the

M. Senthilkumar
Department of Artificial Intelligence and Data Science, Sethu Institute of Technology, Virudhunagar, India
e-mail: senthilkumarsit@sethu.ac.in

K. Suthendran
Department of Information Technology, Kalasalingam Academy of Research and Education, Srivilliputhur, India
e-mail: k.suthendran@klu.ac.in

S. V. S. Aparna · V. N. Kumar (✉)
Department of CSE, CMR Technical Campus, Hyderabad, Telangana, India
e-mail: nareshkumar99890@gmail.com

M. Kotha
Department of CSE (AI&ML), CMR Technical Campus, Hyderabad, Telangana, India

S. Kirubakaran
CMR College of Engineering and Technology, Hyderabad, Telangana, India

S. Dharmireddi
Department of Cybersecurity, MasterCard, St. Louis, Missouri, USA

B. R. Devi et al. (eds.), *Proceedings of Fifth International Conference on Computer and Communication Technologies*, Lecture Notes in Networks and Systems 897,
https://doi.org/10.1007/978-981-99-9704-6_13

proposed method performs using a large number of test images. Numerous tests demonstrate that the suggested approach performs better than traditional methods.

Keywords Medical images · Internet of medical things (IoMT) · Encryption · Cuttlefish particle swarm optimized ciphertext-policy weighted attribute (CPSOCPWA)

1 Introduction

Medical images are essential in today's healthcare since they help diagnose, plan treatments, and monitor different diseases [1]. However, worries concerning their security and privacy are raised by the growing digitalization and sharing of these photographs. A unique encryption system has been created to ease these worries and improve the security of healthcare photographs [2]. The framework considers the particular needs of medical imaging technologies, such as the demand for real-time access, effective storage, and smooth interchange. The photographs may be accessed and used by healthcare providers without jeopardizing their anonymity since it balances privacy and uses [3]. This encryption framework might preserve patient confidentiality, fend against data hacking, and foster confidence in medical imaging technologies [4]. This framework makes an important addition to healthcare cyber Security because as medical technology develops, protecting the Security of medical imaging becomes crucial.

Some security requirements must be completed to send medical photographs securely. These requirements include honesty, validity, and secrecy. By encrypting the medical picture to achieve categorization and using computerized markers to assure truth and uprightness, cryptographic methods may be used to meet the stated security requirements [5]. The pieces displayed in this teaching activity show how encryption methods protect healthcare symbols. The primary goal is to secure medical images during delivery while such cutting-edge information is recorded [6]. User data often contain personal or confidential details; therefore, safeguarding them from vicious assaults, preventing corruption, and ensuring data integrity have all become essential. Steganography, watermarking, and encryption are just a few of the methods that may now be used to ensure that medical photographs are very secure.

This chapter proposes a novel cuttlefish swarm-optimized ciphertext-policy weighted attribute (CSOCPWA) algorithm as an efficient picture encryption technique for the healthcare industry. The remainder of the paper is divided into subsequent parts. Section 3 contains the proposed method explained. Section 4 includes the results, while Sect. 5 discusses the conclusions.

2 Related Work

Researchers used a cutting-edge cryptographic model that utilized optimization techniques to explore the Security of medical pictures in IoT [7]. Most of the time, medical records are kept on a cloud server at the hospital. Therefore, Security is crucial. Therefore, a different framework is needed to efficiently store and safely transfer medical pictures and clinical information. In elliptic curve cryptography, the best key will be selected using hybrid swarm optimization, which combines grasshopper optimization with particle swarm optimization. This will increase the level of safety of the encryption and decoding procedure. Authors in [8] suggested a framework for general medical picture encryption based on a unique combination of dynamic substitution boxes (*S*-boxes) and chaotic maps, two extremely effective constructions. It is demonstrated that the configuration of *S*-box substitution before and after chaotic substitution effectively thwarts selected unencrypted and chosen encrypted text attacks.

Further security measures are implemented to protect against the reset attack on pseudorandom numbers engines. The authors did the medical picture encryption [9] using a dual encryption method. Blowfish Encryption is first considered before using the encryption technique to check the encryption model. The private and public keys are then upgraded via opposition-based flower blooming.

Researchers in [10] have introduced a novel encryption technique for encrypting both grayscale and color medical photos. Picture blocks are the foundation of a novel picture-splitting technology. Then, a zigzag pattern, movement, and random recombination were applied to the picture blocks to create the final mess. A key to decoding the messy image is generated via an unorganized logistic map. A fresh chaos-based encryption technique for photographs of the body is being explored by researchers [11]. It uses a permutation-substitution-diffusion structure and is based on a mix of chaos and DNA computing in the context of 2 encryption sessions followed by a key generation layer. The original secret keys and the SHA-256 hash algorithm create the chaotic systems' private keys. To increase the unpredictability and safety of the encrypted image, the authors introduced a chaotic image encryption scheme that combines two chaotic maps, namely Arnold's Cat Map and 2D Logistic-Sine-Coupling Map. Authors in [12] analyzed the efficiency and safety of our system and compared it to other well-known chaotic image encryption methods.

The investigators addressed the shortcomings of the current chaotic picture encryption system [13] and presented an adaptive medical image encryption algorithm based on enhanced chaotic mapping. The technique first employed Logistic-sine chaos mapping to jumble up the original picture. Then, 2-by-2 sub-blocks of the jumbled picture were created. Authors [14] suggested improving homomorphic Encryption with elliptic curve cryptography. They develop improvements to elliptic curve cryptography since it has several drawbacks in the conventional form. Then, customized elliptic curve cryptography in conjunction with homomorphic Encryption is utilized to encrypt medical images. In the beginning, Authors [15] looked into extremely safe medical pictures with a few subkeys where a few subkeys are provided using chaotic logistics and tent maps. The Security was examined by the messy process, including dissemination and confusion. From chaotic maps, distinct unknowns were created for every map based on the original circumstances.

3 Method

An effective image encryption method for the healthcare sector, this study offers a novel cuttlefish particle swarm-optimized ciphertext-policy weighted attribute (CPSOCPWA) algorithm. Figure 1 depicts the overview of methodology.

3.1 *Cuttlefish Algorithm*

Cuttlefish are mollusks, which means they are marine animals. Cuttlefish may alter their skin's color and design Based on their environment. Cells from chromatophores, leucophores, and iridophores interact to accomplish this. These cells mix with the light that is reflected to produce a variety of colors and patterns. The method is based on visibility and reflection, among the other two primary processes. When light is reflected, it produces a variety of colors and patterns, and visibility adjusts those patterns to fit their surroundings; both, taken together, result in six distinct situations. Thus, adding the two results in a new potential answer. There are four groupings created from the six different examples. Cases 1 and 2 belong to Group 1 and provide a comprehensive solution. Cases 3 and 4 also belong to Group 2 and are in charge of conducting local searches. Case 5 belongs to Group 3 and offers a regional solution. We discover a random solution in Case 6, which stands for Group 4, to symbolize global search. The Cuttlefish Algorithm is shown in Algorithm 1.

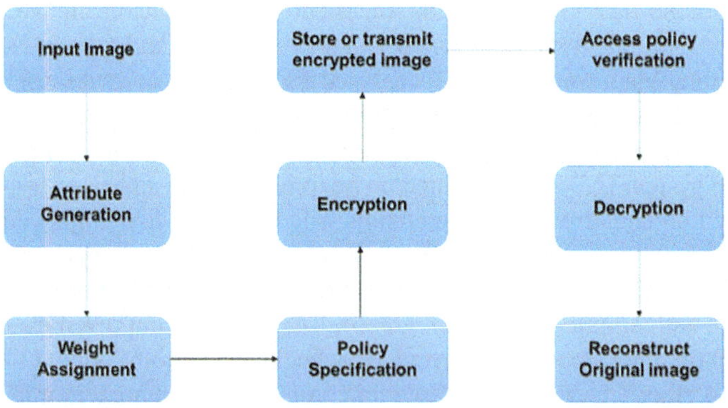

Fig. 1 Overview of methodology

Algorithm 1: Cuttlefish Algorithm

Step 1: Setup population Q using arbitrary M solutions for the c dimensions.

Step 2: Determine the problem domain and assign the values of $s1$, $s2$, $w1$, and $w2$ to the requirements.

Step 3: Measure the population's health.

Step 4: The best option should be kept.

Step 5: Group the cells into $h1$, $h2$, $h3$, and $h4$ groups.

Step 6: Though the halting requirement is not satisfied.

Step 7: Determine the best solution's median amount.

Step 8: Case {1 and 2}.

Do this for each cell in $h1$, and develop fresh solutions utilizing equations

$$\text{reflection} = S * H_1[j] \cdot \text{Points}[i] \tag{1}$$

$$\text{visibility} = W * (\text{BestPoints}[i] - H_1[j] \cdot \text{Points}[i]) \tag{2}$$

Verify whether the new solution is superior to the finest. If the requirement is met, change the Best, and Verify whether the new solution is ideal for the existing one. Substitute the current answer with it if the condition is met.

End

Step 9: Case {3 and 4}.

Do this for each cell in $h2$, and make discoveries using equations.

$$\text{reflection} = V * \text{BestPoint}[i] \tag{3}$$

$$\text{visibility} = W * (\text{BestPoints}[i] - H_1[j] \cdot \text{Points}[i]) \tag{4}$$

Verify whether the new solution is superior to the finest. If the requirement is met, change the Best, and Verify whether the new solution is ideal for the existing one. Replace the current answer with it if the condition is met.

End

Case {5}

Do for each $h3$ cell, make discoveries using equations

$$\text{reflection} = V * \text{BestPoint}[i] \tag{5}$$

$$\text{visibility} = W * (\text{BestPoints}[i] - \text{AVbest}) \tag{6}$$

Verify whether the new solution is superior to the finest. If the requirement is met, change the Best, and Verify whether the new solution is ideal for the existing one. Replace the current answer with it if the condition is met.

End

Case {6}

Do this for each cell in $h4$; Equations can be used to get a randomized solution.

$$P[j] \cdot \text{points}[i] = \text{random} * (\text{upperlimit} - \text{lowerlimit}) + \text{lowerlimit} \quad (7)$$

Verify whether the new solution is superior to the finest. If the requirement is met, change the BestVerify whether the new solution is superior to the existing one. Substitute the existing solution with it if the condition is met.

End

End while

Output and best result

3.2 Particle Swarm Algorithm

In PSO, multiple particles are positioned across the search space of a certain issue, each of which evaluates its intended function in its vicinity. In other words, the positioning of each particle indicates a method for addressing a problem that may be assessed about the objective function. By combining past data on its best (best-fit) locations with that of specific swarm members, each particle decides on its upcoming movement in the search space. The next iteration starts after all particles have been relocated. The swarm continually moves closer to the optimum condition of the fitness function. If the dimension of the investigation space is c, then the vectors y and u, respectively, represent the particle's current position and velocity at time t. The optimal locations of every particle during each previous movement experience and the optimal roles of every particle over the whole region are also memorized. These justifications give the particle motion equations for any c-dimensionality:

$$u_c^{\text{new}} = v \cdot u_c^{\text{old}} + b_1 \cdot s_1(\text{Qbest}_c - y_c) + b_2 \cdot s_2(\text{Hbest}_c - y_c) \quad (8)$$

$$y_c^{\text{new}} = y_c^{\text{old}} + u_c^{\text{new}} \quad (9)$$

where v is the inertia volume, u_c^{old} and y_c^{old} are the particle's initial velocity and place in the city dimension, respectively; u_c^{new} and y_c^{new} are the particle's new velocity and place in its previous dimension, respectively; s_1 and s_2 are random numbers generated uniformly in the range [0, 1] and are intended to add variability to the swarm's flight, and the weighting variables, generally known as the cognitive and social dimensions, are b_1 and b_2, accordingly. The proportional impact of the Pbest and Gbest locations on a particle's velocity is governed by weight coefficients b_1 and b_2. Greater values of these variables enable more extensive searching of regions nearby previously discovered excellent points, even while lower values for b_1 and b_2 allow each particle to investigate sites distant from already discovered good points.

3.3 Cuttlefish Particle Swarm Optimization

In order to improve performance, the hybrid Cuttlefish Algorithm (CA) and Particle Swarm Optimization (PSO) technique integrate the strengths of both methods. PSO emulates a swarm's social behavior, while CA imitates the hunting behavior of cuttlefish. The hybridization tries to make use of each method's advantages by combining CA's capacity for extensive regional exploration with PSO's effective local search abilities. The PSO element sharpens these solutions by iteratively updating their placements in accordance with personal and global best principles, while the CA component directs the search process by providing a variety of potential candidates. This hybrid strategy encourages both exploration and exploitation at the same time, which accelerates convergence and improves solution quality. The hybrid CA-PSO algorithm offers a potential optimization method for effectively resolving complicated issues because of its synergy.

3.4 Ciphertext-Policy Weighted Attribute Encryption (CPWAE)

According to their significance within the system, certain qualities are given varying weights in this structure. The properties do not relate in any way to access management. Each individual has a private key linked to several qualities and weights. A person can decode the ciphertext when these qualities match the weighted access structure in the ciphertext. It is simple to confirm that this architecture does generalize CP-ABE. The concrete building is made as follows:

Setup: The collection of all characteristics v in the system is input as a security parameter λ by the setup method. Each attribute is given a weight based on its significance to the system, and we assume that the highest value for each attribute is $k_1, \ldots, k_V$. The method selects a prime order o group $\mathbb{G}$ and random group members. $h, y, g_{1,0}, \ldots, g_{|V|,0}, g_{1,\text{end}}, \ldots, g_{|V|,\text{end}} \in \mathbb{G}$. In addition, for all $v_i \in V$, it chooses random $g_{v_i,1}, \ldots, g_{v_i,k_i}, \in \mathbb{G}$. At last, $\alpha \in \mathbb{Z}_o$ exponent is selected at random. The public parameters are made available as follows:

$$\text{params} = \left\{ f(h,h)^{\alpha}, h, y, \forall_{v_i \in V} \left\{ g_{v_i,1}, \ldots, g_{v_i,k_i} \right\}, g_{1,\text{end}}, \ldots, g_{|V|,\text{ end}} \right\} \quad (10)$$

$$\text{NTL} = h^{-\alpha} \quad (11)$$

Encryption: The encryption algorithm inputs the public parameters, the access tree Γ and the message $n \in \mathbb{G}_S$. Let Z represent the collection of leaf nodes in Γ. Each leaf node corresponds to a weighted property. E presents an attribute with a weight. Let (λ_j, i) denotes the j-th part of λ_j while x_j denotes the weight of λ_j. The encryption algorithm chooses random $t_{1,0}, \ldots, t_{1,x_j}, \ldots, t_{l,0}, \ldots, t_{l,x_j}, \in \mathbb{Z}_O, t = r_Q(0)$ and creates the ciphertext as follows.

First, it sets $D_n = n,\ f(h, h)^{\alpha \cdot t}$ and

$$D_{(1,0),1} = h^{t_{1,0}}, \ldots, D_{(1,0),1} = h^{t_{1,0}} \tag{12}$$

$$D_{(1,0),2} = \left(g_{1,0}\right)^{t_{1,0}}, \ldots, D_{(1,0),2} = \left(g_{k,0}\right)^{t_{1,0}} \tag{13}$$

Then for $i = 1$ to $x_j - 1$, it sets:

$$D_{(j,i),1} = h^{t_{(j,i)}}, \ldots, D_{(j,i),2} = \left(g_{j,i}\right)^{t_{(j,i)}} y_j^{t_{(j,i-1)}} \tag{14}$$

When $i = x_j$, we let $t_{j,d_i} = r\lambda_j(0)$, it sets:

$$D_{(j,i),1} = h^{r\lambda_j(0)}, \ldots, D_{(j,i),2} = \left(g_{j,i}\right)^{r\lambda_j(0)} y_j^{t_{(j,i-1)}} \tag{15}$$

At last, it calculates: after each attribute has reached its weight

$$\begin{aligned} D_{1,\text{end}1} &= h^{r\lambda_j(0)}, D_{1,\text{end}2} = \left(g_{1,\text{end}}\right)^{r\lambda_j(0)}, \ldots, D_{l,\text{end}1} \\ &= h^{r\lambda_l(0)}, D_{l,\text{end}2} = \left(g_{1,\text{end}}\right)^{r\lambda_l(0)} \end{aligned} \tag{16}$$

$$\text{DS} = \left\{\Gamma, D_n, \forall_{j \in Z, i=0,1,\ldots,x_j}\left(D_{(j,i),1}, D_{(j,i),2}\right), D_{1,\text{end}1}, D_{1,\text{end}2}, \ldots, D_{l,\text{end}1}, D_{l,\text{end}2}\right\} \tag{17}$$

Key Generation: The master key and a collection of attribute S are inputs to the key generation method. The process starts by selecting random group members $C_{j,i}$ connected to a part of the property. Next, it sets $q_{j,i} \in \mathbb{Z}_o$ at random. Making the key starts with:

For $\lambda_j \in T$,

$$L_{(j.0),1} = C_{j,0}\left(g_{j,0}\right)^{s_{j,0}}, L_{(j.0),2} = h^{q_{j,0}}. \tag{18}$$

For each key component that the algorithm generates, it sets:

$$L_{(j.0),1} = C_{(j,i-1)}^{-1} y_j^{q_{(j,i)}}, L_{(j.0),2} = h^{q_{j,i}}, L_{(j.0),3} = C_{(j,i)}\left(g_{j,i}\right)^{q_{j,i}} \tag{19}$$

Finally, it sets:

$$L_{j,\text{end}_w 1} = h^{-\alpha} \cdot C_{(j,w)}\left(g_{j,\text{end}}\right)^{q_{j,\text{end}_w}} \tag{20}$$

The algorithm produces:

$$\text{TL} = \left\{\forall j \in T, L_{j,0}, L_{j,0}, L_{j,\text{end}_w 1}, L_{j,\text{end}_w 2,} \left\{\forall_{j=1,\ldots,x_j'} L_{(j.0),1}, L_{(j.0),2}, L_{(j.0),3}\right\}\right\} \tag{21}$$

Decryption: Assume we are given a private key SK connected to an attribute and a ciphertext CT tied to an access tree. The procedure starts by calculating the following: If $j \in T$, the algorithm begins by computing:

$$A_{j,0} = f(D_{(j,i),1}, L_{(j.0),1}) \cdot f(D_{(j,0),2}, L_{(j.0),2})^{-1} = f(h, D_{j,0})^{t_{j,0}} \tag{22}$$

Then for $i = 1\text{to} : x_j - 1 :$

$$\begin{aligned} A_{j,0} &= A_{j,i-1} \cdot f\left(D_{(j,i-1),1}, L_{(j.i),1}\right) \cdot f\left(D_{(j,i),2}, L_{(j.i),2}\right)^{-1} \cdot f\left(D_{(j,i),1}, L_{(j.i),3}\right) \\ &= f(h, C_{j,i})^{t_{j,i}} \end{aligned} \tag{23}$$

When $i = x_j$:

$$A_{j,\text{end}} = f(h, D_{j,0})^{r\lambda_j(0)} \tag{24}$$

When satisfied $x_j \leq x'_j$ and $A_{j,x_j} = f(h, D_{j,0})^{r\lambda_j(0)}$, it finally computes

$$A_{j,\text{end}} = A_{j,x_j} \cdot f(D_{j,\text{end}1}, L_{j,\text{end}_{x_j}1})^{-1} \cdot f\left(D_{j,\text{end}2}, L_{j,\text{end}_{x_j}2}\right) = f(h, D_{j,0})^{r\lambda_j(0)} \tag{25}$$

Now that we've taken into account the non-leaf node, we compute:

$$E_w = \prod_{d \in T_w} E_d^{\Delta_j, T'_w(0)'} \tag{26}$$

$$= \prod_{d \in T_w} \left(f(h, h)^{\alpha \cdot r_d(0)}\right)^{\Delta_j, T'_w(0)} \tag{27}$$

$$= \prod_{d \in T_w} \left(f(h, h)^{\alpha \cdot r_{\text{parent}(d)}(0)}\right)^{\Delta_j, T'_w(0)} \tag{28}$$

$$= \prod_{d \in T_w} f(h, h)^{\alpha \cdot r_w(j) \cdot \Delta_j, T'_w(0)} = f(h, h)^{\alpha \cdot r_w(0)} \tag{29}$$

If the function is called on the tree Γ *s* root node *R* and the tree is fulfilled by *S*, the following value is set:

$$B = f(h, h)^{\alpha \cdot r_w(0)} = f(h, h)^{\alpha \cdot t} \tag{30}$$

Finally, we can decrypt the message by computing:

$$n = D_n / B \tag{31}$$

4 Result

The study uses an HP laptop with an Intel(R) Core(TM) i6-9000H CPU clocked at 2.60 GHz and 16 G Memory and the proposed CSOCPWA running Python 3.9. Additionally, the procedure is carried out using Windows 10. ChestXray-14 and fcon_1000 were the datasets we used. ChestX-ray13 is a medical imaging dataset comprising 30,805 distinct individuals' 112, 120, 1024*1024*8 bit single channel frontal view X-ray pictures [16]. The COVIDCT-Dataset contains 349 CT scans from 216 individuals that show clinical results for COVID-19. These 8-bit, single-channel grayscale CT scans are of arbitrary size. Numerous brain MRI pictures in nii format may be found in the file Fcon_1000. They are processed as 8-bit grayscale photos in our research. The analysis is done on Security, computation time, Encryption, and Decryption time. TDES [17], AES [18], and RSA [19] are the current methodologies and are compared with the suggested method (CSOCPWA).

To protect the privacy of patients and stop unauthorized access or manipulation, Encryption is crucial when it comes to medical pictures like X-rays, CT scans, or MRI scans. It reduces the chance of data breaches and unauthorized disclosures by ensuring the photos stay private during transport and storage. The Security of the proposed and used procedures are displayed in Table 1 and Fig. 2. While TDES, AES, and RSA only achieve a Security of 90.43, 89.80, and 93.51%, respectively, the suggested technique, CSOCPWA, achieves a Security of 98.74%. The proposed approach by CSOCPWA is more secure than the established ones.

When encrypting medical picture data, computation time means the time needed, considering the encryption technique, image size and complexity, hardware capabilities, and implementation optimizations. It assesses the effectiveness and computational difficulty of the encryption process. It's critical to strike a balance between computing time and Security since more secure encryption methods may take longer to compute, which might slow down high-throughput devices or real-time applications that depend on quick picture processing and transmission. The proposed and authorized processes' computation times are displayed in Table 2 and Fig. 3. The recommended method CSOCPWA can compute in 0.04 s, while only TDES, AES, and RSA can do it in 0.082, 0.078, and 0.07 s, respectively. The CSOCPWA Computation takes less time than existing methods.

In medical picture encryption, the term "encryption time" refers to how long it takes to encrypt the picture data. It reflects the time needed to use cryptographic techniques to transform the initial picture into a safe, encrypted representation. The

Table 1 Security (%)

Methods	Security (%)
TDES	90.43
AES	89.8
RSA	93.51
CSOCPWA (proposed)	98.74

Fig. 2 Security (%)

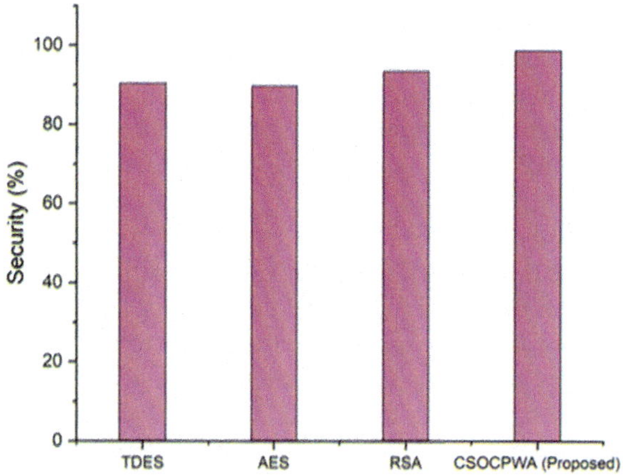

Table 2 Computation time (seconds)

Epochs	Computation time (s)			
	TDES	AES	RSA	CSOCPWA (proposed)
1	0.07	0.078	0.082	0.04
2	0.079	0.084	0.087	0.073
3	0.098	0.102	0.109	0.087
4	0.114	0.119	0.125	0.104
5	0.121	0.129	0.134	0.117

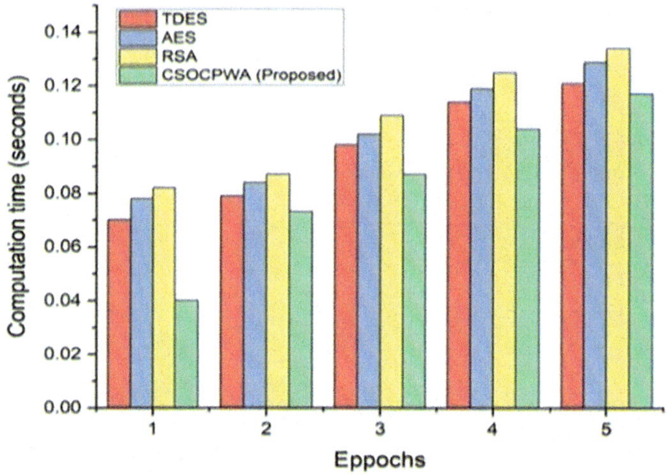

Fig. 3 Computation time (seconds)

encryption process takes time, depending on the encryption technique employed, the complexity and size of the image, the hardware, and any implementation optimizations. For the effective and rapid processing of medical picture data, it is crucial to minimize encryption time while retaining the appropriate degree of privacy. The proposed and existing processes' encryption times are displayed in Table 3 and Fig. 4. The recommended method, CSOCPWA, achieves an encryption time of 0.005 s. However, only TDES, AES, and RSA can obtain encryption times of 0.034, 0.023, and 0.011 s. The CSOCPWA Encryption time is faster than standard methods.

The time required to undo the Encryption procedure and restore the encrypted picture data to its unique readable state is referred to as the decryption time in medical image encryption. It calculates how long it will take. To decode the picture using the right decryption technique and key. The encryption technique employed, picture size and complexity, hardware capabilities, and any optimizations made all affect how quickly data may be decrypted. An effective decryption time is essential for quick access and analysis of medical picture data as well as for the protection and privacy

Table 3 Encryption time (seconds)

Data size (KB)	Encryption time (S)			
	TDES	AES	RSA	CSOCPWA (proposed)
50	0.034	0.023	0.011	0.005
75	0.049	0.036	0.028	0.017
100	0.065	0.057	0.041	0.029
125	0.076	0.069	0.053	0.038
150	0.102	0.093	0.072	0.053

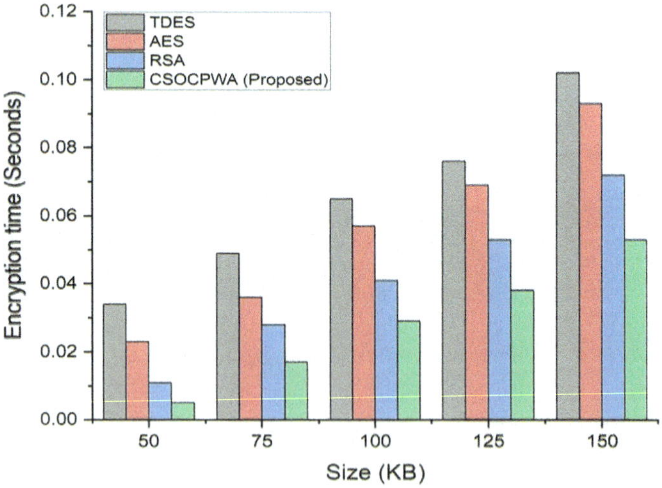

Fig. 4 Encryption time (seconds)

Table 4 Decryption time (seconds)

Data size (KB)	Decryption time (s)			
	TDES	AES	RSA	CSOCPWA (proposed)
50	0.037	0.026	0.015	0.006
75	0.051	0.038	0.031	0.02
100	0.07	0.061	0.045	0.032
125	0.079	0.072	0.064	0.042
150	0.121	0.117	0.102	0.059

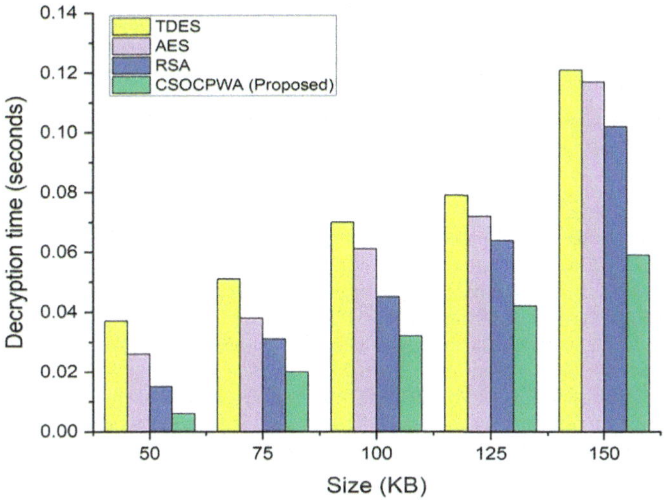

Fig. 5 Decryption time (seconds)

of sensitive information pertaining to patients. The proposed and existing processes' Decryption times are displayed in Table 4 and Fig. 5. The recommended method, CSOCPWA, achieves a Decryption time of 0.006 s. However, only TDES, AES, and RSA can obtain Decryption times of 0.037, 0.026, and 0.015 s. The CSOCPWA Decryption time is faster than standard methods.

5 Conclusion

This paper introduced a unique picture encryption technique named Cuttlefish Swarm Optimized Ciphertext-Policy Weighted Attribute (CSOCPWA) for the healthcare industry. The study recognized the security threats connected to medical picture confidentiality and the shortcomings of the current encryption techniques. It has been proven via rigorous testing and comparison with existing methods that the

CSOCPWA algorithm beats them in terms of computing time and Security. 21,560 distinct individuals' 112, 1024*1024*8 bit single channel frontal view X-ray pictures and 109 CT scans from 156 individuals were used. We analyzed the Security, computation time, Encryption, and Decryption time for the proposed method (CSOCPWA). And the Security was 98.74%, the computation time was 0.04 s, and the Encryption and Decryption time was 0.005 and 0.006 s. The recommended approach improves picture protection in the healthcare industry while solving the unique security concerns associated with medical image encryption. Although the CSOCPWA algorithm outperforms previous techniques in terms of calculation time, it may still need a lot of computational resources, such as processor speed and memory. To make sure the technique can be successfully applied in contexts with limited resources, its scalability should be properly evaluated.

References

1. Kadhim KT, Alsahlany AM, Wadi SM, Kadhum HT (2020) An overview of patients' health status monitoring system based on the Internet of Things (IoT). Wireless Person Commun 114(3):2235–2262
2. Munnangi AK, UdhayaKumar S, Ravi V et al (2023) Survival study on deep learning techniques for IoT enabled smart healthcare system. Health Technol 13:215–228
3. Farsi D (2021) Social media and health care, part I: a literature review of social media use by health care providers. J Med Internet Res 23(4):e23205
4. Durneva P, Cousins K, Chen M (2020) The current research, challenges, and future research directions of blockchain technology in patient care: systematic review. J Med Internet Res 22(7):18619
5. Subburaj T, Shilpa K, Sultana S, Suthendran K (2023) Discover crypto jacker from blockchain using AFS method. In: Proceedings of fourth international conference on computer communication technologies. Lecture notes in networks and systems, pp 145–156
6. Pise AA, Almusaini KK, Ahanger TA, Farouk A, Pareek PK, Nuagah SJ (2022) Enabling artificial intelligence of things (IoT) healthcare architectures and listing security issues. Comput Intell Neurosci 3:12
7. Athish Mon F, Sreeraj M, Prakash P, Suthendran K (2018) Combined cryptography and digital water marking for secure transmission of medical images in EHR systems. Int J Pure Appl Math 13:265–269
8. Khan LS, Hazzazi MM, Khan M, Jamal SS (2021) Novel image encryption based on Rossler map diffusion and particle swarm optimization generated highly non-linear substitution boxes. Chin J Phys 72:558–574
9. Jayahari Prabhu G (2021) Multimodal medical imaging security using hybridization of honey encryption algorithm with binary particle swarm optimization. Turk J Comput Math Educ 12(10):3905–3912
10. Ibrahim S, Alhumyani H, Masud M, Alshamrani SS, Cheikhrouhou O, Muhammad G, Hossain MS, Abbas AM (2020) Framework for efficient medical image encryption using dynamic S-boxes and chaotic maps. IEEE Access 8:160433–160449
11. Athish Mon F, Suthendran K (2017) Arumugam S (2017) A novel reversible data hiding method in Teleradiology to maximize data capacity in Medical images. Lect Notes Netw Syst 10398:55–62
12. Kamal ST, Hosny KM, Elgindy TM, Darwish MM, Fouda MM (2021) A new image encryption algorithm for grey and color medical images. IEEE Access 9:37855–37865

13. Chen X, Hu CJ (2017) Adaptive medical image encryption algorithm based on multiple chaotic mapping. Saudi J Biol Sci 24(8):1821–1827
14. Yin S, Liu J, Teng L (2020) Improved elliptic curve cryptography with homomorphic encryption for medical image encryption. Int J Netw Secur 22(3):419–424
15. Jawahar M, Liu JA, Ravi V et al (2022) CovMnet-deep learning model for classifying coronavirus (COVID-19). Health Technol 12:1009–1024
16. Wu Y, Zhang L, Berretti S, Wan S (2022) Medical image encryption by content-aware DNA computing for secure healthcare. IEEE Trans Ind Inform 19(2):2089–2098
17. Ramachandra MN, Srinivasa Rao M, Lai WC, Parameshachari BD, Ananda Babu J, Hemalatha KL (2022) An efficient and secure big data storage in cloud environment by using triple data encryption standard. Big Data Cognit Comput 6(4):101
18. Aes Abdullah AM (2020) Advanced encryption standard (AES) algorithm to encrypt and decrypt data. Cryptogr Netw Sec 16:1–11
19. Al-Kadei FHMS, Mardan HA, Minas NA (2020) Speed up image encryption by using the RSA algorithm. In: Proceedings of the 6th international conference on advanced computing and communication systems (ICACCS). IEEE, pp 1302–1307

Enhancing DDoS Attack Detection in SDN: A Novel Approach with IG-RFFI Feature Selection

Konda Srikar Goud and Srinivasa Rao Giduturi

Abstract Software-Defined Networking (SDN) is an emerging technology that revolutionizes future network design with its dynamic and efficient capabilities. It introduces a new definition of "network" by allowing network components to be programmable. This centralized, programmable architecture empowers network operators to control and design the entire network from one central location. However, while SDN offers extensive network visibility and robustness, it remains susceptible to security risks such as Distributed Denial of Service (DDoS) attacks, capable of causing network-wide disruptions. To counter such threats, many machine learning (ML) models have been developed to mitigate DDoS attacks on SDN. The effectiveness of these models relies heavily on an appropriate training dataset and the selection of relevant features. In this study, we address this challenge by employing the Information Gain (IG) and Random Forest Feature Importance (RFFI) techniques to identify the most significant features in the dataset. Subsequently, we utilize various Machine Learning algorithms to classify the resulting dataset. Based on our results, the SVM algorithm demonstrated the highest detection rate of 98.14%, the lowest false positive rate, and the shortest detection time.

Keywords SDN · DDoS attacks · Information gain · Random forest feature importance · Feature selection · SVM · BOTNET

K. S. Goud (✉) · S. R. Giduturi
Department of CSE, GITAM School of Technology, GITAM University, Visakhapatnam, Andhra Pradesh, India
e-mail: kondasrikargoud@gmail.com

K. S. Goud
Department of IT, BVRIT HYDERABAD College of Engineering for Women, Hyderabad, India

B. R. Devi et al. (eds.), *Proceedings of Fifth International Conference on Computer and Communication Technologies*, Lecture Notes in Networks and Systems 897,
https://doi.org/10.1007/978-981-99-9704-6_14

1 Introduction

Conventional networks have been struggling to keep up with modern demands for cloud computing, virtualization, high bandwidth, accessibility, and quick connection times. Due to its adaptable, programmable, and dynamic architecture, SDN has emerged as a substitute. Control, data, and application planes are all included in the SDN architecture. Switches and routers are examples of items located on the data plane and maintained and controlled by the control plane [1]. With the controller serving as its central command, the control plane manages the forwarding devices. The application plane interacts with network devices through the controller, whereas devices in the data plane transmit packets by rules specified by the control plane. The controller communicates with the forwarding nodes using the Southbound interface, whereas the Northbound interface makes it easier for the controller to communicate with application-layer services. In this context, OpenFlow is a commonly used protocol [2]. SDN runs from a centralized location and achieves dynamic optimization thanks to the centralization and removal of control logic from local devices [3]. SDN introduces various security problems, including seven separate threat vectors, despite computer networks facing similar security threats [4].

The serious threat to the SDN paradigm is the DDoS attack. Attackers could overwhelm the flow table storage in OpenFlow switches or aim for the controller. Attacker sends voluminous packets to the data plane's OpenFlow switch. The packets that does not have any equivalent flow entries in the flow table is called as Miss-flow. These Miss-flows are sent to the flow buffer and then transmitted to the controller through the Packet-In message to establish new rules. The controller's resources (memory, CPU, bandwidth, etc.) are overloaded by this flood, which breaks down the network and reduces communication line bandwidth [5]. Similarly, they attack network devices' flow tables using DDoS attacks to target the data plane. These attacks include delivering packets to a switch from unknown sources. These packets are subject to rules the controller establishes, which then transmits to the switch's flow table. The flow table capacity can only be exceeded to add the new rules by stopping packet forwarding. Incoming packets are lost due to buffer overflow brought on by the attack [6]. Detecting DDoS attacks in SDN is more complex than in traditional networks because OpenFlow switches only manage packets based on the flow entry provided by the controller.

Machine learning-based methodologies can offer dynamic, effective, and intelligent SDN management, security, and optimization solutions [7]. Network security depends on the quick attack detection mechanism. Self-learning networks with adaptive responses can be created by integrating ML-based frameworks into SDN architecture and processing SDN flow data [8].

We categorized the paper into five sections. The Sect. 2 briefs the literature review. The Sect. 3 outlines the process of applying FS and ML models, along with details of the experimental setup and SDN data collection. The Sect. 4 delves into the analysis of experimental outcomes and draws implications from the findings. Finally, we present the study's concluding remarks in the last section.

2 Literature Review

Several research works have addressed DDoS attacks using various methods. Notably, the authors in [9] introduce a self-adaptive evolutionary extreme learning technique for identifying DDoS attacks. This method encompasses two key elements: detecting the optimal crossover operator and automatically identifying neurons in the hidden layer. Experimental results highlight enhanced accuracy. Additionally, [10] presents a method for detecting DDoS attacks in Software Defined Networks (SDN) utilizing Deep Neural Networks. This technique achieves superior accuracy in less time and with fewer resources than existing approaches. In [11], the author compared various ML methods, revealing that Random Forest (RF) offers superior accuracy in detecting the attacks.

They, furthermore, explored various feature selection methods to enhance intrusion detection. Authors in [12] employ various filter, and wrapper based techniques to select relevant features for web attack detection. Their study concludes that the j48 classifier, coupled with Mutual Information (MI) feature selection, achieves the highest accuracy. The hybrid of PCA + GA are employed as feature selection in [13], leading to improved accuracy in attack detection using a DT classifier. Authors in [14] combine multiple feature selection methods to identify crucial features for intrusion detection, comparing different classifiers.

Other research focuses on reducing computational demands. For instance, [15] identifies the most relevant features using MI selection, achieving high accuracy with reduced feature sets. Ferrag et al. [16] suggested a solution for IoT networks that uses rule-based IDS, improving accuracy and detection rates. The authors of [17] use feature embedding and SVM to classify attacks successfully using publicly accessible datasets.

Authors in [18] propose a model for discovering novel features to detect DDoS attacks. A hybrid machine learning model yields improved accuracy and decrease false positives. The author in [19] integrate various ML techniques with NCA to select an appropriate features and predict traffic as normal or malicious. The proposed method outperforms other algorithms in accuracy. In [20], the author introduced a hybrid approach designed to tackle data imbalance problems in network datasets. This method combines Kernel Principal Component Analysis (KPCA), Weighted Support Vector Machine (WSVM), and Particle Swarm Optimization (PSO) to enhance the efficiency of DDoS attack detection.

3 Proposed Methodology

This section presents the proposed methodology for detecting DDoS attacks in SDN. First the SDN DDoS attack dataset was extracted and the second step involves the data preparation which involves removing noise, normalize features and extract relevant information. In the third step we employ Information Gain and Random Forest

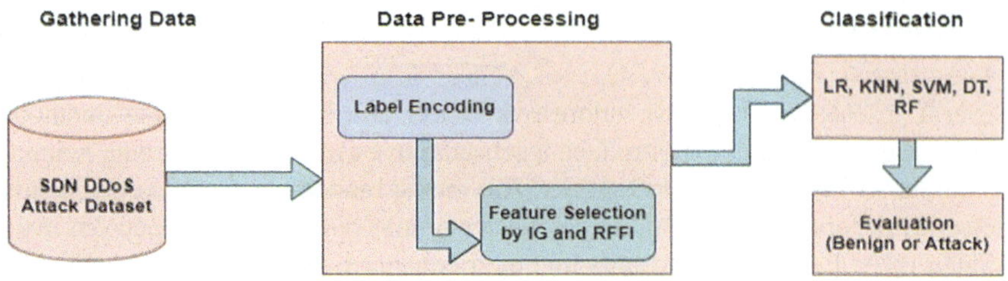

Fig. 1 The proposed model

Feature Importance for selecting the most important features. We apply the IG to rank the extracted features based on their relevance to distinguishing between normal and DDoS traffic. IG measures the amount of information each feature provides in predicting the target variable. Select the top-10 features with the topmost IG values. Further refine the feature selection process using Random Forest Feature Importance (RFFI). Train a Random Forest classifier using the selected features and labelled training data. Determine the feature importance scores by assessing the contribution of each feature to the predictive accuracy of the Random Forest model. Then, pick the ten features with the highest importance scores. During the fourth stage, we proceed to train multiple machine learning models, including LR, KNN, SVM, DT, and RF classifiers. This training is done using the chosen features and labeled training data. Utilize the labelled data to teach the model to differentiate between legitimate and DDoS traffic patterns. Finally, we assess the performance of the trained model using various evaluation metrics. The proposed model is illustrated in Fig. 1.

3.1 Dataset Description

In our research, we made use of the "SDN DDoS attack Dataset," a dataset specifically created for the SDN environment, and it is readily accessible to researchers. This dataset is composed of 23 features and a total of 104,000 network traffic flow records, encompassing various types of traffic, including TCP, ICMP, and UDP. It's worth noting that each record is annotated with both regular traffic class labels and attack traffic class labels. However, before training the machine learning model, the data must undergo preprocessing. During the data preprocessing phase, we managed missing values by replacing them with the median imputation. Categorical features were transformed into numerical format using one-hot encoding. To standardize the values, we applied the standard scalar. Additionally, we addressed data imbalance concerns by implementing the SMOTE technique.

3.2 Feature Selection

The process of feature selection plays a vital role in pinpointing the important characteristics within extensive datasets while also preventing the issue of overfitting. This technique serves to remove irrelevant and noisy attributes, isolating only the pertinent ones. The primary goal of feature selection methods is to reduce computational complexity and improve the effectiveness of machine learning algorithms.

3.3 Information Gain (or) Mutual Information (IG)

Claude E. Shannon first introduced the concept of mutual information. The most basic types of mutual information are entropy and conditional entropy. Mutual information calculates how dependent two variables are on one another. The entropy gauges a random variable's degree of uncertainty. The definitions of the individual entropy of random variable X is:

$$H(X) = -\sum\nolimits_{x \in X} P(x) \cdot \log(x) \tag{1}$$

The conditional entropy is the average uncertainty of X with respect to another random variable Y. It is defined as:

$$H(X|Y) = -\sum\nolimits_{x \in X} \sum\nolimits_{y \in Y} P(x, y) \cdot \log(x|y) \tag{2}$$

The MI between two random variables is formally characterized as:

$$I(X; Y) = H(X) - H(X|Y) \tag{3}$$

where $H(X)$ represents the entropy of random variable X, and $H(X|Y)$ denotes the conditional entropy.

3.4 Random Forest Feature Importance (RFFI)

The Random Forest is a technique that works by constructing a large number of decision trees. Every tree is a collection of internal nodes and leaves. The internal nodes use the chosen features to decide how to split the dataset in to two different sets with related responses. The output of the RF is the class chosen by the majority of the trees. The RFFI can be calculated by evaluating the Gini importance. The main advantage of this technique is its high computational speed with which it calculates the required values throughout the training phase.

4 Results and Discussion

In this section, we evaluate the performance of various machine learning models prior to feature selection and after feature selection. In the first experiment we have trained the models with all the 23 features immediately after data pre-processing. The classification evaluation results of the LR, KNN, SVM, DT, and RF algorithms are recorded. In the second experiment, we have employed IG- and RFFI-based feature selection techniques to find the relevant features. Based on the feature importance, we have selected top correlated features and trained the models with the selected features and recorded the results. In the first experiment the classification results achieved by various ML models before feature selection are displayed in the Fig. 2. The accuracy achieved by SVM was found to be the highest at 97.53%, while the accuracy achieved by RF, KNN, DT, LR are 97, 92, 84, 73.22% respectively. The precision achieved by SVM and RF was found to be highest at 97%, followed by KNN, DT, LR are 96, 78, 75%, respectively. The recall achieved by SVM and RF was found to be highest at 97%, followed by KNN, DT, LR are 92, 81, 71% respectively. The F-Score achieved by SVM and RF was found to be highest at 97%, followed by KNN, DT, LR are 94, 84, 73%, respectively.

In the second experiment the classification results achieved by various ML models after feature selection are displayed in the Fig. 3. We observe, the byteperflow is the most relevant feature followed by bytecount, pktperflow, pktrate and so on. Here we observed that the 10 features are found to be most important features and we have trained the models with those features and recorded the results. The SVM classifier had the highest accuracy with 98.14%, followed by the RF with 98.01%, KNN and DT with 97%, LR with 83.55%. The SVM attained a precision level of 98%. The RF, KNN and DT methods had a precision of 97%, LR with 84%. The SVM model

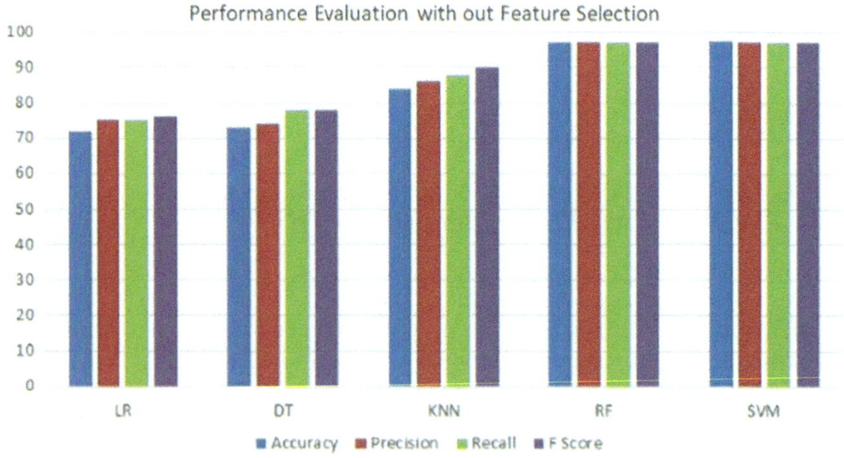

Fig. 2 Depicts the evaluation metrics with all features

had a recall rate of 98%, which was the highest, followed by the RF. KNN and DT with 97%, LR with 84%. The SVM achieved the highest possible F-Score of 98%. On the other hand, the F-Score for the RF, KNN and DT are found to be 97%, LR with 84%.

In Table 1, we conduct a comparison between our novel approach and prior research related to ML-based DDoS attack detection. Various researchers have put forth different methods for identifying DDoS attacks, but a notable limitation is that they relied on conventional datasets originally designed for traditional networks and openly accessible to the public. These datasets include features that may not be suitable for SDN environments.

The results shows that ML models are very effective in classifying the DDoS attack traffic. Based on the results of the proposed work, we can say that, by employing the IG and RFFI based feature selection approaches on the SDN DDoS Dataset, we have achieved highest accuracy of 98.14%.

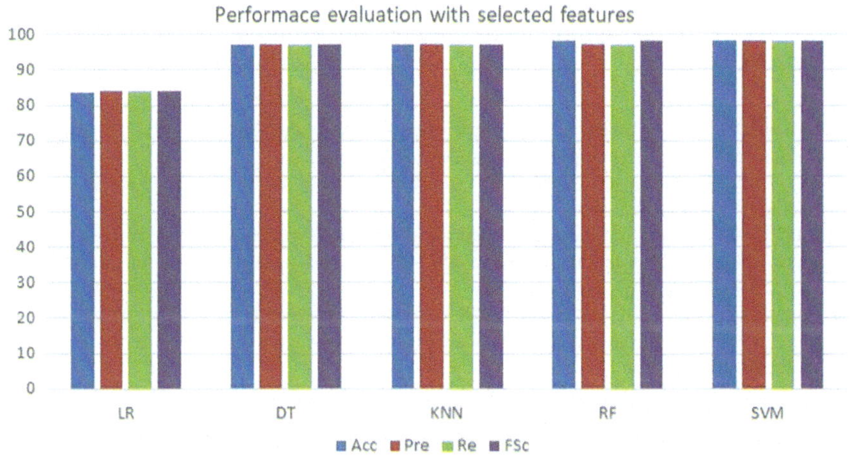

Fig. 3 Depicts the evaluation metrics with features selection

Table 1 Comparison on related studies

Model description	Accuracy in (%)
The author in [9] proposed self-adaptive evolutionary extreme learning technique	97.86
The author in [10] designed a framework with deep neural networks	97.48
The author in [11] proposed a framework with RF and filter based feature selection technique	96
The author in [13] proposed a hybrid framework with DT as classifier and GA + PCA for optimization	96.67
Proposed model	98.14

5 Conclusion and Future Scope

Detecting DDoS attacks in SDN environments poses a significant challenge due to the potential disruptions they cause in services. This study aimed to enhance DDoS attack detection in SDN by leveraging ML techniques. By conducting experiments on the "SDN_DDoS_attack" dataset and employed the IG-RFFI centered feature selection methods to identify the most relevant features. We trained various ML classifiers, including RF, KNN, SVM, DT, and LR, using the selected features. The results highlighted that the SVM classifier attained an exceptional overall prediction accuracy of 98.14% when utilizing the top 12 selected features, outperforming other methods. The SVM classifier with IG and RFFI consistently demonstrated promising outcomes in detecting DDoS attacks. For future research, we can explore wrapper-based feature selection techniques in combination with neural networks for detecting DDoS and other types of attacks.

References

1. Kreutz D, Ramos FM, Verissimo PE, Rothenberg CE, Azodolmolky S, Uhlig S (2014) Software-defined networking: a comprehensive survey. Proceed IEEE 103(1):14–76
2. Xie J, Guo D, Hu Z, Qu T, Lv P (2015) Control plane of software defined networks: a survey. Comput Commun 67:1–10
3. Stancu A, Halunga S, Suciu G, Vulpe A (2015) An overview study of software defined networking. In: Proceedings of the 2015 14th international conference on informatics in economy (IE 2015), Bucharest, pp 50–55
4. Kreutz D, Ramos FM, Verissimo P (2013) Towards secure and dependable software-defined networks. In: Proceedings of the second ACM SIGCOMM workshop on Hot topics in software defined networking, pp 55–60
5. Shu Z, Wan J, Li D, Lin J, Vasilakos AV, Imran M (2016) Security in software-defined networking: threats and countermeasures. Mobile Netw Appl 21:764–776
6. Ali ST, Sivaraman V, Radford A, Jha S (2015) A survey of securing networks using software defined networking. IEEE Trans Reliabil 64(3):1086–1097
7. Goud KS, Gidituri SR (2022) Security challenges and related solutions in software defined networks: a survey. Int J Comput Netw Appl 9(1):22–37
8. Foundation ON (2012) Software-defined networking: the new norm for networks. ONF White Paper 2(2–6):11
9. Kushwah GS, Ranga V (2021) Optimized extreme learning machine for detecting DDoS attacks in cloud computing. Comput Sec 105:102260
10. Makuvaza A, Jat DS, Gamundani AM (2021) Deep neural network (DNN) solution for real-time detection of distributed denial of service (DDoS) attacks in software defined networks (SDNs). SN Comput Sci 2:1–10
11. Bindra N, Sood M (2019) Detecting DDoS attacks using machine learning techniques and contemporary intrusion detection dataset. Autom Control Comput Sci 53:419–428
12. Kshirsagar D, Kumar S (2020) An ensemble feature reduction method for web-attack detection. J Discr Math Sci Cryptogr 23(1):283–291
13. Adhao R, Pachghare V (2020) Feature selection using principal component analysis and genetic algorithm. J Disc Math Sci Cryptogr 23(2):595–602
14. Binbusayyis A, Vaiyapuri T (2019) Identifying and benchmarking key features for cyber intrusion detection: an ensemble approach. IEEE Access 7:106495–106513

15. Stiawan D, Idris MYB, Bamhdi AM, Budiarto R (2020) CICIDS-2017 dataset feature analysis with information gain for anomaly detection. IEEE Access 8:132911–132921
16. Ferrag MA, Maglaras L, Ahmim A, Derdour M, Janicke H (2020) Rdtids: rules and decision tree-based intrusion detection system for internet-of-things networks. Fut Internet 12(3):44
17. Gu J, Lu S (2021) An effective intrusion detection approach using SVM with naïve Bayes feature embedding. Comput Sec 103:102158
18. Ahuja N, Singal G, Mukhopadhyay D, Kumar N (2021) Automated DDOS attack detection in software defined networking. J Netw Comput Appl 187:103108
19. Tonkal Ö, Polat H, Başaran E, Cömert Z, Kocaoğlu R (2021) Machine learning approach equipped with neighbourhood component analysis for DDoS attack detection in software-defined networking. Electronics 10(11):1227
20. Goud KS, Giduturi SR (2023) OML-SDN: Detection of DDoS attacks in SDN using optimized machine learning methods. Int J Intell Syst Appl Eng 11(4):197–208

A Novel Fusion Study on Disease Detection in Cotton Plants Using Embedded Approaches of Neural Networks

Samuel Chepuri and Y. Ramadevi

Abstract If not detected and managed correctly, Cotton diseases can lead to substantial economic losses for farmers. By implementing an automated disease detection system, farmers can save costs by reducing the reliance on manual scouting and labor-intensive inspections. It enables more efficient use of resources by focusing treatments only on affected plants instead of the entire field. Image segmentation methods may need help to generalize well across different datasets or disease types. The performance of segmentation algorithms can be affected by variations in leaf shape, disease symptoms, or even different camera setups. Training and validating the segmentation models on diverse datasets representing various disease severities and environmental conditions can help mitigate this limitation. When trained on large and diverse datasets, deep learning models can generalize well to unseen data. Once trained on a representative dataset, the models can generalize to new cotton leaf images and detect diseases even in different environments or with variations in lighting conditions, leaf shapes, or disease severities. Deep learning approaches promise precision farming. Cotton leaf disease detection in precision farming enables optimized resource management. This paper studies the different approaches for disease detection in cotton plants. Out of the many approaches, most optimization results are obtained by the deep learning approaches.

Keywords Pattern recognition · Image segmentation · Deep learning · Precision farming · Resource management · Optimization techniques

S. Chepuri (✉)
Department of Computer Science and Engineering, Osmania University, Hyderabad, Telangana State, India
e-mail: drsamuelchepuri@gmail.com

Y. Ramadevi
Department of AIML, CBIT, Hyderabad, Telangana State, India
e-mail: yramadevi_cseaiml@cbit.ac.in

B. R. Devi et al. (eds.), *Proceedings of Fifth International Conference on Computer and Communication Technologies*, Lecture Notes in Networks and Systems 897,
https://doi.org/10.1007/978-981-99-9704-6_15

1 Introduction

Cotton leaf disease detection using image segmentation is a popular approach that involves dividing an image into distinct regions or segments based on specific characteristics, such as color, texture, or shape. Image segmentation often involves setting parameters for thresholding, region growing, or edge detection algorithms [22]. Choosing appropriate parameter values can be challenging and often requires manual tuning. The optimal parameters may vary depending on the dataset or disease types, making it time-consuming and subjective. Figure 1 represents the other mechanisms for disease detection mechanisms.

Pattern Recognition techniques to analyze images of cotton leaves and identify disease symptoms. Various image processing methods, such as segmentation, feature extraction, and classification algorithms, are applied to distinguish between healthy and diseased leaves [23]. Features such as color, texture, shape, and lesion characteristics are extracted to differentiate different types of diseases. Machine learning algorithms, such as Support Vector Machines (SVM), Random Forests, and Convolutional Neural Networks (CNN), are commonly used for classification based on these extracted features.

Hyperspectral imaging captures a wide range of electromagnetic spectra from cotton leaves, allowing for detailed spectral analysis [24]. Spectral signatures of healthy and diseased cotton leaves are compared to detect subtle differences indicative of diseases. Classification algorithms, such as Principal Component Analysis (PCA), Linear Discriminant Analysis (LDA), and Deep Learning methods, are applied to classify the hyperspectral data and identify diseased leaves.

1.1 *Leaf Disease Detection Using Different Techniques*

A. **Leaf Disease Detection Using Segmentation Approaches**

Identification of diseases in plants was tuff because it should be detected in the early stage. If not, it may destroy the whole crop. Leaves are affected first, so based on the images, the prediction is considered. The dataset contains fungal-infected leaves and also healthy leaves. While preprocessing the images, five stages are performed

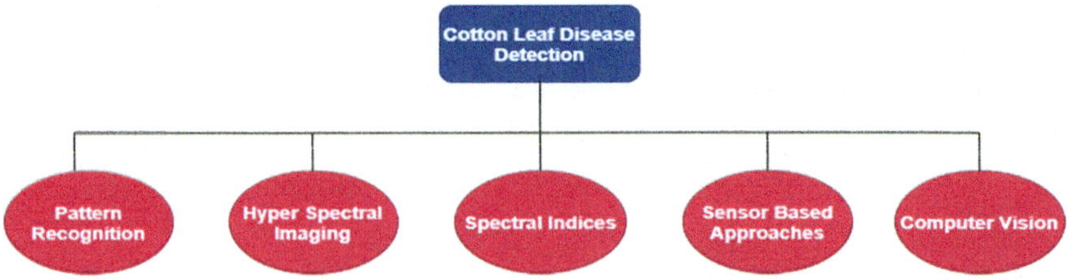

Fig. 1 Leaf disease detection mechanisms

[25]. Removing the background part of the image, converting color images to grey and HSV, those grey are converted to binary, and finally, the images are filtered.

The *K*-means algorithm is commonly employed to determine a photograph's inherent clusters of pixels. Clustering methods provide valuable insights by grouping objects around their respective centroids when constructing vector spaces. Edge detection is a crucial process in image analysis, as it helps identify boundaries and contours in a region. One widely used technique for edge recognition is employing Sobel operators, which estimate the first derivative and offer smoothing capabilities to reduce noise. Image segmentation separates a numeric image into multiple segments, providing meaning and making analysis more understandable. The watershed transform is a technique used for segmentation, where the picture is changed to seem like catchment basins corresponding to the desired objects. The watershed modification technique is used for segmentation, where catchment basins are added to the picture. They correspond to the desired objects. The SIFT method converts images into local feature vectors unaffected by size, translation, or rotation. SVM uses extracted features to build a hyperplane that separates data points and predicts classes for individual inputs. Finally, the prediction was performed, and the results were analyzed with validation techniques (Fig. 2).

B. **Leaf Disease Detection Using GPU's Based CNN**

A deep learning-based approach for identifying diseases in cotton plants is studied based on neural networks. The dataset is gathered using a camera and sensor IoT-based system installed on a crop field and contaminated survey regions. Image preprocessing is performed to enhance the quality and resolution of the images. Image resizing ensures that all images in the dataset are 256 × 256 pixels. A deep convolutional neural network (CNN) is trained using frameworks like Theano, Torch7, or Caffe [26]. Tests are performed on a test set to predict the health status and disease of the leaves. Fine-tuning is conducted to upgrade the prediction validation in the proposed methodology. The CNN model is trained using NVIDIA GPU mode, specifically GTX 1080ti. The method uses programs like Anaconda Python and modules like OpenCV and Caffe, and CudNN. The methodology is trained for 10 k splitting, with an expected accuracy of around 98.0% (Fig. 3).

C. **Detection Using Integrated CNN with ML Techniques**

Images of cotton leaf disease from Xinjiang Agricultural Division seven cotton were utilized in the investigation. The diseases focused on in the study were Anthracnose, Verticillium, and Ascochyta spot, which are prevalent ailments in cotton fields. The image-capturing technique showed excellent generalizability and was unaffected by ambient light [27]. Disease spots were segmented from the picture to erase the composite backdrop. The optimal partition threshold was determined using an iterative method. The initial threshold was set as the median grayscale value, and the goal was to minimize the median difference between scales of average grey values with two segments obtained after partition. Color and texture features were extracted for the SVM model. For both the backdrop and lesion areas, values from the grayscale

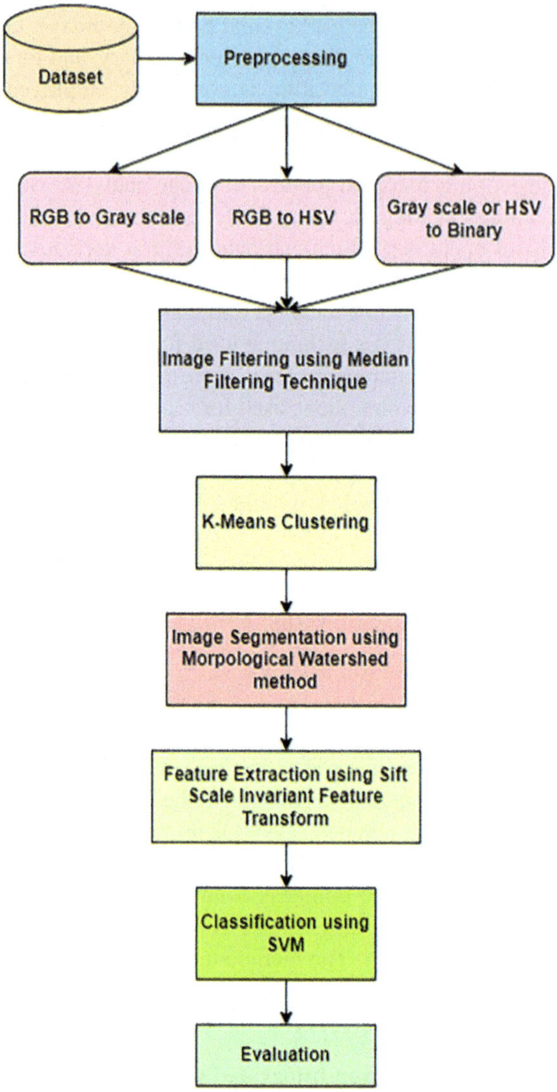

Fig. 2 Segmentation approach for disease detection

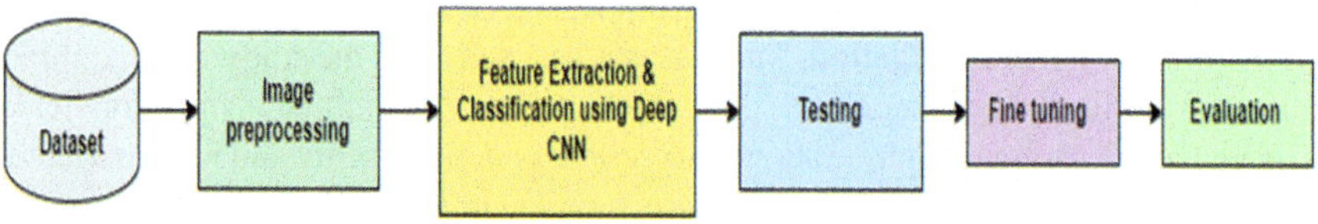

Fig. 3 GPU-based CNN approach

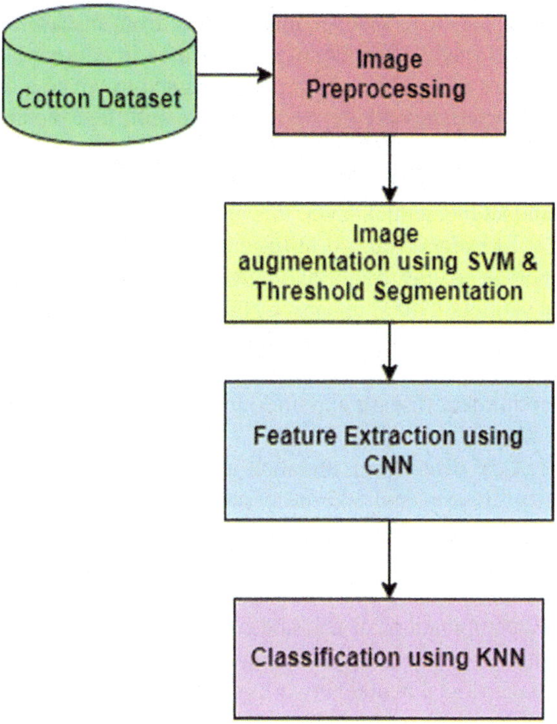

Fig. 4 KNN tuned CNN approach for identification of disease

overlap matrix were incorporated into the texture features. The trained methodology SVM was utilized for labelling the disease. Diseased areas were labelled 1, and non-diseased background areas were labelled 0. The KNN methodology calculates the mean ED metrics to perform the classifying tasks. The feature extractor selection and construction are important in this study (Fig. 4).

2 Literature Review

Patil et al. [1] propose a CNN system to classify cotton plant images into healthy, diseased, and stressed classes. The CNN is trained on a dataset of over 10,000 images. The system also uses an IoT platform to collect data from sensors in the field, such as temperature, humidity, and soil moisture. The system can detect diseases early on, which can help farmers to take preventive measures. The system can recommend appropriate crop management practices, irrigation, fertilization, and pest control.

Zambare et al. [2] Deep learning can be used to accurately and efficiently detect and manage plant diseases. Plant diseases are easier to identify early, which benefits farmers' ability to increase crop yields and economic development. A device that can identify illnesses automatically revolutionizes agricultural surveillance and enable plant leaves to recover immediately from disease discovery. The suggested CNN model method for identifying diseases in cotton plants is effective. The input is a picture of a cotton plant, pre-processed to create a digitized color image. Relevant characteristics are derived from the split leaf picture. The CNN model is used to classify the other leaf picture. Research demonstrates that the suggested CNN model performs better with three layers and 300 training epochs. By flattening the model and adding a thick layer, it is optimized.

Kukadiya et al. [3] utilized a CNN to classify cotton leaf images into four categories: healthy, bacterial blight, powdery mildew, and rust. The CNN was trained on a dataset of 1,000 images. CNN is utilized to detect cotton leaf diseases in real time. They used a webcam to capture images of cotton leaves and then used CNN to classify the images. The study results show that CNNs have been utilized for automatic cotton leaf disease classifier and detection. CNNs are accurate. CNNs are easy to use.

Leaf disease classification is crucial in agriculture as it allows farmers to identify diseases and take necessary precautions. Patil and Burkpalli [4] has demonstrated that color features alone are sufficient for distinguishing between healthy and unhealthy pictures. These color analyses are further utilized as input for the WEKA tool, facilitating analysis with several classifiers. A connection is conducted among 12 combinations of text and color. The findings indicate that color features significantly contribute to enhancing classification accuracy. Additionally, a survey reveals that artificial neural networks outperform other classifiers such as Naïve Bayes, RF, SVM, KNN, and AdaBoost regarding accuracy.

Detecting pests and diseases related to cotton leaves is challenging, especially when the pests are minuscule and difficult to identify. Udawant and Srinath [5] utilized transfer learning to develop a new methodology that can identify pests and diseases related to cotton leaves by ignoring the bg. The model correctly identified the known part of the disease or pest involved in the diseased part. GANs (Generative Adversarial Networks) can be used to augment the data for training, which can improve the model's accuracy.

Kumar et al. [6] Accurate recognition and disease classification of plants can help improve growth and development. Manual identification and detection of plant diseases is time-consuming and costly. Processing the images is a viable tool for plant disease recognition and classification. Computer vision techniques can be used to overcome the challenges of manual identification and detection. A dataset of high-quality images is essential for accurate disease detection. Lighting conditions can affect the quality of images and should be carefully controlled. Researchers should provide detailed information about their methods and tools to enable others to replicate their results.

Liang [7], a small-sample learning framework, is proposed for cotton leaf spot classification. The framework consists of two steps: disease spot extraction and classification. Disease spots are extracted using SVM classification. Classification is performed using S-DenseNet, a convolutional neural network. An SSO is proposed to improve classification accuracy. Experimental results show that the proposed framework attains max accuracy with a small multiple-train specimen. Finding an appropriate technique & a low-disease learning technique with excellent generalization capabilities is the next stage.

Caldeira et al. [8] deep learning models can detect lesions on cotton leaves. GoogleNet and Resnet50 are two deep-learning models that can be used for this task. Deep learning models are highly precise than traditional picture processing methodologies, such as SVM, KNN, ANN and NFC. ResNet50 performed slightly better than GoogleNet, but the difference was not significant. Balancing the data improved the performance of both models. The results of this study suggest that deep learning can be used to develop operational tools for the diagnosis of crop pests and diseases.

Kumar Yadav et al. [9] the detection, location, and destruction or spraying of VC plants are required. The VC identification in a corn field using the YOLOv3 DL at the system was investigated using RGB photos obtained from UAVs. The YOLOv3 could detect VC plants in cornfields with a median detection accuracy of over 80%. The performance metrics of the YOLOv3 behaved when the model was trained with test photos of various frames. The study concluded that YOLOv3 might be utilized to identify VC plants in a corn yield at a rate more than 14% greater in general identification of accuracy and about 29% significant at scaling S1 over the conventional pictures processing methodology employed in prior approaches.

Yadav et al. [10] Volunteer cotton (VC) The pests known as boll weevils (*Anthonomus grandis* L.) may use plants as hosts. The TBWEP employs people to find and eradicate VC yields that are sprouting up beside roadways or in areas used for rotation crops. Using UAS remote sensors pictures, a CV methodology related to YOLOv5 may be used to find VC plants cultivated in the centre of maize fields at 3 distinct growing phases. At the V6 stage of maize, YOLOv5s has to identify VC yields with a max classify precision of 98% and a MAP of 96.3%. The least accurate classification in the VT stage was achieved by YOLOv5s and YOLOv5m, at 85%.

Ramacharan [11] proposed that detecting cotton leaf diseases is of utmost importance to prevent serious outbreaks. The image is then processed using image processing techniques, resulting in a digitized color image of the diseased leaf. Subsequently, a CNN methodology is applied to elevate the root reason for the diseased cotton plants. The algorithm was tested using diseases specific to cotton plants. The experimental results indicate that the proposed approach can effectively distinguish between diseases with minimal computational effort. This technique enables the early-stage recognition of plant diseases, allowing for the timely implementation of pest control measures while minimizing risks to both people and the environment.

Fig. 5 Classification of plants and leaves based on conditions

3 Materials Available

A Kaggle Repository: This dataset contains images of diseased and healthy cotton leaves. It includes four classes: diseased cotton leaf, diseased cotton plant, healthy cotton leaf, and healthy cotton plant. The dataset provides 2694 images for training and testing, and a few cases are presented in Fig. 5.

4 Discussions

Neural networks rely solely on the patterns and information present in the training data. They do not possess explicit domain knowledge about cotton leaf diseases or the underlying biological processes. Incorporating domain knowledge and expert insights into the model design and training process can help enhance its performance and interpretability. Table 1 describes the merits and demerits of different approaches proposed by the existing researchers.

Table 1 Existing approaches evaluation based on different CNN models

Author	Features	Merits	Demerits	Accuracy (%)
Bhushan V. Patil et al	Detects local patterns and features in image Spatial dependencies	It is more accurate, Faster, scalable and cost-effective	The time complexity is high while making it more effective	87
Sheetal S. Zalte-Gaikwad et al	Minimise the feature mappings' temporal dimensionality	The model has high performance with cnn image prediction	Based on the epochs, the results are retrieved	99
Hiren Kukadiya et al	CNNs typically use nonlinear activation functions Enable it to learn complex relationships between features	It can work automatically and classify the images	The machine should connect to the network every time	96.7
Bhagya M. Patil	MLPs use AF to introduce nonlinearity into the network	Any data can be retrieved	Text data cannot be predicted efficiently	96
Pravin Srinath et al	An additional branch in the network called the Mask Head	Detection of pests can easily be identified by ignoring bg	By using GAN, the model performance can be increased	94
Sandeep Kumar et al	ML relies on human-engineered features	Almost every model is analyzed	This model cannot work on mobiles	88.5
Xihuizi Liang	SVM is focused on maximizing the distance between the training and the decision limit samples	The efficiency was increased by 7.7%	The model can only identify images	92
Rafael Faria Caldeira et al	Despite its depth, GoogleNet was designed to be computationally efficient	Using transfer learning, the data can be efficiently identified	ResNet is easy to implement and efficient compared to GoogleNet	89
Pappu Kumar Yadav et al	Effectively detect objects of different sizes, improving its ability to	The model has good efficiency	The model accuracy can be increased	80
J. Alex Thomasson et al	To improve speed and accuracy Reduces computational complexity, allowing YOLOv5	Locating and potential at identification of cotton plants	The method was implemented in areas where it could not find the remaining crop	96
Dr. S. Ramacharan	Leverages the learned features and parameters of the pre-trained model	The diseases are identified efficiently	The process is not automatic	96

5 Conclusion

Disease detection is crucial in promoting sustainable agriculture practices. By identifying diseases early and implementing targeted control measures, farmers can minimize chemical inputs, reduce environmental pollution, and promote the judicious use of resources. Sustainable disease management practices contribute to long-term ecological balance and preserve the health of agricultural ecosystems. Accurate disease detection provides critical information for effective disease management and control. It allows farmers to identify the specific pathogen or disease-causing agent responsible for the infection. This step helps isolate and focus on the leaf area, which is important for disease analysis. In future work, Integrating cotton leaf disease detection systems with decision support systems or mobile applications can provide farmers with comprehensive information and recommendations for disease management. This can include guidance on appropriate treatments, optimized pesticide application schedules, or other agronomic practices based on the detected diseases.

References

1. Patil BV, Patil PS (2021) A computational method for cotton plant disease detection of crop management using deep learning and internet of things platforms. In: Lecture notes on data engineering and communications technologies, vol 53. Springer, New York, pp 875–885. https://doi.org/10.1007/978-981-15-5258-8_81
2. Zambare R, Deshmukh R, Professor A (2022) Deep learning model for disease identification of cotton plants chetan awati* sandeep thorat sheetal zalte. In: Specialists ugdymas/special education
3. Kukadiya H, Meva D (2022) Automatic cotton leaf disease classification and detection by convolutional neural network. Commun Comput Inform Sci 1759:247–266. https://doi.org/10.1007/978-3-031-23092-9_20
4. Patil BM, Burkpalli V (2021) A perspective view of cotton leaf image classification using machine learning algorithms using WEKA. In: Advances in human–computer interaction, vol 2021. Hindawi Limited. https://doi.org/10.1155/2021/9367778
5. Udawant P, Srinath P (2022) Cotton leaf disease detection using instance segmentation. J Cases Inform Technol 24(4):721. https://doi.org/10.4018/JCIT.296721
6. Kumar S, Jain A, Shukla AP, Singh S, Raja R, Rani S, Harshitha G, Alzain MA, Masud M (2021) A comparative analysis of machine learning algorithms for detection of organic and nonorganic cotton diseases. Math Probl Eng 27:171. https://doi.org/10.1155/2021/1790171
7. Liang X (2021) Few-shot cotton leaf spot disease classification based on metric learning. Plant Methods 17(1):813. https://doi.org/10.1186/s13007-021-00813-7
8. Caldeira RF, Santiago WE, Teruel B (2021) Identification of cotton leaf lesions using deep learning techniques. Sensors 21(9):3169. https://doi.org/10.3390/s21093169
9. Kumar Yadav P, Alex Thomasson J, Hardin R, Searcy SW, Braga-Neto U, Popescu SC, Martin DE, Rodriguez R, Meza K, Enciso J, Solorzano Diaz J, Wang T (2023) Detecting volunteer cotton plants in a corn field with deep learning on UAV remote-sensing imagery. Comput Electr Agricult 204:7551. https://doi.org/10.1016/j.compag.2022.107551
10. Yadav PK, Thomasson JA, Searcy SW, Hardin RG, Braga-Neto U, Popescu SC, Martin DE, Rodriguez R, Meza K, Enciso J, Diaz JS, Wang T (2022) Assessing the performance of YOLOv5

algorithm for detecting volunteer cotton plants in corn fields at three different growth stages. Artif Intell Agricult 6:292–303. https://doi.org/10.1016/j.aiia.2022.11.005
11. Ramacharan S (2021) A 3-stage method for disease detection of cotton plant leaf using deep learning CNN algorithm. Int J Res Appl Sci Eng Technol 9:2503–2510

A Security Framework for the Detection of Targeted Attacks Using Honeypot

P. Subhash, Mohammed Qayyum, C. Likhitha Varsha, K. Mehernadh, J. Sruthi, and A. Nithin

Abstract The reliance on the Internet is growing steadily day by day, making susceptible to various security risks such as code injection, session hijacking, Denial-of-Service attacks, etc. These attacks threaten the CIA triad, that is, Confidentiality, Integrity, and Availability. As a result, ensuring uninterrupted security has become a demanding undertaking. Of all the options available, a honeypot is one of the best security mechanisms an organization can rely on. It is a system used as a trap for threat actors to believe it is a real system. The study of tricks and their attack vectors enables an understanding of potential security vulnerabilities, allowing for the implementation of measures to safeguard assets before any compromise occurs. This work presents the development of a real organizational network on the AWS Cloud, with a focus on enhancing cyber security measures. The network includes an all-in-one honeypot, *TPOT*, and vulnerable *web servers* on one server, while a secure web server and *database server* are deployed on another. The system aims to detect nine different types of attacks, such as DoS, brute force, and XSS, leveraging the T-Pot framework to analyze attack parameters. The crucial aspect of log monitoring is addressed through AWS Cloud Watch, which logs all processes on the connected instances. Additionally, Route 53 health checks are used to analyze traffic levels and

P. Subhash (✉) · C. Likhitha Varsha · K. Mehernadh · J. Sruthi · A. Nithin
CSE-(CyS, DS) and AI&DS, VNR VJIET, Pragathi Nagar, Hyderabad, Telangana 500090, India
e-mail: subhash.parimalla@gmail.com

C. Likhitha Varsha
e-mail: likhithavarsha555@gmail.com

K. Mehernadh
e-mail: karumurimehernadh02@gmail.com

J. Sruthi
e-mail: sruthi.j603@gmail.com

A. Nithin
e-mail: nithinaavula@gmail.com

M. Qayyum
Computer Science Engineering, King Khalid University, Abha 62529, Saudi Arabia
e-mail: mdqayyum.se@gmail.com

B. R. Devi et al. (eds.), *Proceedings of Fifth International Conference on Computer and Communication Technologies*, Lecture Notes in Networks and Systems 897,
https://doi.org/10.1007/978-981-99-9704-6_16

implement necessary mitigation strategies. This comprehensive network setup offers a robust defense against potential cyber threats, ensuring the organization's security and enabling proactive measures to safeguard its digital assets. Proposed security framework exhibits the significant results in detecting multiple targeted attacks.

Keywords CIA triad · AWS Cloud · TPOT · AWS Cloud Watch · Route 53 health checks · Web servers · Database server

1 Introduction

In this world of technology, it is very important to secure software, data in use, transit and rest, and applications as they are being widely used all over the world and are related to every field. As the methods and tactics used by hackers are improving, it is also essential for security professionals to improve methods to safeguard the data. As a result, advanced security measures are being implemented and found everywhere. Honeypot is one of the advanced security mechanisms used by companies in order to not only distract the attackers from the real server but also to understand their attack vectors and vulnerabilities found. Honeypots mimic the real target server and lure the attackers by making them believe that it is the target. A server that is the same as the real one is created and intentionally compromised. When attackers hack these, security administrators can study the methods and tactics used by the hackers, to harden security of their software or application.

Different types of honeypots can be configured depending on the requirements and end goals of the company or organization. Some honeypots are attack-based while some are based on protocols. Results shown by the honeypots also differ. They can show the location, the IP address of the attacker, commands or scripts used by them, attack vectors, the number of attacks performed, and many more insights that data can be represented visually through graphs and charts. Various types of attacks [1–4] that can be logged by the honeypot include DoS attacks, DDoS attacks, Brute force attacks, and many more. DoS and DDoS attacks are one of the most dangerous attacks that have no direct prevention. Honeypot is one of the best ways to detect and mitigate DoS and DDoS attacks where multiple requests are sent to the target server in such a way that it can no longer respond to requests even if they are from legitimate users. Our proposed work mainly aims to understand and demonstrate the usage of honeypots. A cloud platform is used to configure the honeypot as it is scalable and isolated from other products. Different types of attacks are performed in order to understand the efficiency of the honeypot and to understand how dangerous an attack could be. CloudWatch log groups collect honeypot data and are used to detect and understand the activity happening in the server, which in turn centralizes data to make the analysis and management of logs easier. The outcome of the proposed work is analyzing the attacks performed on the honeypot, analyzing the network traffic using Route 53 Health checks and cloud watch by performing various attacks on the server. It is all about the attacker and target sides of the process. The main objectives of the work investigated are:

1. Configure the Real Organization Network using AWS and Deploy TPOT in one of the servers in the network.
2. Implement CloudWatch to analyze the logs and monitor the traffic levels using Route 53 Health Checks.
3. Perform Attacks like DoS, Brute force, XSS, SQL Injection, File upload, some protocol-based attacks by performing the attacks

2 Literature Survey

The work proposed using a Honeypot framework uses the LIH, HIH for the detection DDoS [5, 6] and it has the two levels of filtering which only included few parameters like threshold value and using the routers. Found that these parameters were insufficient to prevent the DDoS attack and there was a need to add some more parameters to detect attack in a more efficient manner. This experiment only focused on DDoS attack [7]; rest attacks were not taken under consideration to prevent them. In the paper, Catching the Insider Threats, honeytokens were used to detect the insider threats. In this experiment they used unwanted entries in the database, unwanted entry to the system, etc. It was very useful to detect the insider threats but, in some scenarios, if the insider has more privileges and knows about the honeytokens, in that case, the implementation of these honeytokens fails [8]. The insider tries to create a false assumption to the organization by misleading them. The paper, Detecting Ransomware with Honeypot techniques, mentioned that he would detect the ransomware based on the File Screening service and Event Sentry [9]. The methods proposed by this work were able to detect the ransomware with a great accuracy but when new signature ransomwares are used to penetrate or other wrapping techniques, the method was not able to identify ransomware with a high level of precision.

The work proposed in the paper predication attacks based on intelligent honeypot technique for predication of intelligent honeypot based model which is used and make the machine learning methods to detect the intrusion Algorithm like DT decision tree used for classification tasks and identifying whether network activities are normal or indicative of intrusion detection improves the accuracy of the intrusion which also helps in identifying the True positives and false positives of the intrusion By using DT and ML algorithm, we can improve the overall accuracy in detecting and false alarm rate enhanced by this intelligent model based honeypot technique [10]. Using MTD and SDN—grounded Honeypots to Defend DDoS Attacks in IoT, authors Xupeng Luo, Qiao Yan, Mingde Wang, Wenyao Huang developed a moving target defense (MTD) armature using software-defined networking (SDN), which increases query due to the constantly changing attack face. Also, used SDN-grounded honeypots to imitate IoT bias and lure malware and bushwhackers. A combination of MTD and SDN-grounded honeypots can successfully hide network means from scanners and defend against DDoS attacks in the Internet of effects, according to testing results [11, 12].

In the paper, Roaming honeypots for mitigating service level DoS attack, roaming honeypots were proposed, a mechanism that allows the locations of honeypots to be unpredictable, continuously changing, and disguised within a server pool. It identifies and blocks harmful traffic coming from outside a firewall, and it also reduces attacks from within the firewall by cutting off all connections when a server switches from being a honeypot to being in active use. Effectiveness of the roaming honeypots scheme is observed through ns-2 simulators [13].

The work is proposed on Detecting and Analyzing Zero-day Attacks using honeypots. This network security system comprises a honeypot system with two components: Honeyd, a low-interaction honeypot on a customized Linux machine, and Metasploitable, a real system. These honeypots lure potential attackers and monitor their actions. A framework generates attack signatures used by an intrusion detection/prevention system like Snort to block threats. Metasploitable's data, lacking native logging, is securely transferred for analysis. This setup captures a wide range of attack patterns and behaviors, enhancing network security by effectively detecting and responding to potential threats [14].

All the above papers just tried to implement a method to secure using a honeypot but none of them performed the various testing on the honeypot and came to analysis about how much damage and impact it can cause and various ways to prevent or mitigate them.

3 Proposed Methodology

This experiment encompassed a diverse range of tools essential for conducting a comprehensive analysis. Notably, AWS, renowned for its cost-effective cloud computing services, played a pivotal role. PuTTY, as an open-source terminal emulator, facilitates secure connections to instances via SSH or Telnet. To simplify attack simulations, Kali Linux, a Debian-based distribution, was employed, streamlining the execution of various attacks. In addition, Burp Suite, a specialized software security testing application, was utilized by penetration testers to conduct web-based attacks efficiently. The integration of T-Pot, an "all-in-one honeypot" comprising more than 20 honeypots, allowed for detailed attack analysis. For web server configuration or Java code, Apache Tomcat, an open-source web server, proved invaluable. Lastly, MySQL, a reliable relational database management system, was used to establish well-organized databases in the form of tables. The amalgamation of these powerful tools provided a robust foundation for the experiment, enabling thorough exploration and comprehension of attack methodologies, thereby contributing to the formulation of effective security measures to protect valuable assets from potential compromise.

The first step involved developing the virtual network infrastructure, which included launching instances in both public and private regions. A web server was then deployed on one of the instances and connected to the database server. Subsequently, another instance was set up to configure the honeypot and vulnerable web server. Finally, multiple attacks were executed to assess the web server's security. The analysis of attack results was conducted through the implementation of the T-Pot honeypot framework, Route 53 Health Checks, and CloudWatch Logs. The outcomes revealed that the web server exhibited vulnerabilities to a diverse array of attacks. Detection of these attacks was successfully accomplished using the honeypot framework, while the Route 53 Health Checks and CloudWatch Logs provided valuable insights into the traffic patterns and attack logs. This experiment underscores the significance of employing multiple tools to assess web server security comprehensively. The utilization of diverse tools allows for the identification and mitigation of a broader spectrum of vulnerabilities, enhancing the overall resilience of the web server against potential threats. As shown in Fig. 1, using the AWS console, a virtual private cloud (VPC) was created to establish a virtual network. The VPC was configured with both public and private subnets. Inside the VPC, a honeypot server and a web server were launched separately on different ports. Additionally, two additional instances were deployed, one for the web server and the other for the database server. These instances were interconnected through a bastion server. To enable internet access for the private subnet without directly connecting it to the internet via the public subnet, a NAT gateway was set up. Additionally, route tables and network access control lists (NACLs) were appropriately configured to manage traffic flow and control access within the VPC.

Figure 2 shows our proposed system deals with the analysis of the output obtained, from which necessary measures can be taken. First, a network is created in AWS using VPC. This network is connected to a Debian OS instance, on which a honeypot and a

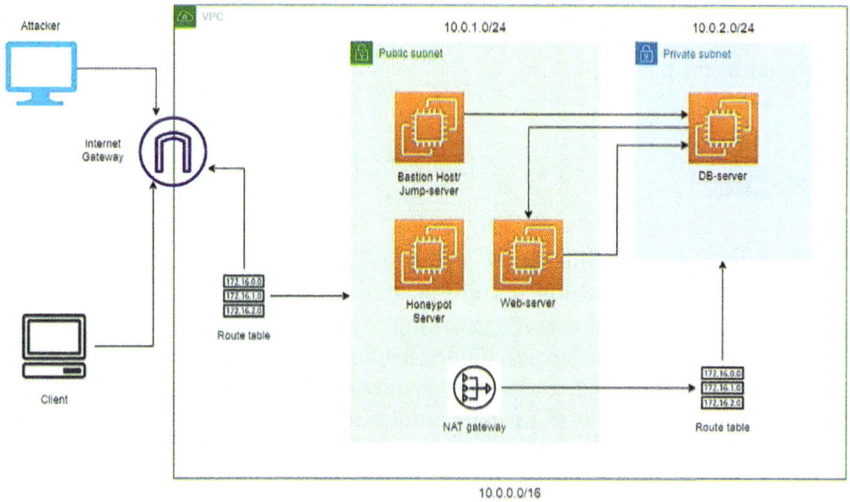

Fig. 1 Proposed system architecture diagram

Fig. 2 Flowgraph of proposed system

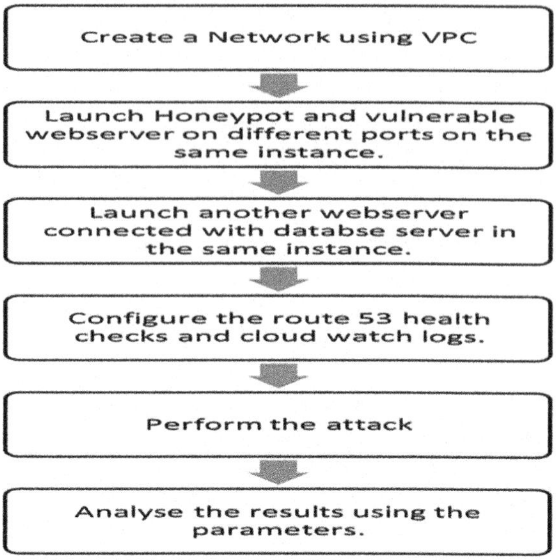

vulnerable web server are launched on different ports. Another web server connected to the database server is launched. For any organization, examining and analyzing logs is very crucial. So, cloud watch logs and health checks are configured on Route 53 for monitoring. Now, once the required target server is created successfully, it will be tested by performing various attacks like DoS, Brute-force attacks, XSS, SQL injection attacks, etc. Honeypot shows the results in the form of graphs, pie-charts and some of them even show commands used to compromise the server, from which we can understand the vulnerabilities exploited. Some of the other features that will be shown are the location and IP address of the attacker and, the number of attacks performed.

4 Results

In Fig. 3, the analysis of the attack on the web server included an examination of the source IP addresses and their corresponding counts. This analysis revealed the number of unique IP addresses engaged in the attack and the frequency of attacks from each IP address. The acquired information serves as a critical tool for implementing targeted filters and necessary precautions to defend against these attacks. By comprehending the origins and patterns of the attacks, proactive security measures can be implemented to strengthen the web server's defenses and safeguard it against potential threats.

Figure 4 from the honeypot framework reveals the origins of attacks, including those from major sources like India, where the attacks were performed from our end.

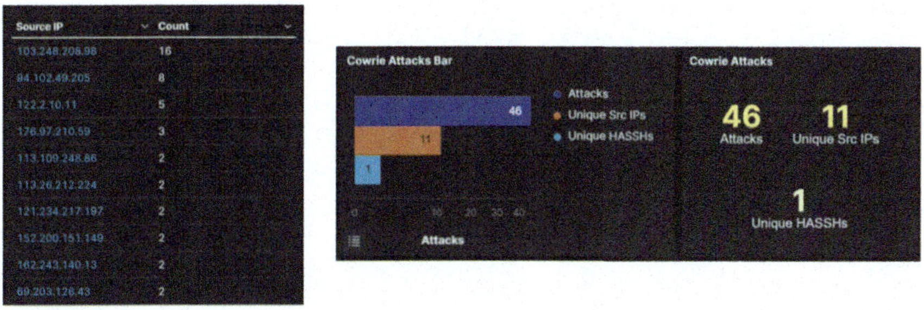

Fig. 3 Source IP and attack count in T-Pot framework

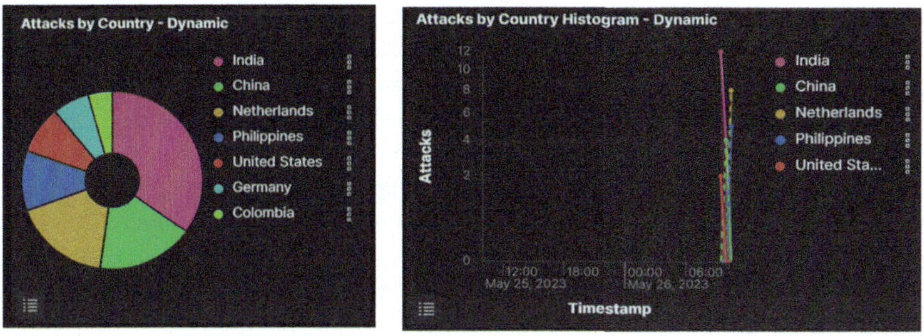

Fig. 4 Attacking country in T-Pot framework

In certain scenarios, identifying the attacking country becomes vital to implement appropriate countermeasures. Understanding the root cause of attacks, influenced by political or other factors, allows us to develop targeted strategies to mitigate future occurrences effectively.

By closely monitoring and recording each attack's port number, gathered data to construct Fig. 5, which displayed the percentage of attacks targeting specific ports. This graphical representation highlighted the most frequently targeted ports, with Telnet and SSH attacks being notably common and posing significant risks. Valuable insights led to critical security measures, including disabling unnecessary services on vulnerable ports and enforcing robust authentication. The strengthened defenses reduced Telnet and SSH attack exposure, elevating overall security.

The analysis of CloudWatch logs brought to light significant vulnerabilities in the honeypot, raising serious security concerns for the system. The successful execution of malicious scripts without error prompts exposed potential weaknesses in the honeypot's defences, granting effortless access to the user portal and revealing vulnerabilities during automated injections. Furthermore, the honeypot's unrestricted acceptance of all file types and malicious code added to the security risks. The bypassing of password prompts allowed unauthorized access to sensitive areas, leading to the exposure of usernames and passwords through brute force attacks. Notably, trial and error methods even facilitated access to the web server's shell

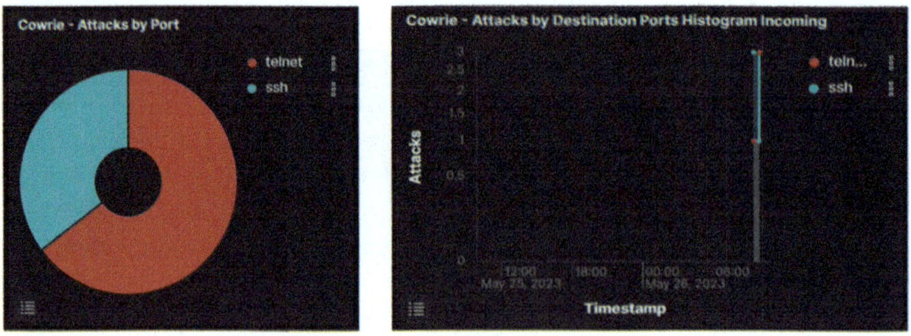

Fig. 5 Ports attacked in T-pot framework

and enabled privilege escalation. The FTP server's lax controls further compounded the problem by permitting the posting of malicious files. Additionally, the observed slowdown during attacks highlighted the system's vulnerability to potential Denial of Service (DoS) threats. Strengthening the honeypot's security, conducting a comprehensive review and fortification of script handling, enforcing stringent file upload restrictions, and enhancing password security, shell access controls, and FTP server configurations are crucial steps to bolster the system's integrity and availability against adversarial attacks. The analysis of the attack outcomes demonstrated the efficacy of Route 53 Health Checks in providing valuable insights into the traffic patterns generated during the attacks. The accompanying Fig. 6 depicted the health check status as initially healthy under normal conditions. However, as the intensity of the attack increased, there was a significant surge in traffic. Consequently, the health check status transitioned to unhealthy, activating alarms.

This research paper underscores the importance of employing Route 53 Health Checks as a proactive measure for monitoring and responding to unusual traffic levels, thereby bolstering the network's overall security and resilience against potential attacks.

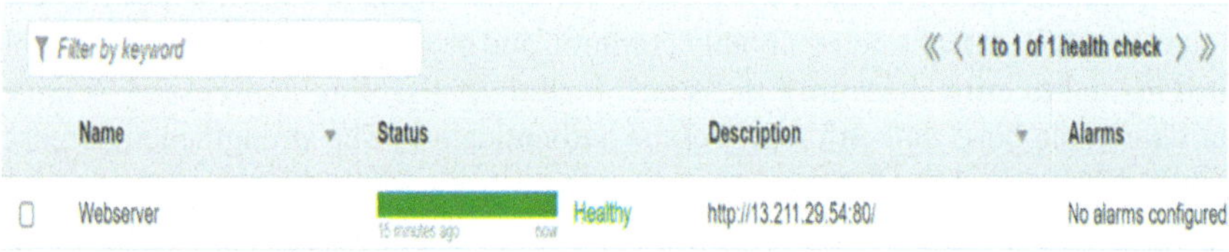

Fig. 6 Route 53 health check status

5 Conclusions

The primary aim of this experiment was to create a network similar to the ones utilized by organizations on the AWS Cloud. The network was then tested for security vulnerabilities. A honeypot was launched on one of the vulnerable web servers, Route 53 health checks were created to analyze traffic intensity, and Cloud Watch logs were used to analyze all incoming logs into the honeypot server. The goal was to analyze the impact of the attacks and provide measures to mitigate them. The experiment found that the honeypot server was vulnerable to common web-based attacks. The attackers were able to escalate privileges, inject payloads, and execute scripts. Cloud Watch logs provided a lot of information about each attack, including the scripts that were executed, the IP of the attackers, and the countries from which the attacks stemmed. Route 53 health checks were only useful for detecting DoS attacks. The findings of this experiment can be used to rectify vulnerabilities and configure settings in a more secure manner. In the future, honeytokens could be implemented to detect insider threats. Investigate the use of other tools, such as SIEMs, to analyze the collected data. More rigorous testing with DDoS attacks could be performed to analyze how auto-scaling helps to mitigate these attacks. Additionally, develop a more comprehensive plan for mitigating the vulnerabilities that were identified.

References

1. Das VV (2009) Honeypot scheme for distributed denial-of-service. In: 2009 International conference on advanced computer control. IEEE, pp 497–501
2. Pashaei A, Akbari ME, Lighvan MZ, Charmin A (2022) Early intrusion detection system using honeypot for industrial control networks. Results Eng 16:100576
3. Leaden G, Zimmermann M, DeCusatis C, Labouseur AG (2017) An API honeypot for DDoS and XSS analysis. In: 2017 IEEE MIT undergraduate research technology conference (URTC). IEEE, pp 1–4
4. Kambow N, Passi LK (2014) Honeypots: the need of network security. Int J Comput Sci Inf Technol 5(5):6098–6101
5. Sembiring I (2016) Implementation of honeypot to detect and prevent distributed denial of service attack. In: 2016 3rd international conference on information technology, computer, and electrical engineering (ICITACEE). IEEE, pp 345–350
6. Melhem H, Dayoub Y (2022) A hybrid honeypot framework for DDOS attacks detection and mitigation
7. Weiler N (2002) Honeypots for distributed denial-of-service attacks. In: Proceedings on eleventh IEEE international workshops on enabling technologies: infrastructure for collaborative enterprises. IEEE, pp 109–114
8. Spitzner L (2003) Honeypots: catching the insider threat. In: Proceedings on 19th annual computer security applications conference, 2003. IEEE, pp 170–179
9. Moore C (2016) Detecting ransomware with honeypot techniques. In: 2016 cybersecurity and cyberforensics conference (CCC). IEEE, pp 77–81
10. Alshahrani A (2023) Predication attacks based on intelligent honeypot technique
11. Hakim MA, Aksu H, Uluagac AS, Akkaya K (2018) U-pot: a honeypot framework for upnp-based iot devices. In: 2018 IEEE 37th international performance computing and communications conference (IPCCC)). IEEE, pp 1–8

12. Luo X, Yan Q, Wang M, Huang W (2019) Using MTD andSDN-based honeypots to defend DDoS attacks in IoT. In: 2019 Computing, communications and IoT applications (ComComAp). IEEE, pp 392–395
13. Khattab SM, Sangpachatanaruk C, Mossé D, Melhem R, Znati T (2004) Roaming honeypots for mitigating service-level denial-of-service attacks. In: Proceedings on 24th international conference on distributed computing systems, 2004. IEEE, pp 328–337
14. Musca C, Mirica E, Deaconescu R (2013) Detecting and analyzing zero-day attacks using honeypots. In: 2013 19th international conference on control systems and computer science. IEEE, pp 543–548

SNMP Watcher: A Model to Detect DoS and DDoS Attacks Using SNMP Management Information Base Analysis

V. Surya Narayana Reddy, P. Subhash, Ch. Vinay Kumar, M. Samuel, G. Vishal Reddy, and G. Tharun

Abstract Internet has become daily routine of the world in the present generation, there is no individual who has no internet access. The increase in usage of internet has also increased varieties of vulnerabilities which empowered the attackers to make use of these vulnerabilities. Denial of Service (DoS) and Distributed Denial of Service (DDoS) are attacks which make large impact on many reputed organizations which results in monetary loss and defamation. Addressing these, we propose a model that makes use of Simple Network Management Protocol Management Information Base (SNMP-MIB) data to effectively detect DoS and DDoS attacks and provide accurate results using Machine Learning Algorithms. SNMP-MIB data is used to monitor flow of network and also identify abnormal behaviour in the traffic. This data makes it less difficult to detect malicious behaviour, if occurred in the network. Machine Learning algorithms such as Extreme Gradient Boosting, CatBoost (Categorical Boosting) and LightGBM generated results with astonishing accuracies of 99.67%, 99.78% and 99.78% respectively. Usage of broad dataset and strict evaluation metrics, including accuracy, recall and $F1$-score increased the robustness and dependability of our

V. Surya Narayana Reddy · P. Subhash (✉) · Ch. Vinay Kumar · M. Samuel · G. Vishal Reddy · G. Tharun
CSE-(CyS, DS) and AI&DS, VNR VJIET, Pragathi Nagar, Hyderabad, Telangana 500090, India
e-mail: subhash.parimalla@gmail.com

V. Surya Narayana Reddy
e-mail: suryanarayanareddy_v@vnrvjiet.in

Ch. Vinay Kumar
e-mail: chvinay126@gmail.com

M. Samuel
e-mail: vishalsam149@gmail.com

G. Vishal Reddy
e-mail: ismartvishal2002@gmail.com

G. Tharun
e-mail: tharungurram28@gmail.com

B. R. Devi et al. (eds.), *Proceedings of Fifth International Conference on Computer and Communication Technologies*, Lecture Notes in Networks and Systems 897,
https://doi.org/10.1007/978-981-99-9704-6_17

models. As a whole the primary goal of our research is to provide innovative and cost-effective solution to build robust defence system that helps Enterprises in protecting their assets and resume their operations in the phase of evolving cyberthreats.

Keywords Denial of service (DoS) · Distributed denial of service (DDoS) · Simple network management protocol (SNMP) · Management information base (MIB) · Machine learning (ML) · Object identifiers (OIDs)

1 Introduction

Technology in present generation is highly advanced and is evolving to digital. This evolution of technology has raised many cyber threats that can shut down the organizations. Among those cyber threats DoS and DDoS are the attacks which tend to halt the functioning of the organizations. The potentiality of these attacks ranges from temporarily denying the service of any website, online services, and servers to make resources unavailable, making it a serious impact on enterprises including government services. In order to deal with these potential attacks SNMP-MIB and ML technology would be equally potential. The diversity and robustness of ML made it suitable to integrate with SNMP-MIB data. These technologies are effective and robust to protect online services.

DoS is a type of attack that halts online services of any organization to prevent its access to legitimate users. The primary goal is to make the service unavailable to its users. This causes monetary loss to organizations and defames the reputation of organizations. DDoS is an alike to DoS attack. Unlike DoS attack methodology DDoS uses multiple systems to flood the traffic. DDoS intend to shut down the services of the organizations causing catastrophic damage to the organizations. The targets of these attacks can be websites, servers, and applications. These attacks can be executed by an individual attacker or a group of attackers depending on their intention. The major attacks take place in the day-to-day operations are UDP Flood, TCP SYN Flood, ICMP-Echo Flood, Slowloris, Slowpost, HTTP Flood.

The primary goal of this research is to minimise the impact of DoS/DDoS attacks. To find an effective way we make use of SNMP-MIB data, beneficial network traffic information and suitable ML Algorithms. The integration of ML with SNMP-MIB data builds a robust system that effectively and accurately detects several types of DoS and DDoS attacks. In addition, this system helps enterprises to protect themselves from these attacks and continue their services. To be more clear, network topology contains devices such as Routers and Switches which helps in routing packets to the destination. These devices can be analysed and managed in network using SNMP. SNMP is designed to monitor and control these devices in the network. The main components of SNMP are SNMP manager and SNMP agent. SNMP agents are installed in a variety of devices, including routers, computer systems, and servers and they serve to store vital device data which can be accessed by the SNMP Manager. SNMP manager sends a query to these agents to retrieve information. SNMP Agent software contains MIB database, this database contains vast range of OIDs. Each OID represents a variable in the SNMP Agent's system and stores associated values. SNMP Manager queries these OIDs to retrieve information such as Agent System performance, CPU utilization, transmitted packets i.e., number of packets received, and number of packets transmitted and much more.

MIB data contains information such as statistics of network traffic, system uptime, contact information, etc. under regular conditions where the behaviour of network is normal. This information can be used to inspect the pattern of network in the regular conditions. The behaviour of the network traffic is unusual such as immediate raise in received packets, uncommon traffic pattern during any of the DoS or DDoS attacks. However, these unusual patterns can be identified by using appropriate Machine Learning Algorithms. When the MIB data is fed to ML algorithms, the algorithms analyse the traffic pattern during normal behaviour and abnormal behaviour, which allows us to be able to detect DoS or DDoS attacks effectively.

Feature selection is the preparatory step before using ML algorithms to detect DoS and DDoS attacks. This step involves selecting most related and impactful OIDs in the MIB, which increase the probability of detecting DoS and DDoS attacks. This process helps in removing unwanted OIDs and makes the ML algorithm focus on beneficial data points that improve the accuracy in identifying normal traffic and abnormal traffic. OIDs can be grouped in different categories such as TCP, UDP, IP, Interface and ICMPgroups to effectively organize and process the selected OIDs.

2 Related Work

The work proposed on network anomaly detection model [1] focuses on machine learning algorithms like Random Forest, REP tree, and decision tree (J48), which have accuracy rates of 99.98%, 99.98%, and 99.88%, respectively. However, due to the model's construction relying solely on the IP group of the SNMP-MIB, it cannot be used to identify and categorise a specific form of DoS/DDoS attack.

Al-Harwari et al. [2, 3] proposed models on anomaly detection in the network using ML. The accuracy achieved with IP group and Interface group are 100% and 99.93% respectively. The accuracy achieved with only Interface group using Random Forest 99.93%, with Adaboost 99.60% and with MLP 98.86%. These models do not focus on detecting and classifying types of DoS or DDoS attacks.

The methodologies [4, 5] introduced in the literature to detect DoS and DDoS attacks efficiently with minimal consumption of hardware resources. Results shows that accuracy of 99.93% was achieved by Lazy. IBK classifier with Interface group. The model developed by Namvarasal et al. [6] achieved an accuracy of 99.03% with neural networks classifiers. A variety of DoS and DDoS attack types cannot be detected by or classified using the models.

Brahanyaa et al. [7] constructed a model using ML algorithms like Gradient Boosting, Decision Trees, Perceptron and Random Forest with SNMP-MIB dataset for DDoS attack prevention. Results show that Gradient Boosting classifier achieved

99.60% accuracy with ICMP group, TCP group and UDP group. Alhaidari et al. [8] applied HMM based on Viterbi algorithm and recorded 99.86%, Adaboost M1 recorded 93.57% with Interface group of SNMP-MIB dataset. These articles didn't focus on recognising and sorting out multiple attack types that are there in DoS and DDoS attacks.

Gayathri et al. [9] developed a model using Linear Regression and SNMP-MIB variables focused on TCP group. The methodology investigated on detecting DoS attacks using ML algorithms [10] such as Random Forest and Logistic regression achieved 100%, 99.9% accuracies on Slowloris attack. These articles does not detect other types of DoS/DDoS attacks.

Hwoij et al. [11] developed a network anomaly detection using SNMP-MIB dataset with Rule-based ML algorithms focussed on detecting ICMP, Slowloris, HTTP attacks and achieved an accuracy of 99.7% and Shaheen et al. [12] achived 99.6%. They do not focus on finding and labelling of all possible types of DoS/DDoS attack categories.

The literature consists of all the required novelties but none of the above focussed on developing a model that can deeply investigate to detect and classify various types of DoS attacks. In view of the incompleteness in the aforementioned literature, the developed model in the current research uses the SNMP-MIB database and ML algorithms to detect and categorise various types of network attacks.

3 Proposed Methodology

The proposed strategy for delivering DoS as well as DDoS detection framework is followed by using ML algorithms and SNMP-MIB data. The dataset of SNMP-MIB is generated from real network traffic [13]. The proposed model able to detect and classify several DoS/DDoS kinds of attacks. Detecting and identifying the type of attack performed helps to reduce the cause and take appropriate precaution. The process was initiated by collecting and process SNMP-MIB dataset. This dataset contains useful traffic information such as network performance and behaviour. We feed this dataset to ML algorithms for classification of traffic behaviour. For building the models on MIB dataset XGBoost, CatBoost, LightGBM are the ML algorithms opted based upon their adaptability and performance in classification problems. The collected SNMP-MIB data is categorized into different groups such as Interface, UDP, TCP, IP and ICMP. This division helps in increasing the performance of classifiers to detect network attacks using MIB data. The proposed research focuses on integrating all these groups for analysis.

3.1 Dataset

SNMP-MIB dataset is used to test our framework which was developed to spot attacks involving DDoS/DoS. The collected MIB dataset is categorized into 34 MIB variables which contains 4998 records. The information of network performance and behaviour is provided by these MIB variables. TCP-SYN flood, Slowloris, ICMP-ECHO, UDP flood, Slowpost, and HTTP flood are few categories of attacks including instances of brute force attacks in the collected MIB dataset. The fine documentation of network attack features and its patterns in [14], was helpful in providing necessary insights for evaluation and analysis of various DoS attacks. We intended to evaluate the accuracy and performance of our built framework to efficiently determine and label several DoS attack using the gathered MIB dataset.

3.2 Classifiers

Detecting network attacks is mainly based on Classification technique in ML. These classification techniques are known as Classifiers which work effectively in classifying and identifying various types of network attacks. The primary goal of Algorithms is to categorize traffic into malicious and regular network traffic. Based on the collected labelled data a ML model is constructed and implemented to classify network traffic. In this research, we opted three ML algorithms, such as CatBoost, XGBoost, and LighGBM based on their excellent performance and adaptability in classification problems. These algorithms are known for their accuracy and robustness. The collected SNMP-MIB dataset is fed to these algorithms to differentiate network traffic into malicious or regular traffic. After this, we evaluated the results of each classifier to regulate the effectiveness and ability of using ML algorithms to figure out several DOS/DDoS utilising SNMP-MIB data. The ML algorithms used to detect attacks are XGBoost, CatBoost, LightGBM.

In the execution phase of the experiment, SNMP-MIB data collected is fed to ML algorithms for the identification and classification of DoS and DDoS attacks. Malicious traffic and regular traffic are present in the MIB dataset that is furtherly divided into two parts i.e., for training 80% of the data is fed to Classifiers and the remaining 20% data is used to test the trained classifiers. In the implementation, to identify and differentiate the network traffic we used XGBoost, LightGBM and CatBoost classifiers. These algorithms perform well in classification problems by generating accurate results and handling errors.

3.3 Metrics for Assessing Performance

SNMP-MIB dataset is used to test our framework which was developed to spot attacks involving DDoS/DoS. The collected MIB dataset is categorized into 34 MIB variables which contains 4998 records. We examine the proposed model using the performance assessment metrics such as Accuracy, Precision, Recall and $F1$-Score. The examination reveals the significant results and evidence for the better model performance.

4 Results and Discussion

This section examines the results received after the experiment conducted on SNMP-MIB dataset using ML algorithms such as XGBoost, CatBoost and LightGBM. The experiment is carried out with different algorithms on various types of attacks. Measures of performance including $F1$-Score, recall and precision are steadily high for each type of attack. This depicts that applying these ML algorithms to MIB dataset can effectively detect and classify different type of network attacks very accurately.

The recall and precision values are either 1.00 or close to it, this means all type of attacks (HTTP Flood, Slowloris, TCP-SYNC Flood, ICMP-ECHO Flood, Slowpost and UDP Flood) including normal traffic are correctly classified with both negative and positive instances with exciting accuracy using ML algorithms. Figure 1 shows the precision comparison of algorithms for various attacks where the results are accurate. Figure 2 demonstrates the Recall comparison of algorithms used in various attacks. $F1$-score represents ability of algorithms to perform well in classification tasks which is also very high for each type of attack. The contrasting view of the $F1$-score for various algorithms and attacks is shown in Fig. 3.

All three classifiers have accuracy scores that are very high, indicating that they make tremendously accurate predictions. The accuracy score is an overall indicator of the proportion of accurate forecasts among all predictions. Additionally, considering the class imbalance in the dataset, the weighted average $F1$-Score and precision are both near to 1.00, demonstrating that the classifiers are performing well across all classes.

The algorithms such as XGBoost, CatBoost and LightGBM are used for detection and classification of various types of DoS attacks and has produced significant results with accuracies 99.67%, 99.78% and 99.78% respectively, which is represented in Fig. 7. The detailed result of the XGBoost, LightGBM and CatBoost algorithms is cleary depicted in the Figs. 4, 5 and 6, respectively. The effectiveness and robustness of the used ML algorithms in precisely detecting and categorizing network behaviour is grounded on MIB dataset. The potentiality of these algorithms is proved by achieving accurate results without employing any hybrid algorithms and using all MIB groups such as TCP, UDP, IP, Interface and ICMP in security of real-time networking applications.

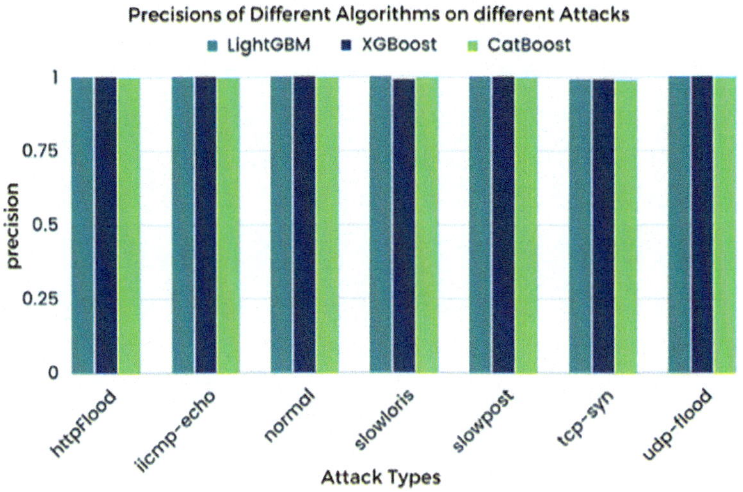

Fig. 1 Precision comparison

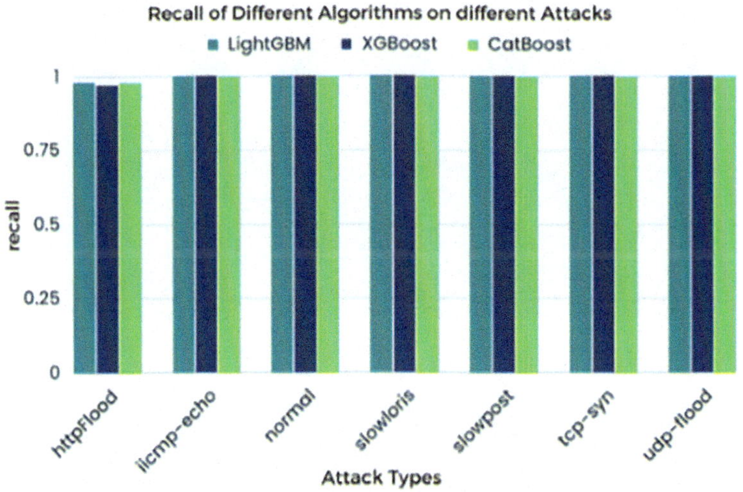

Fig. 2 Recall comparison

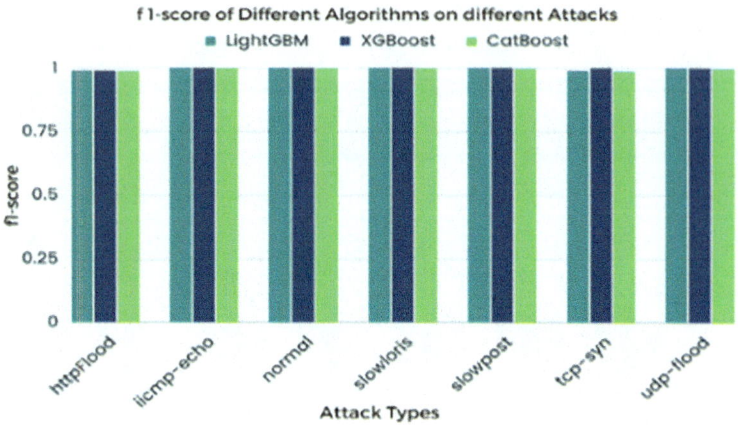

Fig. 3 $F1$-score comparison

Classification Report:

	precision	recall	f1-score	support
httpFlood	1.00	0.97	0.99	115
icmp-echo	1.00	1.00	1.00	127
normal	1.00	1.00	1.00	130
slowloris	0.99	1.00	1.00	154
slowpost	1.00	1.00	1.00	96
tcp-syn	0.99	1.00	0.99	160
udp-flood	1.00	1.00	1.00	152
accuracy			1.00	934
macro avg	1.00	1.00	1.00	934
weighted avg	1.00	1.00	1.00	934

Fig. 4 XGBoost algorithm report

Classification Report:

	precision	recall	f1-score	support
httpFlood	1.00	0.98	0.99	115
icmp-echo	1.00	1.00	1.00	127
normal	1.00	1.00	1.00	130
slowloris	1.00	1.00	1.00	154
slowpost	1.00	1.00	1.00	96
tcp-syn	0.99	1.00	0.99	160
udp-flood	1.00	1.00	1.00	152
accuracy			1.00	934
macro avg	1.00	1.00	1.00	934
weighted avg	1.00	1.00	1.00	934

Fig. 5 LightGBM algorithm report

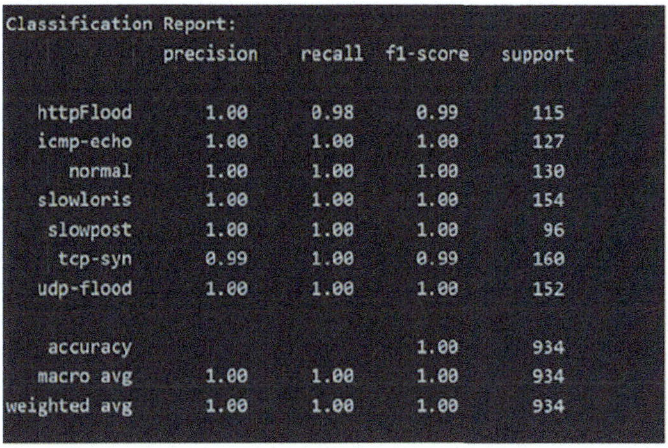
Classification Report:

	precision	recall	f1-score	support
httpFlood	1.00	0.98	0.99	115
icmp-echo	1.00	1.00	1.00	127
normal	1.00	1.00	1.00	130
slowloris	1.00	1.00	1.00	154
slowpost	1.00	1.00	1.00	96
tcp-syn	0.99	1.00	0.99	160
udp-flood	1.00	1.00	1.00	152
accuracy			1.00	934
macro avg	1.00	1.00	1.00	934
weighted avg	1.00	1.00	1.00	934

Fig. 6 CatBoost algorithm report

Fig. 7 Accuracies of algorithms

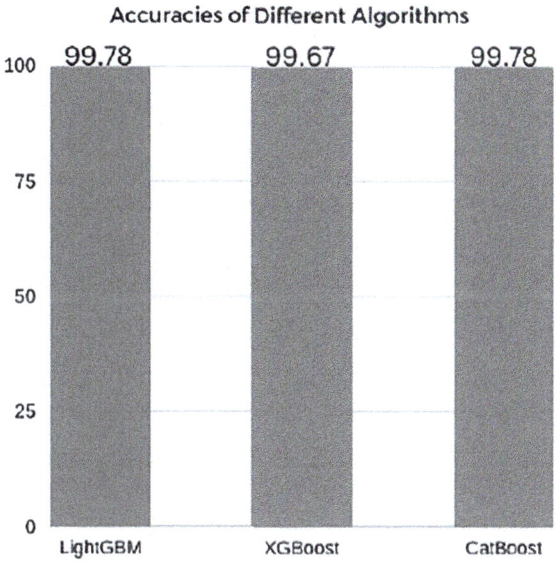

5 Conclusion and Future Work

In this work, the SNMP-MIB dataset has been successfully assessed for the effectiveness using various machine learning techniques. The research revealed that the results of various algorithms for each attack is investigated and found the future opportunities in the same line. Three machine learning algorithms are used to analyse the patterns in the SNMP-MIB data, and the detection of each attack is taken place with maximum accuracy. These algorithms are capable to detect and classify various type of DoS and DDoS attacks efficiently considering selected MIB groups such as TCP, UDP, IP, Interface and ICMP. As part of future research, the work can be extended to be able to classify and detect other than DoS and DDoS attacks with more accuracy and with a different composition of algorithms to make sure that the organizations are secure and optimal utilization of their resources.

References

1. Manna A, Alkasassbeh M (2019) Detecting network anomalies using machine learning and SNMP-MIB dataset with IP group. In: 2019 2nd international conference on new trends in computing sciences (ICTCS). IEEE
2. Al-Naymat G, Al-Kasassbeh M, Al-Harwari E (2018) Using machine learning methods for detecting network anomalies within SNMP-MIB dataset. Int J Wirel Mob Comput 15(1):67–76
3. Al-Naymat G, Al-Kasassbeh M, Al-Hawari E (2019) Exploiting snmp-mib data to detect network anomalies using machine learning techniques. In: Proceedings of the 2018 intelligent systems conference (IntelliSys) on intelligent systems and applications, vol 2. Springer
4. Al-Naymat G, Hambouz A, Al-Kasassbeh M (2019) Evaluating the impact of feature selection methods on snmp-mib interface parameters to accurately detect network anomalies. In: 2019 IEEE international symposium on signal processing and information technology (ISSPIT), pp 1–6. IEEE
5. Al-Naymatm G, Hambouz A, Alkasassbeh M (2019) Network attacks anomaly detection using SNMP MIB interface parameters. arXiv preprint arXiv:1906.00865
6. Namvarasl S, Ahmadzadeh M (2014) A dynamic flooding attack detection system based on different classification techniques and using SNMP MIB data. Int J Comput Netw Commun Secur 2(9):279–284
7. Brahanyaa S, Jani Anbarasi L (2018) Classification of SNMP network dataset for DDoS attack prevention. In: 2018 IEEE international conference on computational intelligence and computing research (ICCIC). IEEE
8. Alhaidari S et al (2019) Network traffic anomaly detection based on Viterbi algorithm using SNMP MIB data. In: Proceedings of the 2019 3rd international conference on information system and data mining
9. Gayathri R, Neelanarayanan V (2019) Identification of regression function and distribution model for denial of service attack in second life online community using simple network management protocol. Int J Web Based Commun 15(3):225–237
10. Abushwereb M et al (2020) Attack based DoS attack detection using multiple classifier. arXiv preprint arXiv:2001.05707
11. Hwoij A, Al-Kasassbeh M, Al-Fayoumi M (2020) Detecting network anomalies using rule-based machine learning within SNMP-MIB dataset. arXiv preprint arXiv:2002.02368
12. Shaheen YK, Al Kasassbeh M (2019) A proactive design to detect denial of service attacks using SNMP-MIB ICMP variables. In: 2019 2nd international conference on new trends in computing sciences (ICTCS). IEEE

13. https://www.kaggle.com/datasets/malkasasbeh/network-anomaly-detection-dataset
14. Al-Kasassbeh M, Al-Naymat G, Al-Hawari E (2016) Towards generating realistic SNMP-MIB dataset for network anomaly detection. Int J Comput Sci Inf Secur 14(9):1162

Comparative Analysis for Loan Approval Prediction System Using Machine Learning Algorithms

Ashok Reddy Kandula, K. Divya, Sri Rama Krishna Datta Movva, Mohan Sai Motineni, Harika Pappala, and Karthika Venkata Satish Reddy Jakkireddy

Abstract In the banking sector, predicting loan acceptance is a crucial task, and machine learning techniques may be used to create models that can do this. To identify which machine learning method produces the best results for this problem, we compare a variety of techniques, which include decision trees, random forests, support vector machines, Gaussian Naive Bayes, XGBoost, and K-Nearest Neighbor. The analysis takes a number of criteria into account, including each algorithm's computational cost and scalability as well as accuracy, precision, recall, and $F1$ score. Based on their unique requirements and dataset, financial institutions may use the findings of this investigation to determine the best suitable algorithm for loan approval prediction.

Keywords Loan approval prediction · Machine learning · Algorithms · Comparative analysis · Decision tree · Random forest · Support vector machines · XGBoost · K-Nearest neighbors · Naive Bayes

1 Introduction

Predicting loan approval is a critical challenge in the finance sector since it enables financial firms to manage risks and make educated decisions [1]. Previously, a borrower's credit score and other financial data were used to make decisions on whether to approve a loan. But, with the expansion of data sources and improvements in machine learning algorithms, these algorithms are now being utilized to create models that can accurately forecast loan acceptance.

A. R. Kandula (✉) · K. Divya · S. R. K. D. Movva · M. S. Motineni · H. Pappala · K. V. S. R. Jakkireddy
Department of AI&DS, Seshadri Rao Gudlavalleru Engineering College, Gudlavalleru, Andhra Pradesh, India
e-mail: ashokreddy.gec@gmail.com

B. R. Devi et al. (eds.), *Proceedings of Fifth International Conference on Computer and Communication Technologies*, Lecture Notes in Networks and Systems 897,
https://doi.org/10.1007/978-981-99-9704-6_18

In this comparative study [2], we compare several machine learning methods for predicting loan acceptance in this comparative study. We take into account well-known algorithms such as Naive Bayes, XGBoost, K-Nearest Neighbors, Random Forests, Decision Trees, and Support Vector Machines [7]. The many features connected for analysis are Loan ID, Gender, Married, Dependents, Education, Self-Employed, Applicant Income, Co-Applicant Income, Loan Amount Term, Credit History, Property Area, and Loan Status [3]. These are all analyzed by these algorithms to determine how well they perform. These attributes are then used by the algorithms to forecast whether or not the loan will be authorized. We take into account a number of criteria, including accuracy, precision, recall, and the $F1$ score, to assess the effectiveness of these algorithms. The proportion of properly anticipated loan approvals or rejections is measured by accuracy, whereas the percentage of accurately predicted loan approvals across all predicted loan approvals is measured by precision [5]. The proportion of accurately anticipated loan approvals out of all actual loan approvals is measured by recall, and the $F1$ score is the harmonic mean of accuracy and recall. Also, we take each algorithm's scalability and computational complexity into account. While alternative techniques could be simpler to scale to bigger datasets, some may require more computing than others.

The findings of this investigation can assist financial firms in choosing the best algorithm for their unique dataset and requirements. Financial institutions may better manage risks and make educated decisions, which will eventually result in improved financial performance, by properly anticipating loan approval.

2 Literature Review

The banking industry has been witnessing significant growth in recent years, resulting in an increased demand for bank loans. Banks and other financial institutions need to make loans to maintain their primary source of income, which is necessary for them to generate interest. However, granting loans involves a considerable amount of risk, and banks need to be able to identify borrowers who will be unable to repay the loan. To tackle this problem, machine learning-based strategies have been developed to decrease risk by identifying borrowers who are more likely to default on their loans [4]. Logistic regression, XGBoost, random forest, and Naive Bayes models are some of the machine learning algorithms used to create more accurate credit risk analysis models. One of the primary objectives of the banking industry is to limit the risk of financial loss by accepting loan applications.

Machine learning models have been developed to determine loan eligibility and assist consumers in maintaining a good credit score [6]. These models are trained and evaluated to predict whether or not to accept a loan application. Baseline modeling and K-Fold cross-validation procedures are employed to validate the models, and the model with the best accuracy is chosen. The trained model is then used to collect input from the user and produce the projected outcomes. To anticipate customer loan defaults, various machine learning techniques have been employed to develop loan default prediction models. Logistic regression, decision trees, random forests, and other machine learning techniques have been used to predict loan defaults [8]. The random forest approach has been shown to be more effective than logistic regression and decision trees for forecasting default samples. However, class imbalances in the dataset pose a challenge, and the SMOTE method is used to deal with this problem [9]. Additionally, data cleaning and dimensional reduction techniques are employed.

The credit approval process involves the consideration of several aspects before granting permission. Machine learning methods such as decision trees, random forests, multilayer perceptron, support vector machines, and logistic regression [10] have been used to analyze credit approval datasets. Attribute selection approaches such as CFS Subset Eval Best First, Consistency Subset Eval Best First, and exhaustive search have been employed to improve the performance of credit approval models [11]. These techniques have been found to be more effective than previous strategies, resulting in superior performance.

Priscilla et al. [12] the logistic regression model is a useful tool in predictive analytic that uses variables other than checking account information to determine the likelihood of loan default [13]. The model indicates that banks should target wealthy clients for loans and evaluate their other characteristics when making credit choices.

3 Data Presentation

Information from loan applications submitted to banks is included in the dataset utilized for this research. There are a total of 614 rows in the dataset, and its 13 characteristics include the following variables: loan status; credit history; gender; marital status; number of dependents; number of years after high school graduation; loan amount; loan term; occupation; location of primary residence; and loan ID [14]. In order to get rid of the blanks To conduct EDA, a dataset with no missing values and one-hot encoding for all categorical variables is used. We were able to compile the following summary of EDA procedures as a result. This is how the dataset breaks out in terms of loan approval status: After 192 declined loans, 42 were approved for a total of 422. Here is how the dataset is divided by gender: There are 112 women and 489 men. In the data set, marital status appears to be distributed as follows: There were 398 married applications and 216 single ones. Here's how the dataset stacks up in terms of average years spent in school: Only 134 out of 480 applicants did not have a degree.

Jena and Vasantha [15] The following describes the demographics of the self-employed people in the sample: Only 32 out of 582 applicants had self-employment listed as an occupation. The dataset has a credit history distribution that looks like this: While 475 applicants had established credit histories, 139 did not. The dataset contains the following breakdown of property sizes: The application pool represented 233 city dwellers, 179 suburbanites, and 202 country folk. The range of loan amounts in the dataset is rather wide, from $9000 to $700,000. The median loan amount is

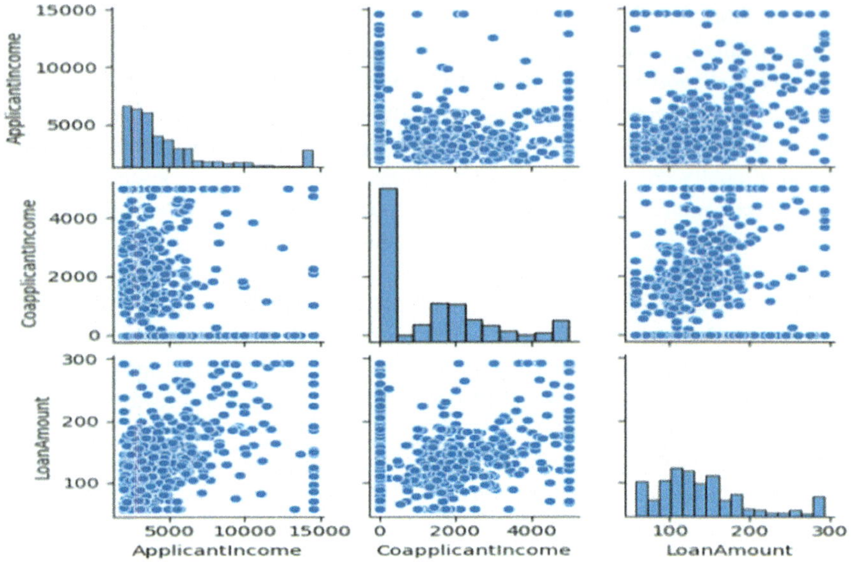

Fig. 1 Pair plot relation for features applicant income, loan amount and co-applicant income

$128,000. The median income in the dataset is $38,000, with a range from $1500 to $81,000. The income distribution of co-applicants in this dataset has a median of $0 and a range of $0 to $41,667.

Figure 1 shows the link between applicant income, loan amount, and co-applicant income, which may play a role in the loan approval process. Understanding the underlying trends and making educated loan decisions in financial institutions requires this kind of visual research. With co-applicants in the picture, it's easier for banks to calculate the level of risk associated with loans of varying sizes and income levels.

Figures 2 and 3 allows us to a significant tool for lenders, borrowers, and financial analysts is to examine the density of distributions of applicant income in relation to loan status. It offers a graphical depiction of the income-related elements that are considered in deciding whether or not to provide a loan. Financial institutions may make more educated lending decisions when income limits, inequalities, and risk levels are all clearly identified and borrowers can better understand their prospects of acceptance. As a result of this investigation, the loan process will be more open and equitable.

These descriptive statistics provide insight on the characteristics of the dataset and may aid in selecting the machine learning technique to be used for loan approval prediction.

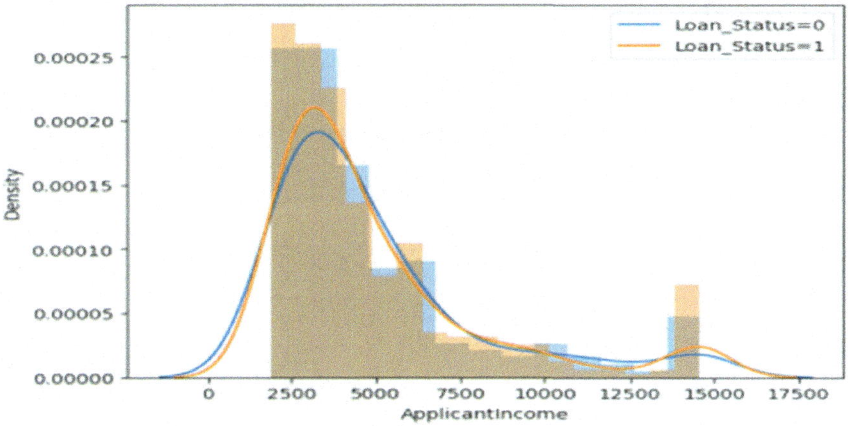

Fig. 2 Loan status based on relation between density versus applicant income

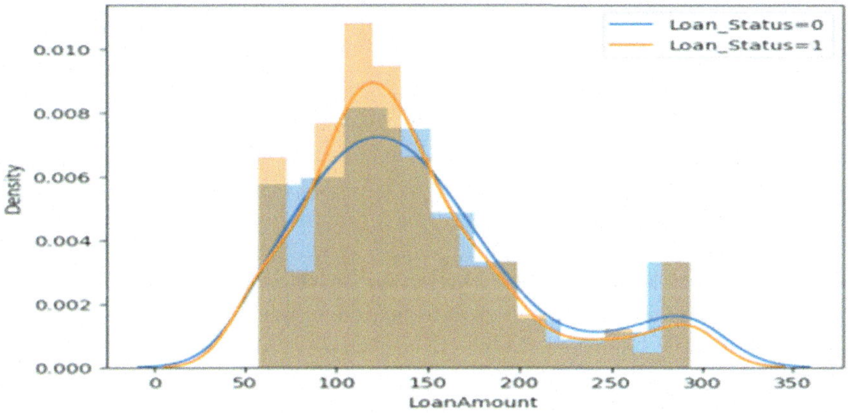

Fig. 3 Loan status based on relation between density versus loan amount

4 Methodology

See Fig. 4.

4.1 Exploratory Data Analysis

Data is evaluated and displayed visually in what is called the EDA Data Analysis. This step involves validating the data type, or shape, of the features, such as object for 8, int64 for 1, and float64 for 4. Checking for missing data in the rows under characteristics revealed 149 blanks, with 13 for age, 3 for marital status, 15 for

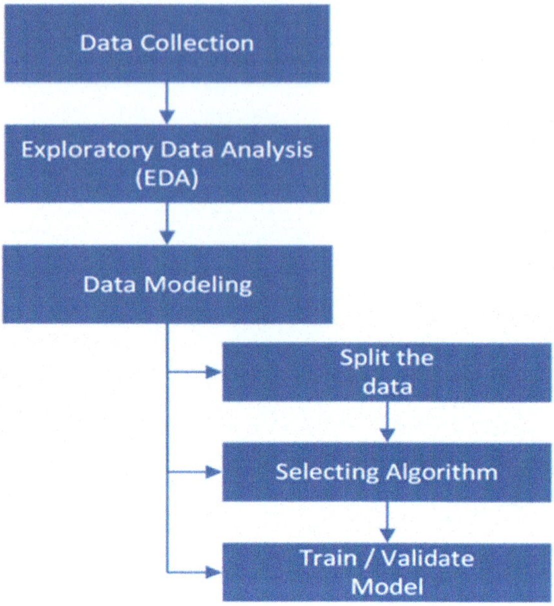

Fig. 4 Flow chart representing the process

number of dependents, 32 for number of years spent self-employed, 22 for loan amount, 14 for loan_amount_term, and 50 for credit history all having no entries. The mean was used on the numerical data, while the max count was used on the categorical data, both of which decreased the number of missing values. There were no occurrences of duplicate values in the dataset, marking this stage as complete data cleaning. In order to comprehend the information, visual representations were used for each field. Data visualization refers to the process of making data visible through visual means. The data transformation is complete after the data is normalized using standard scalars.

4.2 *Data Modeling*

The process of data modeling may be broken down into three distinct stages: segmenting the data, deciding which algorithm to use, and either training or verifying the model.

4.3 *Splitting the Data*

The dataset is divided into four parameters: x_train, y_train, x_test, and y_test; the train size is 70% and the test size is 30%; the random state or iterative count is

100; and the split test data is divided into dependent and independent variables, also known as input variables and output variables, commonly referred to as x and y variables, respectively. The features of the x variable include Dependents, Education, Self-Employed, and Applicant Income.

4.4 Selecting the Algorithm

Algorithms are picked out depending on factors including the difficulty of the task, the amount of data available, how easily they can be understood, how quickly they can be trained, and how well they perform. Six algorithms (RF, DT, NB, XGBoost, KNN, and SVM) are utilized in comparison to all these factors.

4.5 Training or Validating Model

The model is trained with different algorithms, like RF, DT, NB, XGBoost, KNN, and SVM, using the parameters as x and y variables (input and output variables, respectively) and the random state (verified by prediction and accuracy scores) for 100 iteration count runs.

5 Results and Discussion

Different algorithms yielded varying degrees of accuracy, precision, $F1$ score, and recall when used in the loan acceptance prediction system. An $F1$ score of 85%, a recall of 98.0%, and an accuracy of 80.7% were all achieved by NB on the dataset. The DT results on the dataset were 69.3% accuracy, 74% precision, 70% $F1$ score, and 66.0% recall. When applied to the dataset, RF achieved a 76.8% accuracy rate, a 77% precision rate, an 83% $F1$ score, and an 89% recall rate. After applying KNN to the dataset, the accuracy, precision, $F1$ score, and recall all improved to 79.3%, 72.0%, 83.0%, and 99%, respectively. When applied to the dataset, XGBoost achieved 74.1% accuracy, 74% precision, a 77% $F1$ score, and 80% recall. An $F1$ score of 84%, a precision of 75.1%, an accuracy of 80.8%, and a recall of 96.0% were all achieved by applying SVM to the dataset.

Incorporating a box plot, a visual approach to evaluation Fig. 5 plots algorithm types along the x-axis and accuracy percentages along the y-axis. By contrasting results, shows how well different algorithms performed.

Making use of the bar plot as a means of visualizing a comparison Fig. 6 plots algorithm types along the x-axis and accuracy percent along the y-axis. A comparison of each of these methods can reveal how well each algorithm performs.

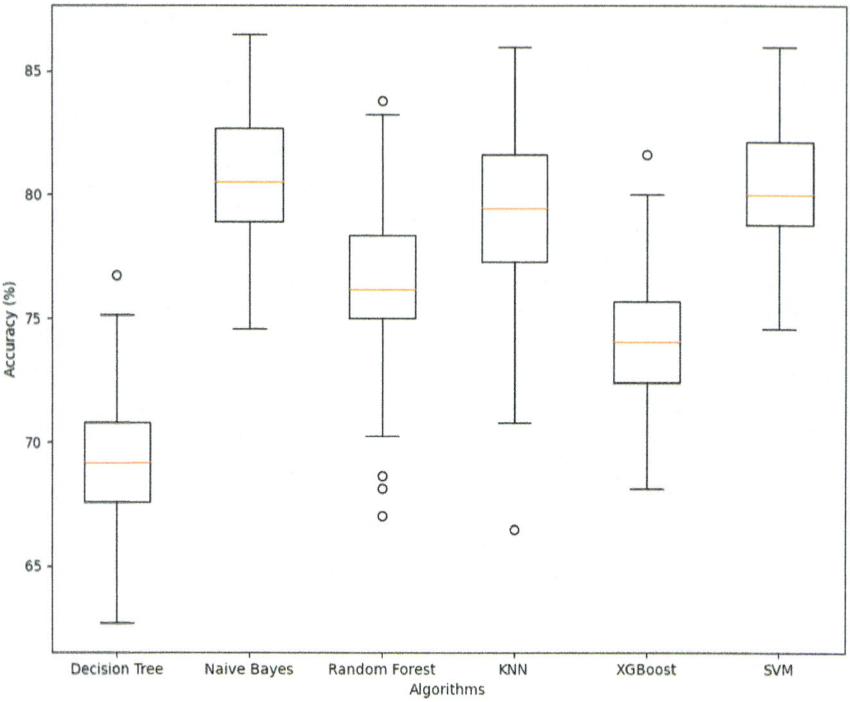

Fig. 5 Accuracy comparison using box plot

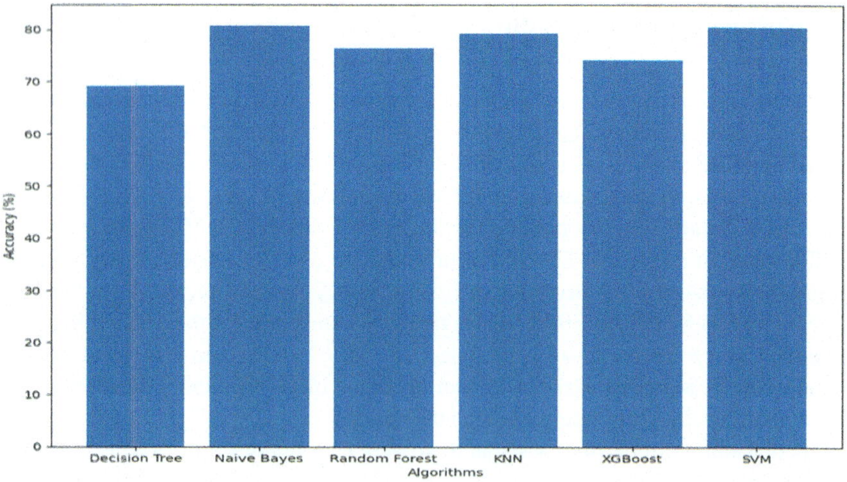

Fig. 6 Accuracy comparison using bar plot

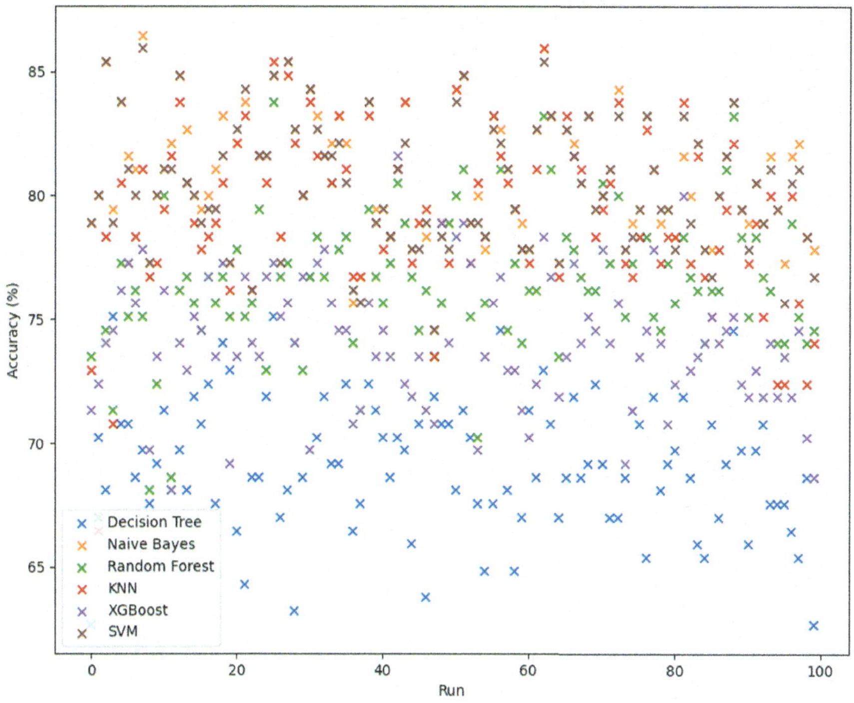

Fig. 7 Accuracy comparison using scatter plot for 100 iterations

Figure 7 uses a scatter plot to compare the different methods, with the number of iterations shown by the x-axis and the accuracy (in percent) represented by the y-axis, for a total of 100 iterations.

Table 1 results indicate that the Gaussian Naive Bayes and SVM algorithms are more effective in predicting whether or not a loan application will be approved. This suggests that a sizable fraction of the loan applications in the dataset were correctly predicted to have been approved by using these techniques. The KNN method also did well, classifying loan applications accurately as either approved or declined (with an accuracy of 79.31%). The DT, RF, and XGBoost algorithms all have an accuracy of between 69.33 and 76.84%. Data over-fitting, in which a model becomes overly complex and memorizes the training data instead of learning generalization patterns, might be to blame for the dwindling accuracy. Because of their flexibility in processing numeric, categorical, and missing data, NB and SVM algorithms achieve high accuracy. In addition to outperforming DT, RF, and XGBoost on this loan approval dataset, these methods also exhibit less overfitting than those previously mentioned.

Table 1 Performances of all machine learning algorithms

	Precision	Recall	$F1$-score	Accuracy
Random forest	0.77	0.89	0.83	76.84
Decision tree	0.74	0.66	0.70	69.33
Gaussian Naive Bayes	0.76	0.98	0.85	80.75
K-nearest neighbor	0.72	0.99	0.83	79.31
Support vector machines	0.75	0.96	0.84	80.83
XGBoost	0.74	0.80	0.77	74.15

6 Conclusion

In conclusion, the SVM algorithm's 80.83% accuracy in predicting loan acceptance status was the highest of any approach tested. Strong generalization performance was shown on the loan application dataset, which included numeric, categorical, and missing data types. Based on the results of the research, the SVM algorithm may help banks and other financial organizations make better loan approval choices.

References

1. Arun K, Ishan G, Sanmeet K (2022) Loan approval prediction based on machine learning approach. FUDMA J Sci 6:41–50. https://doi.org/10.33003/fjs-2022-0603-830
2. Sivaniah R, Pushpalatha D (2020) Comparative analysis of machine learning and attribute selection techniques for credit approval data. Int J Pure Appl Math 118:305–311
3. Sheikh MA, Goel AK, Kumar T (2020) An approach for prediction of loan approval using machine learning algorithm. In: 2020 international conference on electronics and sustainable communication systems (ICESC), Coimbatore, India, pp 490–494. https://doi.org/10.1109/ICESC48915.2020.9155614
4. Kandula AR, Sathya R, Narayana S (2023) Multivariate analysis on personalized cancer data using a hybrid classification model using voting classifier. Int J Intell Syst Appl Eng 11(1):354–362
5. Sharma A, Kumar V (2022) An exploratory study-based analysis on loan prediction. Inventive Commun Comput Technol Proc ICICCT 2022:423–433
6. Hegde SK et al (2023) Performance analysis of machine learning algorithm for the credit risk analysis in the banking sector. In: 2023 7th international conference on computing methodologies and communication (ICCMC). IEEE
7. Kandula AR, Sathya R, Narayana S (2022) Comparative analysis of machine learning techniques on genetic mutation based cancer diagnosis data. J Theor Appl Inf Technol 100(6)
8. Zhu L et al (2019) A study on predicting loan default based on the random forest algorithm. Procedia Comput Sci 162:503–513
9. Masmoudi K, Abid L, Masmoudi A (2019) Credit risk modeling using Bayesian network with a latent variable. Expert Syst Appl 127:157–166
10. Kandula AR, Tamilarasi K, Maan S (2022) Deep neural network for image recognition in medical diagnosis. J Pharm Negative Results 13
11. Dutta P (2021) A study on machine learning algorithm for enhancement of loan prediction. Int Res J Modernization Eng Technol Sci 3

12. Priscilla R et al (2023) Baseline modeling for early prediction of loan approval system. In: 2023 International conference on artificial intelligence and knowledge discovery in concurrent engineering (ICECONF). IEEE
13. Ndayisenga T (2021) Bank loan approval prediction using machine learning techniques. Diss
14. Gupta A et al (2020) Bank loan prediction system using machine learning. In: 2020 9th international conference system modeling and advancement in research trends (SMART). IEEE
15. Jena SR, Vasantha S. Prediction of modernized loan approval system based on machine learning approach

Classification of Credit Card Delinquent Customers Using Random Forest Algorithm

Vadapally Praveen Kumar, Satyanarayana Nimmala, Burri Naresh, and Ch Ravikumar

Abstract Customers receive a little plastic payment card called a credit card. It allows the holder access to the market to purchase goods and services mostly on the promise of payment from the holder. The card issuer sets up a revolving account and offers the cardholder a credit line from which they can borrow money for either personal use or to pay for goods or services. To control risks and keep a strong credit portfolio, financial institutions must foresee credit card defaults. In this study, we investigate how to categorize clients who have not paid their credit card balances using the random forest algorithm. We examine the effectiveness of random forest in precisely detecting probable defaulters using a dataset comprising credit card payment history. The study highlights the benefits and drawbacks of the random forest technique through data preprocessing, model application, and performance evaluation. For credit card issuers to make wise judgments and reduce potential financial losses, the findings offer insightful information. The experiment in this study uses real-time data collection obtained from a third-party agency and a random forest technique with tenfold cross-validation. The findings demonstrated that random forest is the best method for categorizing credit card delinquent clients. The classifier outperforms itself in predicting the most likely defaulters with an accuracy of 97.73%, while the classifier's overall accuracy is 84.703%, according to the experimental data.

V. Praveen Kumar (✉) · S. Nimmala
Department of CSE (Data Science), CVR College of Engineering, Hyderabad, India
e-mail: micro091983@gmail.com

S. Nimmala
e-mail: satyauce234@gmail.com

B. Naresh
Department of CSE (CS), CVR College of Engineering, Hyderabad, India
e-mail: naresh.burri@gmail.com

C. Ravikumar
Department of Artificial Intelligence and Data Science, Chaitanya Bharathi Institute of Technology, Hyderabad, India
e-mail: chrk5814@gmail.com

B. R. Devi et al. (eds.), *Proceedings of Fifth International Conference on Computer and Communication Technologies*, Lecture Notes in Networks and Systems 897,
https://doi.org/10.1007/978-981-99-9704-6_19

Keywords Random forest algorithm · Classification · Risk management · Credit portfolio · Data preprocessing · Efficacy · Accuracy

1 Introduction

The credit card industry has evolved significantly, offering consumers convenient access to goods and services with the promise of repayment. Credit card issuers provide revolving accounts, granting users a line of credit. However, credit card defaults, where users fail to make timely payments, pose risks to financial institutions. Such defaults can lead to substantial financial losses and affect the broader financial ecosystem. Given these challenges, predicting credit card defaults has become vital for financial stability. In this study, we explore the random forest algorithm's application for classifying credit card delinquencies. Its ability to handle complex data and offer insights into feature importance makes it a promising tool for accurate prediction [1–3] (Fig. 1).

Our investigation utilizes a real-time credit card payment history dataset obtained from an independent third-party source. This dataset enables us to analyze credit card default patterns. We evaluate the random forest method's effectiveness in identifying likely defaulters after data preprocessing and hyperparameter tuning. We employ performance metrics like accuracy, precision, recall, $F1$-score, and ROC-AUC to assess the model. Our study aims to provide valuable insights to credit card issuers for informed decision-making and risk mitigation. It also discusses the random forest approach's strengths and limitations in credit card delinquency classification, offering avenues for improvement. Ultimately, this research contributes to credit card risk management and financial stability, empowering institutions to proactively address defaults [4–6].

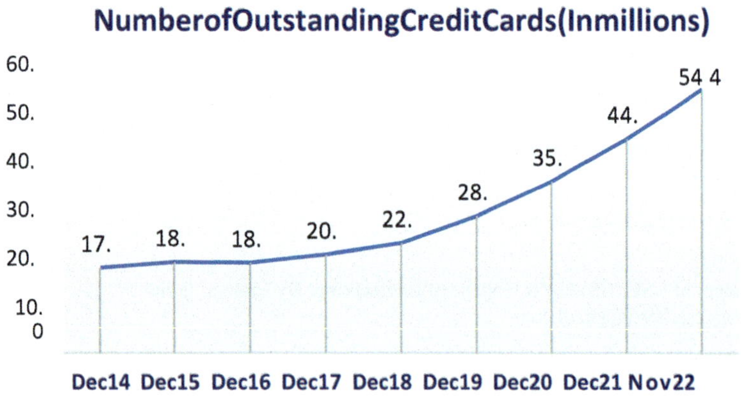

Fig. 1 The increase in outstanding credit cards in the year 2022

1.1 Contributions

(a) This article addresses credit card defaults as a significant risk, showcasing machine learning expertise using the random forest algorithm for client categorization.
(b) Real-time data collection, practical data preprocessing, and hands-on implementation of random forest resulted in impressive prediction accuracy of 97.73% for likely defaulters and an overall accuracy of 84.703%.
(c) Valuable insights for credit card issuers empower informed decisions, reducing potential financial losses, and emphasizing the research's practical relevance.

2 Literature Review

Credit card default prediction is essential for managing financial risks and a stable credit portfolio. Default instances can lead to significant losses and industry instability. Precise prediction empowers issuers to take proactive measures, mitigate financial losses, and make informed credit risk decisions [7]. Credit cards have revolutionized payments but introduced default risks, necessitating effective prediction methods. A research article explores random forest's application for classifying delinquent cardholders, known for handling complex, imbalanced data, and for offering feature importance insights. The study uses a real-time dataset, enhancing the research's relevance and applicability to current credit risk management scenarios [8, 9].

The research paper "Credit Card Default Prediction as a Classification Problem" addresses credit card default prediction, emphasizing its classification nature and the importance of precise forecasting for financial institutions. Likely utilizing client demographics, payment history, and transaction data, it explores various classification methods like logistic regression, decision trees, and support vector machines to predict defaulters. Its goal is early high-risk customer identification and credit risk management optimization. However, the study's performance evaluation does not comprehensively compare these methods, and generalizability to different datasets or financial institutions is not examined. Despite credit card usage growth, this research highlights the need for improved default prediction in a dynamic financial landscape, driven by factors like EMI options and e-commerce incentives [1, 10, 11].

The research paper "Credit Card Default Prediction as a Classification Problem" highlights the importance of precise credit card default forecasting for financial institutions. It likely employs client demographics, payment history, and transaction data to create prediction models using classification methods like logistic regression, decision trees, and support vector machines. However, the study may not comprehensively compare these methods or examine generalizability across datasets. Nevertheless, the rising popularity of credit cards, as indicated by RBI data, underscores the significance of proactive credit risk management in an evolving financial landscape (Fig. 2).

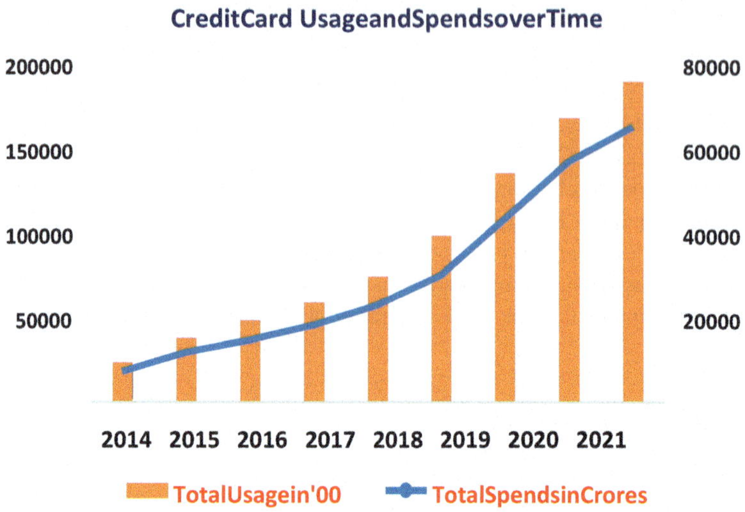

Fig. 2 The increase in outstanding credit cards in the year 2022

The paper "Research on Credit Card Default Prediction Based on k-Means SMOTE and BP Neural Network" aims to improve credit card default prediction by addressing imbalanced data issues. It combines k-means SMOTE to balance the dataset and a backpropagation neural network (BPNN) for classification. The research likely involves parameter experimentation and performance evaluation. However, the paper lacks a thorough discussion of evaluation metrics. A comparison with existing methods could enhance its value. Moreover, the computational overhead of k-means SMOTE and neural networks may limit scalability in large datasets or real-time scenarios, as discussed in the study [12].

The research paper "Prediction Of Credit Card Defaulters Using Machine Learning" explores various machine learning techniques for credit card default prediction. It likely employs gradient boosting, random forest, and support vector machines to classify potential defaulters using a comprehensive dataset. The study aims to identify the most effective predictive model by comparing machine learning methods. However, it lacks in-depth coverage of feature engineering, a critical aspect of model accuracy. Handling imbalanced data is also not addressed, potentially skewing results. External validation on diverse datasets or time periods is absent, raising concerns about real-world applicability. Lastly, interpretability of complex models is not discussed, hindering understanding of credit default determinants [13].

3 Experimental Setup

In this section, we outline the experimental setup for conducting the credit card default prediction using the random forest algorithm. The real-time dataset of 438 customers belonging to ICICI Bank is collected. All the customers are located in Hyderabad. The dataset used in this study is obtained from a reliable third-party agency, ensuring that it contains comprehensive credit card payment history and relevant customer information. The dataset comprises features such as credit utilization, payment history, credit limit, age, income, city, type of credit card, postal code of the customer; amount to be paid, number of days pending, principal balance, history of payment as pattern, and other demographic information. The data are preprocessed to handle missing values, and outliers, and to select the input features. Feature scaling and normalization techniques are applied to ensure consistent scales across different features. The random forest algorithm with tenfold cross-validation is implemented using a machine learning library, scikit-learn in Python.

3.1 The Sample Pseudocode of the Proposed Algorithm is Given Below

```
Import numpy as np
from sklearn. ensemble import RandomForestClassifier
rf_classifier=RandomForestClassifier (n_estimators=100, max_depth=10,
random_state=42)
rf_classifier.fit (X_train, y_train)
predictions=rf_classifier. predict(X_test)
```

The dataset is divided into training and testing sets in the split ratio 65:30, ensuring that the model is trained on a majority of the data while being tested on unseen data for evaluation. The random forest model is trained on the training dataset using the optimized hyperparameters obtained from the tuning process. The model learns from the training data and constructs multiple decision trees through the bagging process to form the random forest ensemble. The performance of the random forest model is assessed using classification metrics, including accuracy, precision, recall, $F1$-score, and the receiver operating characteristic (ROC) curve. As part of the evaluation, the performance of the random forest model is compared with baseline models such as logistic regression, decision trees, and support vector machines (Table 1).

Table 1 Categorical values of input parameters

S. No	Attribute name	Categorical values based on a heuristic approach		
1	Age	≤ 30 (young)	> 30 && ≤ 45 (middle-aged)	> 46 (old)
2	Principal balance	< 50 K (low)	> 50 && < 2lakhs (medium)	> 2 Lakhs (high)
3	Amount to be paid	< 50 K (low)	> 50 && < 2lakhs (medium)	> 2 Lakhs (high)
4	Number of days pending	≤ 300 (Class1) least likely defaulters		> 300 (Class2) Most likely defaulters

4 Results and Discussion

Various accuracy metrics help assess a classifier's performance across different scenarios. These metrics provide insights into its strengths and limitations. In this analysis, the classifier's performance is evaluated using a dataset divided into class 1 (173 records, least likely to default) and class 2 (265 records, most likely to default). The classifier achieved an 84.703% accuracy in predicting unseen test records, employing the random forest algorithm and tenfold cross-validation. This method is widely used to evaluate model performance and make data-driven decisions, especially valuable for small or unbalanced datasets. The experimental findings are summarized in Table 2, shedding light on the classifier's predictive capabilities across various decision types (Figs. 3 and 4).

Overall, the classifier shows good performance with relatively high accuracy and F-measure. However, there is some imbalance in the classification, as indicated by the differences in TP rates and precision between the two classes. The ROC areas suggest reasonable discriminative power for both classes. As per the experimental results, the

Table 2 Class-wise performance details of the classifier

TP rate	FP rate	Precision	Recall	*F*-measure	ROC area	Class
0.977	0.353	0.809	0.977	0.885	0.829	Class 2
0.647	0.023	0.949	0.647	0.770	0.829	Class 1

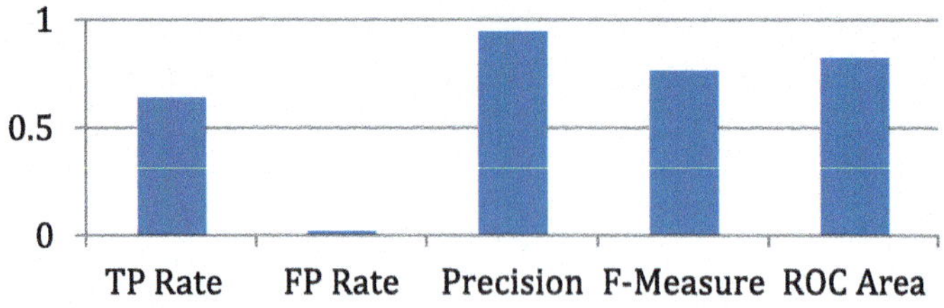

Fig. 3 Performance measures of class 1

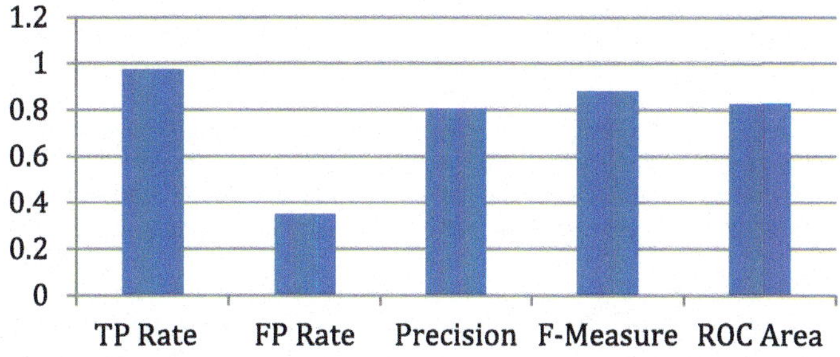

Fig. 4 Performance measures of class 2

classifier is outperformed in predicting the most likely defaulters with an accuracy of 97.73%; however, the classifier's accuracy in predicting least likely defaulters is 65%. The experiments are aimed at catching the most likely defaulters so that banks can initiate the necessary steps to recover the credit bill from the customers.

5 Conclusion and Future Direction

The classifier's experimental evaluation, using various accuracy measures, revealed insights into its performance. It achieved an overall accuracy of 84.703%, indicating its proficiency in predicting unseen test records. Employing the random forest algorithm with tenfold cross-validation, the evaluation divided the dataset into class 1 (173 records, least likely to default) and class 2 (265 records, most likely to default). The classifier excelled at predicting class 2 with 97.73% accuracy but showed lower accuracy (65%) for class 1. This imbalance warrants further investigation to enhance predictions for class 1. Despite this, the classifier is valuable for credit recovery efforts, with potential for improvement through tuning and adjustments for more accurate credit risk assessment and recovery.

References

1. Durkin T, Elliehausen G (2008) The impact of legislation on credit card pricing, costs, and availability. J Econ Perspect 22(3):75–98
2. Thomas LC (2000) A survey of credit and behavioral scoring: forecasting financial risk of lending to consumers. Int J Forecast 16(2):149–172
3. Baesens B, Setiono R, Mues C, Viaene S (2003) Using neural network rule extraction and decision tables for credit-risk evaluation. Manage Sci 49(3):312–329
4. Hand DJ, Henley WE (1997) Statistical classification methods in consumer credit scoring: a review. J R Stat Soc A Stat Soc 160(3):523–541
5. Thomas LC, Edelman DB, Crook JN (2002) Credit scoring and its applications, vol 21. SIAM
6. Breiman L (2001) Random forests. Mach Learn 45(1):5–32

7. Chen T, Guestrin C (2016) XGBoost: a scalable tree boosting system. In: Proceedings of the 22nd ACM SIGKDD international conference on knowledge discovery and data mining, pp 785–794
8. Huang J, Ling CX, Su Y (2006) Credit risk assessment with a simple random forest model. Int J Inf Technol Decis Mak 5(3):451–460
9. Liaw A, Wiener M (2002) Classification and regression by randomForest. R News 2(3):18–22
10. Yu H, Kim K (2015) Credit card fraud detection using random forests. Inf Sci 301:250–261
11. Bhattacharyya A, Bose I (2017) Credit card default prediction using random forests. Expert Syst Appl 79:93–103
12. Bai YH, Wang H (2018) Research on credit card default prediction based on k-means SMOTE and BP neural network. In: Proceedings of the 2018 4th international conference on economics, management engineering and education technology (ICEMEET 2018), pp 110–115
13. Wang J, Lu Y, Wang F, Feng M (2016) Prediction of credit card defaulters using machine learning. Int J Comput Appl 135(10):11–15

Evaluating E-learning Engagement Through EEG Signal Analysis with Convolutional Neural Networks

Dharmendra Pathak and Ramgopal Kashyap

Abstract Amidst the COVID-19 pandemic, online e-learning platforms have experienced a surge in popularity, serving a diverse audience across multiple subjects. Nevertheless, the absence of learner-centric approaches and efficient learning validation on these platforms has become a concern within the online learning community. This has led to an increased demand for personalized learning recommendations for both learners and educational materials. In response to these challenges, we propose an innovative solution that utilizes real-time EEG data acquired from individuals wearing EEG headsets while participating in online courses. Our approach centers on a deep learning convolutional neural network (CNN) model, which adeptly classifies these EEG signals with an impressive accuracy rate of 70%. This remarkable performance underscores the speed and precision of our developed models in handling e-learning EEG signals, presenting a promising resolution to prevailing e-learning validation issues. Beyond classification, our work introduces an automated framework that continuously tracks users' learning progress and provides invaluable recommendations for e-learning materials, effectively elevating the overall learning experience. By leveraging this framework, we envision a transformative impact on e-learning platforms, as it not only validates user learning more effectively but also offers personalized recommendations, creating a more efficient and user-centric e-learning environment. As a result, our research contributes significantly to the advancement of e-learning practices, especially during challenging times like the pandemic, where remote learning has become increasingly prevalent.

Keywords Convolution neural network · EEG signals · e-learning · e-learning framework · LSTM

D. Pathak (✉) · R. Kashyap
Amity University, Chhattisgarh, India
e-mail: dharmendra.pathak@s.amity.edu

B. R. Devi et al. (eds.), *Proceedings of Fifth International Conference on Computer and Communication Technologies*, Lecture Notes in Networks and Systems 897,
https://doi.org/10.1007/978-981-99-9704-6_20

1 Introduction

In the current educational landscape, technology plays a pivotal role, encompassing various aspects like online courses, virtual laboratories, and e-tutoring. Especially during the COVID-19 pandemic, online educational platforms have demonstrated high effectiveness and cost-efficiency in terms of infrastructure. Recent studies have shown that e-learning outperforms conventional classroom teaching with a user retention rate surpassing by 25–60% [1]. E-learning provides users with time flexibility, portability, location independence, resourcefulness, and ease of access.

However, despite its effectiveness, e-learning has its limitations, with users facing challenges in maintaining focus and spending extended periods on teaching sessions [2]. Therefore, there is a growing demand for a comprehensive framework that not only validates users' real-time learning progress and customizes e-learning content based on their needs but also visualizes their learning curve.

Electroencephalographic (EEG) signals represent digital recordings of the brain's electrical activities, measured in microvolts (μV). These signals exhibit distinct frequencies associated with various mental states:

- Delta frequency (0–4 Hz) corresponds to sleep and healing.
- Theta frequency (4–8 Hz) indicates creativity, and emotional states.
- Alpha frequency (8–12 Hz) is linked to focus and relaxation.
- Beta frequency (12–40 Hz) is associated with problem-solving ability and conscious focus.
- Gamma frequency (40–100 Hz) represents acute senses, cognition, and learning.

To capture and analyze EEG signals, researchers employ EEG headsets available for consumers, subsequently applying machine learning or deep learning methods.

Deep learning, a specialized branch of machine learning, implements neural networks to simulate human brain functions for the purpose of classification [3].

Figure 1 illustrates the structure of a deep neural network, consisting of input, hidden, and output layers. Neurons within the network serve as processing units, managing feature extraction and classification through parameters. Deep learning's feature to process complex datasets and dense parameters through nonlinear functions, depicted in Fig. 2, makes it a popular choice for tasks like classification, pattern recognition, and identifying unknown patterns [4].

This paper presents the results of EEG classification using an CNN deep learning network, successfully validating users' real-time e-learning EEG data and tracking attention levels while showcasing learning patterns. Additionally, the proposed framework offers valuable customization recommendations for e-learning materials and improved feedback for both individual and collaborative learning experiences. Furthermore, this framework has the potential to be extended for identifying learning disabilities in students.

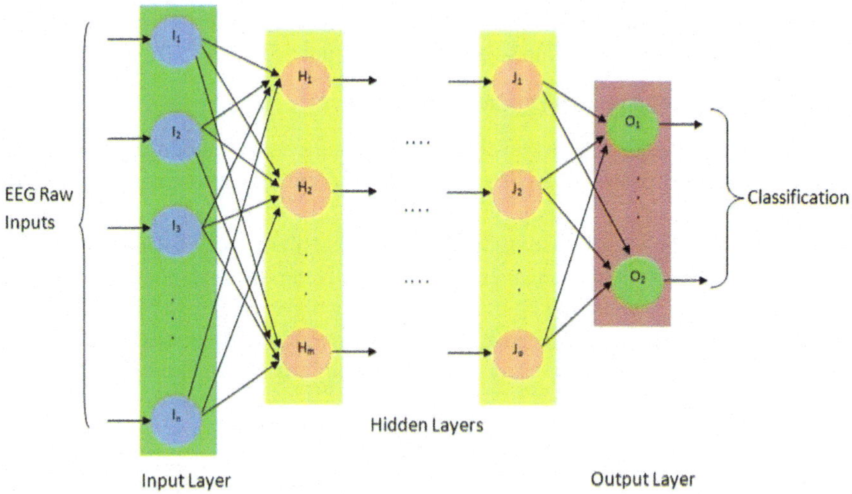

Fig. 1 Working diagram of deep learning

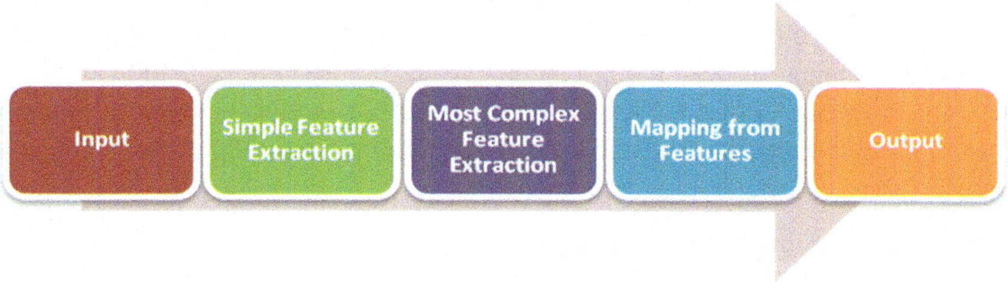

Fig. 2 Components of deep learning

2 Challenges Facing in E-Learning Approaches

The impact of e-learning on a student's career path is undeniable. In recent years, the digital revolution has transformed the way education is delivered, making it more accessible and flexible than ever before. E-learning platforms offer students the opportunity to acquire knowledge and skills at their own pace, often from the comfort of their homes. This accessibility allows learners to explore a wide range of subjects and specialties, enhancing their career prospects by providing a diverse skill set. Additionally, e-learning fosters self-discipline, time management, and technological proficiency, all of which are valuable attributes in today's job market. Furthermore, online courses and certifications are increasingly recognized by employers, enabling students to demonstrate their expertise and commitment to continuous learning, which can significantly impact their career progression and open doors to new opportunities in the ever-evolving professional landscape. The ongoing global pandemic has resulted in the widespread adoption of online educational platforms. Despite their advantages in terms of flexibility in time, location, and resources, e-education faces critical challenges [5]. These challenges include:

1. Limited student attention period:

Numerous parameters are responsible for this challenge, including platform user interface and user experience (UI/UX), modes of communication, media consumption, expert accessibility, timing, internet bandwidth, student engagement, subject proficiency, and real-world case studies [6]. To tackle this issue, we need an approach capable of gauging students' engagement and attentiveness during electronic lectures. Previous endeavors that relied on machine learning algorithms faced limitations due to small EEG datasets and reduced accuracy. The integration of diverse deep learning methods, as examined in recent research contributions [7–9], holds promise as a prospective remedy.

2. Ineffective outcomes despite good content and experts:

Research findings indicate that electronic learning platforms frequently encounter challenges in achieving their intended outcomes, even when delivering high-quality digital resources and expert guidance. The quality of learning experiences is notably influenced by attributes such as student engagement, interest in the subject matter, and instructional methods. To enhance educational outcomes, there is a demand for an approach that suggests personalized e-content adaptations tailored to individual user needs. Furthermore, offering supplementary remedial sessions for students who may be struggling to keep pace with the average learning curve could prove advantageous [10].

3. Lack of participant learning tracking:

Numerous electronic learning platforms do not possess the capacity to monitor and exhibit the progress of participants, which can impede the assessment of enhancements and outcomes. A viable strategy to address this issue is to introduce periodic quizzes and assessments following each instructional session, as suggested in [11]. It is crucial to maintain comprehensive feedback and performance records by conducting session-based evaluations through tests, quizzes, and similar assessment methods, as emphasized in [12].

4. Teaching methodologies:

Enhancing the effectiveness of e-learning can be achieved by integrating animations, graphics, and two-way communication, such as live one-to-one communication, chatbots, and forums. While replicating conventional off-line teaching exercises can be challenging online, group-based exercises can promote collaborative learning and validate e-learning [13].

5. Identifying individual learning problems:

E-learning materials are typically crafted for a broad audience, making it challenging to pinpoint the specific reasons for suboptimal learning outcomes among individual students. To address this challenge, it is essential to conduct a comprehensive assessment of students based on various criteria, including their prior academic performance, feedback, level of engagement, achievements, and learning trajectory [14].

While there is currently no existing framework that integrates EEG data with e-learning, recent research in the fields of EEG classification has achieved accuracy rates of up to 65% using machine learning techniques on limited EEG datasets [15]. Particularly noteworthy is the application of deep learning algorithms for EEG data classification, which has demonstrated substantial performance improvements, achieving accuracy rates of up to 88%. This improvement is attributed to auto-feature selection and the handling of extensive EEG signals [16]. In our own research, we have implemented 2-D convolutional neural network (CNN) algorithm, leveraging substantial datasets and considering parameters such as attention levels, focus, and data processing. This approach has led to enhanced classification performance compared to prior methodologies.

3 Methodology

3.1 EEG Signal Classification in Real-Time

In Fig. 3, we illustrate the live classification of EEG signals, highlighting the predominant frequency bands: 1. alpha, 2. beta, 3. gamma, 4. delta, and 5. theta. Each of these frequency bands is associated with distinct cognitive states and holds significant importance in monitoring learning and information-processing activities, with their ideal values indicating stable brain conditions.

3.2 Summary of Real-Time E-Learning EEG Data Collection

In our research, we enrolled over 300 participants aged between 18 and 35 years, with a particular focus on engineering students spanning from their first to final years and educators, primarily from the computer science domain. Our investigation encompasses multiple e-learning platforms. To capture real-time raw EEG signals, participants wore the Muse Neuro Headband during online classes on the selected e-platforms. On average, participants attended sessions lasting 10–20 min, concentrating on specific subject matter. Following each session, a multiple-choice test was administered to assess the participants' learning progress, which contributed to the grading process for training our deep learning model.

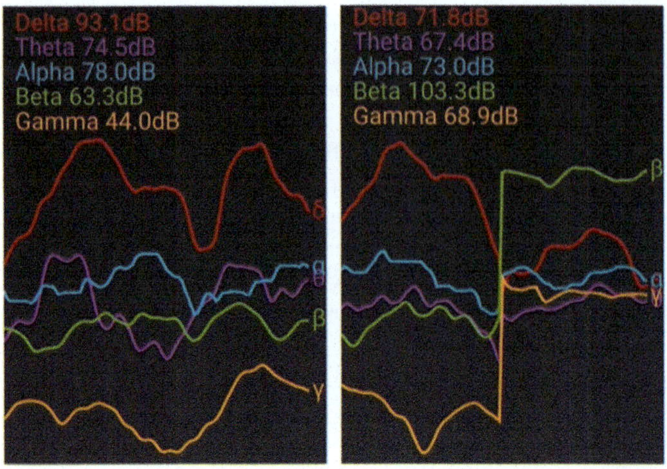

Fig. 3 EEG signal classification

3.3 EEG Data Discretization

Our research encompasses a vast dataset of over 300 h of raw EEG data from e-learning sessions, divided into three classes: CL A (excellent), CL B (good), and CL C (poor), mentioning different levels of learning for model training and testing. The Muse Neuro Headband, equipped with four channels (TP9, AF7, AF8, TP10), was used to simultaneously capture EEG frequencies [17].

To process the captured EEG data, we applied fast Fourier transformation (FFT) for discretization, converting the EEG signals into their respective frequency domain. The FFT algorithm optimally reduces computational complexity to O(NlogN), an improvement from the original $O(N^2)$, with N representing the data size [18, 19]. We utilized the ML library to execute the fast Fourier transformation operation on the collected EEG signals.

The FFT operation produced log-transformed spectra measured in 0.1 Hz intervals and averaged across EEG signals. Figure 4 visually presents the output of the FFT implementation; wherein, EEG frequencies have been transformed into discrete values.

3.4 EEG Signal Preprocessing Methods

Following EEG data transformation with FFT, we perform various data preprocessing operations, as shown in Fig. 5:

1. Artifact Handling: Specific noise, particularly ocular noise, was omitted from the collected EEG signals using the signal amplitude thresholding method.

Delta_TP9	Delta_AF7	Delta_AF8	Delta_TP10	Theta_TP9	Theta_AF7	Theta_AF8	Theta_TP10	Alpha_TP9	Alpha_AF7	Alpha_AF8	Alpha_TP10	Beta_TP9	Beta_AF7	Beta_AF8	Beta_TP10	Gamma_TP9	Gamma_AF7	Gamma_AF8	Gamma_TP10
0.9055106	0.19767758	0.8386295	1.0337056	0.4692589	0.72792876	0.46537602	0.77046466	0.7840342	0.78058875	0.34648076	1.026552	0.7588374	1.086706	0.6162969	1.0472555	0.4699146	0.8259294	0.39030132	0.7199756
0.77826667	0.19767758	0.8386295	0.8780114	0.63993365	0.72792876	0.46537602	0.8950842	0.5183032	0.78058875	0.34648076	1.0337733	0.70964783	1.086706	0.6162969	1.1255846	0.5639925	0.8259294	0.39030132	0.8045895
0.9123004	0.19767758	0.8386295	0.85046464	0.6987383	0.72792876	0.46537602	0.64798725	0.4513754	0.78058875	0.34648076	0.8255265	0.7171922	1.086706	0.6162969	1.0278713	0.67221045	0.8259294	0.39030132	0.9378289
0.44915062	0.19767758	0.8386295	0.3787943	0.5966056	0.72792876	0.46537602	0.2902317	0.71246403	0.78058875	0.34648076	0.6085298	0.8074056	1.086706	0.6162969	1.0871301	0.9264966	0.8259294	0.39030132	0.7188052
0.24834666	0.19767758	0.8386295	0.26443508	0.80577415	0.72792876	0.46537602	0.55870456	0.97423154	0.78058875	0.34648076	1.0408574	0.7826683	1.086706	0.6162969	1.0792127	0.71322656	0.8259294	0.39030132	0.71220046
0.436004	0.19767758	0.8386295	0.20432952	0.96212244	0.72792876	0.46537602	0.48247537	0.9193652	0.78058875	0.34648076	1.0294033	0.9067077	1.086706	0.6162969	0.9994916	0.81724304	0.8259294	0.39030132	0.84803545
0.49236685	0.19767758	0.8386295	0.2971314	1.0029414	0.72792876	0.46537602	0.42442834	0.67316854	0.78058875	0.34648076	0.7279316	0.6903894	1.086706	0.6162969	0.7108619	0.83723754	0.8259294	0.39030132	0.9275011
0.5234939	0.19767758	0.8386295	0.38062817	0.31854466	0.72792876	0.46537602	0.4985653	0.6256748	0.78058875	0.34648076	0.7804015	0.7570167	1.086706	0.6162969	0.88742276	0.82038885	0.8259294	0.39030132	0.5347242
0.43493807	0.19767758	0.8386295	0.7649537	0.3819946	0.72792876	0.46537602	0.36634436	0.60445255	0.78058875	0.34648076	0.96280585	0.6425202	1.086706	0.6162969	0.9018058	0.73320484	0.8259294	0.39030132	0.799957
0.43493807	0.19767758	0.8386295	0.9714576	0.3819946	0.72792876	0.46537602	0.4474628	0.60445255	0.78058875	0.34648076	0.97406346	0.6425202	1.086706	0.6162969	0.9597668	0.73320484	0.8259294	0.39030132	0.7056993
1.558127	0.19767758	0.8386295	0.980611	0.984605	0.72792876	0.46537602	0.52448565	0.76857283	0.78058875	0.34648076	0.9505767	1.0697614	1.086706	0.6162969	0.9747036	0.6839995	0.8259294	0.39030132	0.65348065
0.8862661	0.19767758	0.8386295	0.8717031	0.7638964	0.72792876	0.46537602	0.69592094	0.90901273	0.78058875	0.34648076	1.0267582	0.69636714	1.086706	0.6162969	0.9578463	0.82316348	0.8259294	0.39030132	0.44407147
0.7834249	0.19767758	0.8386295	0.7800393	0.5619764	0.72792876	0.46537602	0.6424843	0.70094305	0.78058875	0.34648076	0.8210941	0.6346358	1.086706	0.6162969	0.77823805	0.29166165	0.8259294	0.39030132	0.58934706
0.5667798	0.19767758	0.8386295	0.59804106	0.38694555	0.72792876	0.46537602	0.45316213	0.49774334	0.78058875	0.34648076	0.74885184	0.6342003	1.086706	0.6162969	0.7703722	0.34733307	0.8259294	0.39030132	0.69270396
0.7951799	0.19767758	0.8386295	0.59804106	0.26972795	0.72792876	0.46537602	0.45316213	0.98746615	0.78058875	0.34648076	0.74885184	0.74817127	1.086706	0.6162969	0.7763722	0.25457144	0.8259294	0.39030132	0.69270396
0.84997225	0.19767758	1.054837	0.7676231	0.20163286	0.72792876	0.41822973	0.10395945	1.0661039	0.78058875	0.47643256	0.9054631	0.67953277	1.086706	0.7427959	0.7349431	0.3177234	0.8259294	0.11150793	0.5342817
0.68057746	0.19767758	0.78297335	0.657666	0.3244071	0.72792876	0.3161362	0.42748472	0.848771	0.78058875	0.57018644	1.1109436	0.7198851	1.086706	0.84132695	0.74624485	0.2140836	0.8259294	0.23534752	0.4487436
0.5726646	0.19767758	0.6609062	0.5167335	0.4856427	0.72792876	0.76633006	0.5431934	0.32008493	0.78058875	0.89215076	1.0737444	0.6194331	1.086706	0.802303	0.7601717	0.23885375	0.8259294	0.36386418	0.5720658
0.46654797	0.19767758	0.7006276	0.27791566	0.5226879	0.72792876	0.78601795	0.44815797	-0.0231574	0.78058875	0.93369914	0.7559319	0.3671221	1.086706	0.770892	0.7840309	0.3122602	0.8259294	0.32998496	0.62347937
0.20820203	0.19767758	0.8902158	0.5146091	0.14628313	0.72792876	0.84879416	0.28604602	0.48703132	0.78058875	0.6878394	0.30717030	0.7141685	1.086706	0.7553137	0.693923	0.4621957	0.8259294	0.15567324	0.4925703

Fig. 4 Fast Fourier transformation applied on EEG signal

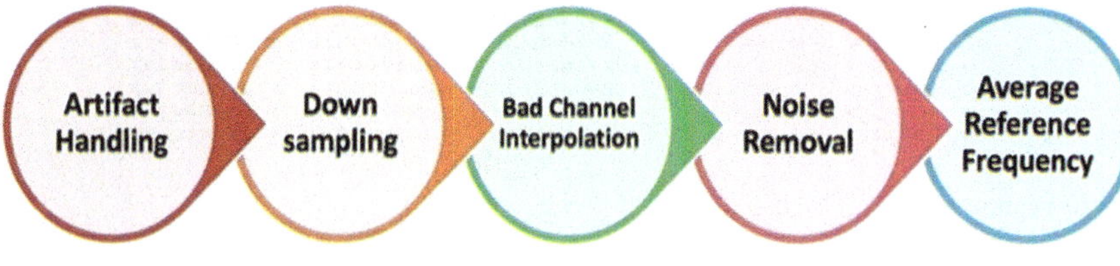

Fig. 5 Preprocessing methods applied on EEG data

2. Down Sampling: We reduced the sampling rate of digital signals by down sampling, accomplished by lowering the bit rate of the four headband's channels.
3. Bad Channel Interpolation: Signals inadequately captured because of poor EEG band channel were excluded.
4. Noise Removal: Empty rows and columns were eliminated.
5. Average Reference Signal Frequency: Ultimately, we computed average reference frequencies from the four channels, which were then converted into five reference frequencies [20, 21].

Figure 6 provides a glimpse of the preprocessed signals after the completion of these data preprocessing implementations, which will later serve as input for various deep learning methods for the model building.

4 Implementation Approach

4.1 *Label Encoding*

In the context of deep learning, we need data in a numerical form. Consequently, categorical values necessitate conversion into numerical representations before carrying out subsequent phases [22]. Figure 7 visually indicates the label encoding procedure

CON_ALPHA	CON_BETA	CON_GAMMA	CON_DELTA	CON_THETA
0.55139975	0.814039385	0.660967125	0.18791766	0.279385525
0.59346208	0.845765468	0.641883833	0.20038046	0.295844023
0.56223008	0.818177417	0.640633443	0.34257286	0.38895329
0.619125805	0.838558255	0.538445555	0.25161285	0.360469445
0.604041758	0.862075743	0.514852715	0.120514738	0.38615584
0.572642575	0.708282695	0.29900759	-0.018859392	0.079699711
0.552921365	0.744020895	0.249295511	0.181566283	0.245259383
0.617140298	0.727487478	0.304086623	0.188645403	0.327845573
0.545739675	0.71455214	0.317504583	0.232302293	0.446424746
0.5532	0.687677315	0.327950806	0.176330548	0.470582969
0.407381263	0.692725918	0.360047196	0.116102998	0.311765399
0.52323906	0.694090215	0.312320023	-0.011428558	0.311570583
0.52339049	0.68467627	0.428327572	-0.038429465	0.026987553
0.543172294	0.683254108	0.392220183	0.031279049	0.16768824
0.588203605	0.691280548	0.341186638	0.09000275	0.217233066
0.45346238	0.767320388	0.329087881	0.11687905	0.172201523
0.403981533	0.741319455	0.287995771	0.054325773	0.171086024
0.582234485	0.769646155	0.29636118	0.064499669	0.275171149
0.505379393	0.746199628	0.231248871	-0.040982848	0.184988443
0.443124083	0.715251373	0.280112329	-0.094896977	0.220259175
0.43649122	0.721385603	0.372681803	-0.115484462	0.186906018
0.346804143	0.726757078	0.411428298	-0.040857703	0.112408004
0.48187858	0.72439197	0.34234393	0.250062812	0.285358085
0.5831911	0.79528548	0.324690545	0.24230338	0.229296253
0.680045335	0.813952128	0.371607872	0.22284483	0.399735547
0.59777517	0.715617693	0.352480163	0.03065755	0.298594585
0.532153208	0.66803264	0.3485437	0.033485613	0.154997312
0.44180947	0.663160963	0.372421025	0.131204433	0.33606337
0.55957743	0.61714795	0.350933473	0.001086394	0.24024383
0.64016585	0.659553981	0.357373691	0.130738288	0.154647264
0.593472838	0.70108049	0.384043309	0.247095513	0.157449668

Fig. 6 Feature-wise EEG Signals

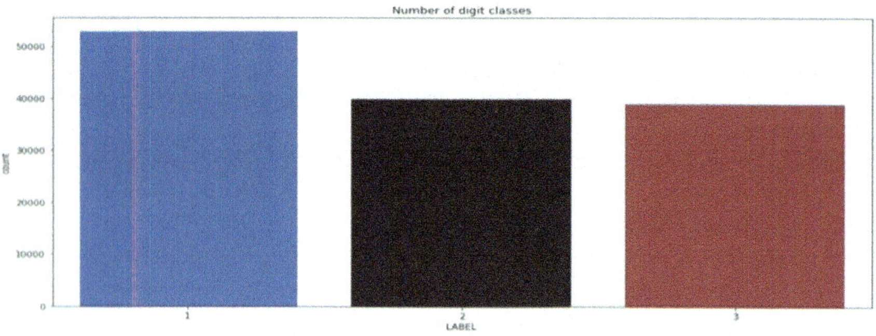

Fig. 7 Label encoding on EEG preprocessed signals

applied to EEG samples collected. All EEG samples are classified into three grade categories: CL A, CL B, and CL C.

4.2 *Representation Learning*

Within our research, we employed 2-D CNN model for classifying e-learning EEG datasets, as illustrated in Fig. 8. Out of 131,604 EEG samples, 118,443 samples were dedicated for the purpose of model training, while 13,161 EEG samples were reserved.

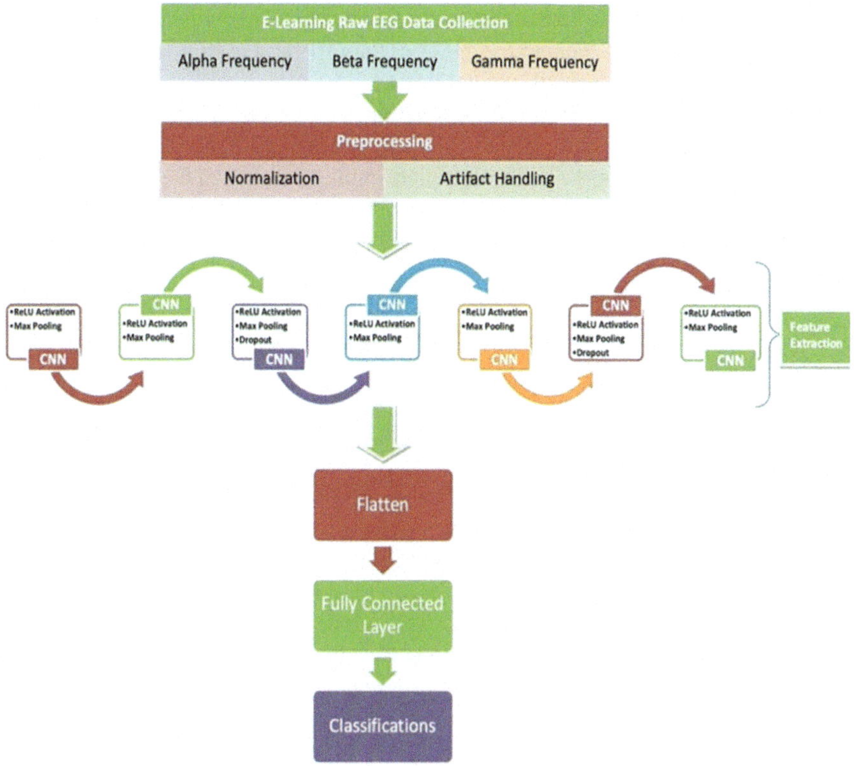

Fig. 8 Convolution neural network (CNN) working flow diagram

The mathematical formula for the implemented process is described by Eq. (1):

$$a(t) = (x * w)(t) = \int_{-\infty}^{\infty} x(b)w(t - b)\mathrm{d}b \tag{1}$$

Model was built using ML Keras Packages. Table 1 outlines the hyperparameters employed for the 2-D CNN model:

Table 1 Hyperparameters for the 2-D CNN

Hyperparameters	Values
Optimizer	ADAM
Loss function	Categorical cross Entropy
Metrics	Accuracy
Batch size	200
Epochs	100
Dropout	0.25

Table 1 indicates the utilization of two convolution layers with filter sizes, along with the ReLU function, denoted as $f(x) = \max(0, x)$. Pooling is applied for the dimension reduction for the subsequent layer.

For classification purpose, SoftMax function is employed. To combat overfitting, the dropout technique is incorporated. Additionally, we utilized the Adam optimizer, to address sparse gradients [23]. The complete implementation of the aforementioned CNN algorithm was carried out in Python within the Google Colab platform using Jupyter Notebook.

5 Results

The loss and accuracy curves are illustrated in Fig. 9, representing the test loss over the number of epochs.

Figure 10 displays the loss details during model training, showing that the model's accuracy increases with each repetition, while the loss decreases correspondingly.

In Fig. 11, we display the confusion matrix, to represent the performance of our implemented algorithm. This matrix offers a visual representation of the classifier's errors. Following the training of our 2-D CNN model, it underwent testing using the designated dataset, and we evaluated its performance using several key statistical metrics, including (TP), (FP), (FN), and F-measure [24].

We utilized the following formulas to calculate these performance metrics:

- Accuracy : $((\mathrm{TP} + \mathrm{TN})/(\mathrm{FP} + \mathrm{FN}))$
- Precision : $(\mathrm{TP}/(\mathrm{TP} + \mathrm{FP}))$

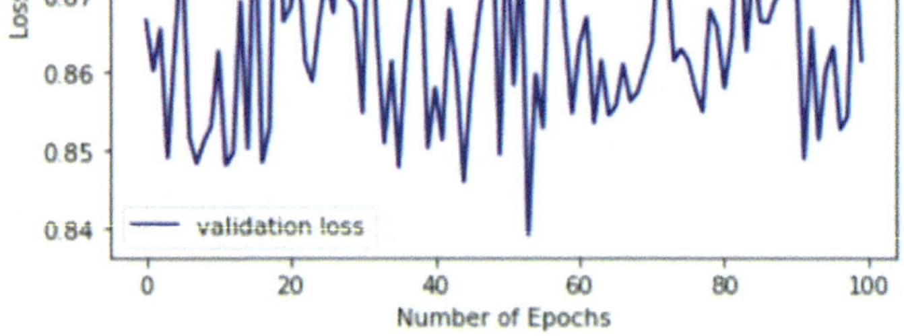

Fig. 9 Epochs versus loss curve

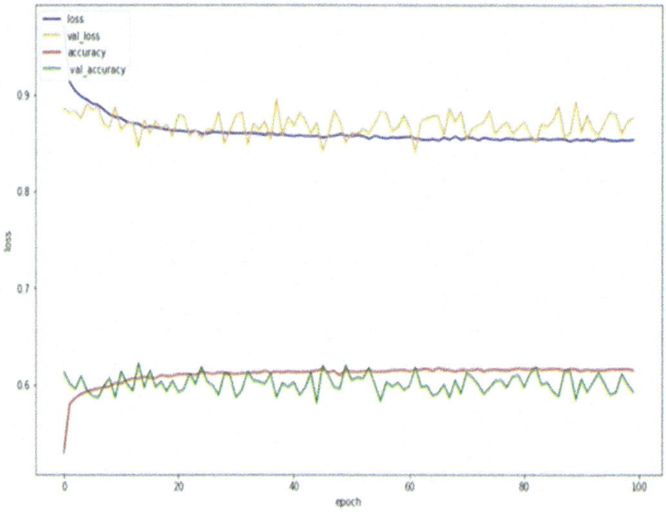

Fig. 10 Model loss during training

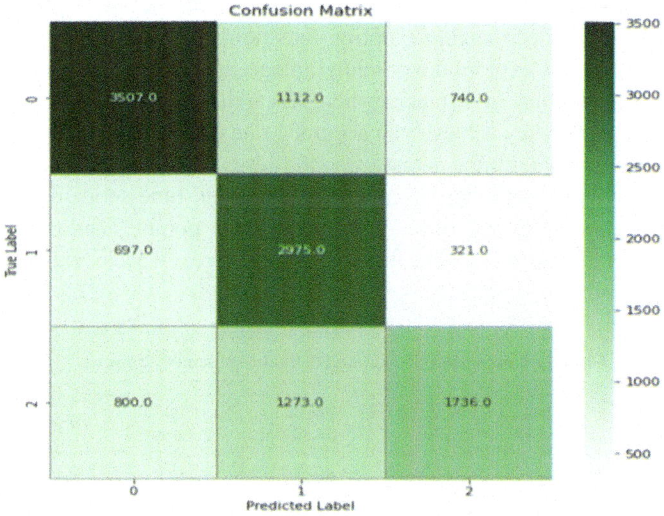

Fig. 11 Confusion matrix representation

- Recall : (TP/(TP + FN))
- F − measure : ((2 ∗ Recall ∗ Precision)/(Recall + Precision))

Table 2 presents the computed values based on performance attributes for implemented model:

Table 2 Performance evaluation on trained model

Performance attributes	Values (%)
Accuracy	70
Precision	68
Recall	70
F-measure	0.68

As illustrated in Table 2, following 2-D CNN model building, our model attained a satisfactory accuracy of 70.34%, a good precision of 68.2%, a recall of 69.8%, and an F-measure of 0.689. The implemented model based on EEG signals employing the 2-D CNN architecture has demonstrated superior performance in comparison with traditional machine learning algorithms.

6 Discussion

Although there has not been a specific investigation applied to EEG with e-learning, recent research, as depicted in Fig. 12, has delved into various aspects, including focus and attention, sleep staging, emotion detection, workload detection, and mental stages. These studies explore the potential integration of EEG data into e-learning for diverse applications. In instances where researchers have focused on classifying focus and attention, they achieved an accuracy rate of 77% using machine learning methods, albeit with a limited dataset.

It is important to note that EEG signals constitute time-series data, where the signals at evert time points depend on fellow time points. While the 2-D CNN model excels in capturing spatial features from images, it lacks explicit modeling

Parameter	**Reviewed recent papers in various EEG BCI Deep Learning Categories**				
Objective (Classification)	Focus and Attention	Temporal Sleep Staging	Emotion Detection	Workload Detection	Mental stages
Dataset	Limited	Limited	Limited	Limited	Limited
Method	ML Algorithms	CNN	LSTM	CNN	ML Algorithms
Model Validation Technique	Not used	Not used	Not used	Not used	K-fold cross validation
Result	Accuracy= 77%	Accuracy= 87%	Accuracy= 88%	Accuracy= 96%	Accuracy= 75%

Fig. 12 Comparative analysis based on reviewed EEG papers

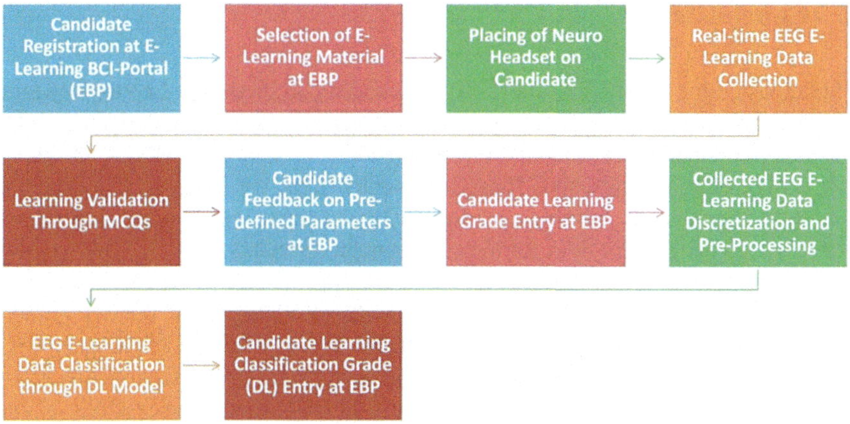

Fig. 13 Recommended framework for e-learning classification

of temporal dependencies among input data. The significant contribution of our work lies in effectively classifying participants' learning using real-time EEG data, surpassing conventional historical approaches. Furthermore, our study introduces an automated approach that offers personalized recommendations to both learners and e-learning authorities, fostering overall improvement. By monitoring participants' learning curves, our approach also aids in identifying potential learning challenges among candidates. The functioning of the automated approach is depicted in Fig. 13.

In the e-learning BCI portal (EBP) process, candidates initially register and select their learning topics. Subsequently, they wear a neuro-headset to collect real-time EEG data during online sessions. Their learning performance is assessed through multiple-choice questions (MCQs), and feedback is provided on a scale of 1–10. Learning grades are recorded, and the collected EEG data are preprocessed and used to train deep learning models for classification. Ultimately, the calculated grades are stored in the approach for tracking of learning and personalized recommendations, offering a comprehensive data-driven approach to online education.

Although our study primarily based on engineering students in the computer science domain, it does have certain limitations. Future work will aim to encompass other academic disciplines and courses. Additionally, while the EEG dataset collected (300 + hours) is substantial, deep learning algorithms generally perform better with larger datasets. Therefore, further research in this domain should explore diverse academic disciplines with extensive e-learning EEG data collection. In future, this research may extend to identifying learning difficulties in the studied subjects and suggesting appropriate interventions [25].

7 Conclusion

Numerous online educational platforms provide a broad spectrum of courses encompassing various subjects. Our research introduces an innovative framework tailored to validate participants' learning and furnish customized recommendations within the current educational landscape. These recommendations are intended to elevate the quality of digital study materials, efficiently monitor learners' progress, and furnish personalized insights for both learners and content providers.

The paper proposes the utilization of deep learning models for the classification of EEG employed e-learning data collected in real time. Specifically, the CNN model exhibits superior performance compared to traditional machine learning models, owing to its capability for automatic feature extraction from extensive e-learning EEG datasets. In essence, our implemented 2-D CNN model successfully classifies candidates' learning based on collected EEG signals and holds several practical applications across domains such as education, psychology, and neurorehabilitation. These benefits encompass personalized learning experiences, objective evaluation of learning progress, early identification of learning challenges, support for rehabilitation efforts, and opportunities for research applications. Future research endeavors may delve into uncovering the underlying determinants of specific learning outcomes, exploring potential interventions, and validating the effectiveness of these interventions.

References

1. Keskin S, Yurdugül H (2020) Factors affecting students' preferences for online and blended learning: motivational vs. cognitive. Euro J Open Distance E-Learn 22(2):72–86. https://doi.org/10.2478/eurodl-2019-0011
2. Shahzad A, Hassan R, Aremu A, Hussain A, Lodhi R (2020) Effects of COVID-19 in E-learning on higher education institution students: the group comparison between male and female. Qual Quant. https://doi.org/10.1007/s11135-020-01028-z
3. Zhao F, Wu Z, Li G (2023) Deep learning in cortical surface-based neuroimage analysis: a systematic review. Intell Med 3(1):46–58. https://doi.org/10.1016/j.imed.2022.06.002
4. Valverde-Berrocoso J, Garrido-Arroyo M, Burgos-Videla C, Morales-Cevallos M (2020) Trends in educational research about e-learning: a systematic literature review (2009–2018). Sustainability 12(12):5153. https://doi.org/10.3390/su12125153
5. Pathak D, Kashyap R, Rahamatkar S (2022) A study of deep learning approach for the classification of Electroencephalogram (EEG) brain signals. In: Artificial intelligence and machine learning for EDGE computing, pp 133–144. https://doi.org/10.1016/b978-0-12-824054-0.00009-5
6. Obpaet J, Paoprasert N (2020) Factors affecting engineering program performance using structural equation modeling technique. In: 2020 International symposium on educational technology (ISET). https://doi.org/10.1109/iset49818.2020.00051
7. Arnicane A, Souza AS (2022) Tracking attentional states: assessing the relationship between sustained and selective focused attention in visual working memory. https://doi.org/10.3758/s13414-021-02394-y

8. Peng C, Chen Y, Chen C, Chen S, Cagneau B, Chassagne L (2019) An EEG-based attentiveness recognition system using Hilbert-Huang transform and support vector machine. J Med Biol Eng 40(2):230–238. https://doi.org/10.1007/s40846-019-00500-y
9. Yoshida K et al (2020) Focused attention meditation training modifies neural activity and attention: longitudinal EEG data in non-meditators. Soc Cogn Affect Neurosci 15(2):215–224. https://doi.org/10.1093/scan/nsaa020
10. Salloum S, Alhamad AQM, Al-Emran M, Abdel Monem A, Shaalan K (2019) Exploring students' acceptance of E-learning through the development of a comprehensive technology acceptance model. IEEE Access 7:128445–128462. https://doi.org/10.1109/access.2019.2939467
11. Mousavi A, Mohammadi A, Mojtahedzadeh R, Shirazi M, Rashidi H (2020) E-learning educational atmosphere measure (EEAM): A new instrument for assessing e-students' perception of educational environment. Res Learn Technol 28(0). https://doi.org/10.25304/rlt.v28.2308
12. Singh N, Gunjan VK, Zurada JM (2022) Building seistutor intelligent tutoring system for experimental learning domain. In: Advanced technologies and societal change, pp 61–78. https://doi.org/10.1007/978-981-19-5197-8_4
13. Onah D, Pang E, Sinclair J, Uhomoibhi J (2021) An innovative MOOC platform: the implications of self-directed learning abilities to improve motivation in learning and to support self-regulation. Int J Inf Learn Technol 38(3):283–298. https://doi.org/10.1108/ijilt-03-2020-0040
14. Takeuchi N (2022) Perspectives on rehabilitation using non-invasive brain stimulation based on second-person neuroscience of teaching-learning interactions. Front Psychol 12. https://doi.org/10.3389/su12125153
15. Lu D, Le H, Vu T (2020) The factors affecting acceptance of E-learning: a machine learning algorithm approach. Edu Sci 10(10):270. https://doi.org/10.3390/educsci10100270
16. Peterson V, Galván C, Hernández H, Spies R (2020) A feasibility study of a complete low-cost consumer-grade brain-computer interface system. Heliyon 6(3):e03425. https://doi.org/10.1016/j.heliyon.2020.e03425
17. Pathak D, Kashyap R (2022) Electroencephalogram-based deep learning framework for the proposed solution of e-learning challenges and limitations. Int J Intell Inf Database Syst 15(3):295. https://doi.org/10.1504/ijiids.2022.124081
18. Xie Z, Yu X, Gao X, Li K, Shen S (2022) Recent advances in conventional and deep learning-based depth completion: a survey. IEEE Trans Neural Netw Learn Syst 1–21. https://doi.org/10.1109/tnnls.2022.3201534
19. Li J (2021) Editorial: recent developments of deep learning in analyzing, decoding, and understanding neuroimaging signals. Front Neurosci 15. https://doi.org/10.3389/fnins.2021.652073
20. Ko W, Jeon E, Jeong S, Phyo J, Suk H-I (2021) A survey on deep learning-based short/zero-calibration approaches for EEG-based brain–computer interfaces. Front Hum Neurosci 15. https://doi.org/10.3389/fnhum.2021.643386
21. Ghimire A, Sekeroglu K (2022) Classification of EEG motor imagery tasks utilizing 2D temporal patterns with deep learning. In: Proceedings of the 2nd international conference on image processing and vision engineering. https://doi.org/10.5220/0011069400003209
22. Das N, Zegers J, Hamme H, Francart T, Bertrand A (2020) Linear versus deep learning methods for noisy speech separation for EEG-informed attention decoding. J Neural Eng 17. https://doi.org/10.1101/2020.01.22.915181
23. Mansoor A, Usman M, Jamil N, Naeem M (2020) Deep learning algorithm for brain-computer interface. Sci Program 2020:5762149. https://doi.org/10.1155/2020/5762149
24. Edwards BI (2021) Emerging trends in education: envisioning future learning spaces and classroom interaction. In: Emerging technologies for next generation learning spaces, pp 7–18. https://doi.org/10.1007/978-981-16-3521-2_2
25. Ramya R, Srinivasan K, Dharshini V, Johnson T, Varsha N (2020) An overview of non-invasive detection of fetal abnormalities using deep learning techniques. J Adv Res Dyn Control Syst 12(8):01–13. https://doi.org/10.5373/jardcs/v12i8/20202442

Food Item Recognition and Calories Estimation Using YOLOv5

Mahmoud Y. Shams, Ahmed Hussien, Aya Atiya, Logyn Medhat, and Roheet Bhatnagar

Abstract This paper explores the application of the YOLOv5 algorithm for food recognition and calorie estimation. The study focuses on the specific context of Egyptian cuisine, utilizing a dataset dedicated to detecting Egyptian food. The paper outlines the data collection process, including techniques for labeling food items in images and data preprocessing steps. It provides a detailed explanation of the YOLOv5 architecture and its components, along with insights into the training process, transfer learning, and model tuning. The segmentation of the dataset into training and validation sets is discussed, along with the importance of selecting appropriate evaluation metrics for food detection. A step-by-step guide for training YOLOv5 on the Egyptian food dataset is presented, along with strategies for monitoring and improving the training process. Various metrics, such as accuracy, recall, and the interpretation of the confusion matrix, were used in the study for evaluating the performance of the trained model. The paper discusses the obtained results, identifies limitations and challenges encountered during the training and evaluation process, and suggests future improvements and areas for further research. Real-world applications of food detection using YOLOv5, including nutritional analysis and personalized nutrition tracking, are highlighted. The chapter concludes by summarizing the main points discussed and emphasizing the importance of food disclosure and the potential impact of YOLOv5 training on specific culinary traditions. Additionally, the paper introduces a separate chapter focused on calorie estimation. It covers the methodology for estimating calorie intake based on individual characteristics, such as age, gender, height, weight, and activity level. The integration of food recognition using YOLOv5 for adjusting calories is explained, emphasizing the creation of a comprehensive calorie database. This research contributes to the field of food recognition and calorie estimation, showcasing the potential of YOLOv5 and its applicability in various real-world scenarios.

M. Y. Shams · A. Hussien · A. Atiya · L. Medhat
Faculty of Artificial Intelligence, Kafrelsheikh University, Kafrelsheikh 33516, Egypt
e-mail: mahmoud.yasin@ai.kfs.edu.eg

R. Bhatnagar (✉)
Department of Computer Science and Engineering, Manipal University Jaipur, Rajasthan, India
e-mail: roheet.bhatnagar@jaipur.manipal.edu

B. R. Devi et al. (eds.), *Proceedings of Fifth International Conference on Computer and Communication Technologies*, Lecture Notes in Networks and Systems 897,
https://doi.org/10.1007/978-981-99-9704-6_21

Keywords Food recognition · Calorie estimation · Egyptian cuisine

1 Introduction

Healthy lifestyle is of prime importance in today's modern world, and many apps and devices are gaining popularity for monitoring the overall health of an individual. With the outbreak of diseases like Covid-19, the significance of a healthy diet in boosting our immune system and overall well-being cannot be overstated [1]. Food plays a crucial role in shaping our health, and understanding its nutritional composition is essential for making informed dietary choices. In this paper, we delve into the realm of food identification and calorie calculation, with a focus on the integration of these processes to provide users with a convenient and powerful tool for maintaining a healthy lifestyle [2].

One of the key challenges in achieving a healthy lifestyle is accurately tracking our calorie intake. Traditional methods are time-consuming, cumbersome, and prone to errors. Therefore, there is a growing need for automated systems that can streamline this process and provide users with real-time and accurate information about the nutritional content of their meals. This is where the concept of food identification and automatic calorie calculation comes into play [3].

Our proposed model addresses this challenge by leveraging advanced technologies, particularly the power of deep learning and computer vision. By utilizing state-of-the-art object detection algorithms like YOLO (You Only Look Once), we enable users to effortlessly capture an image of their food, which is then processed by the model to identify the different dishes present. This ability to recognize multiple dishes within a single picture is a significant advantage, as it eliminates the need for users to manually input each food item separately.

Furthermore, our model goes beyond just identifying the dishes. It also incorporates automatic calorie calculation, allowing users to obtain accurate information about the caloric content of their meals directly from the image. By integrating trusted databases and reliable sources of nutritional information, the model provides real-time calorie calculations for each identified dish. This approach not only saves time and effort but also decreases the chances of inaccuracies that can occur during manual calculations.

One of the remarkable aspects of our model is its adaptability to the diverse range of food types and variations. Egyptian cuisine, in particular, presents a unique challenge due to its rich variety of dishes, each with its distinct ingredients, cooking methods, and calorie compositions. Our model's ability to recognize and calculate calories for multiple dishes in a single picture is a significant breakthrough in addressing this challenge. Users can now capture an image of their entire meal, and our model will accurately calculate the total calorie count by analyzing the different dishes present.

The integration of food identification and automatic calorie calculation is a groundbreaking development in the field of dietary analysis and nutrition tracking. It empowers users to make more informed choices about their diet, promotes healthier eating habits, and assists in achieving personal health goals. By simplifying the process of calorie tracking and eliminating the need for manual input, our model offers a convenient solution for individuals seeking a healthier lifestyle.

In this paper, we will delve into the intricacies of food identification and automatic calorie calculation, focusing on the significance of these technologies in the context of a healthy lifestyle. We will explore the challenges associated with recognizing and calculating calories for various food types, with a specific emphasis on the complexities of Egyptian cuisine. By leveraging the power of deep learning and computer vision, we aim to provide a robust and accurate system that enables users to effortlessly track their caloric intake and make informed dietary decisions.

Overall, this paper is targeted to throw light on the importance of food identification and calorie calculation in promoting healthy life, particularly in the face of prevailing health concerns like Covid-19. By introducing our model's capabilities in recognizing multiple dishes within a single image and automatically calculating their caloric content, we strive to offer a valuable tool that supports individuals on their journey toward a healthier and more balanced lifestyle.

2 Related Works

Numerous studies have explored the field of food recognition and calorie estimation, utilizing various algorithms and methodologies. Here, we provide an overview of the existing literature that has contributed to the advancement of this research area.

In the realm of food recognition, convolutional neural networks (CNNs) have demonstrated remarkable performance. Early approaches, such as Food-101 [4], employed CNN architectures to classify food images into different categories. However, these methods primarily focused on general food recognition without specific consideration for cultural or regional cuisines. To address the need for cuisine-specific food recognition, recent studies have investigated the use of transfer learning techniques. Zhang et al. [5] applied transfer learning using a pre-trained VGGNet to recognize Chinese dishes. Similarly, Ung et al. [6] utilized transfer learning with Inception-V3 for Vietnamese food recognition. These studies showcase the effectiveness of transfer learning in adapting models to specific culinary traditions.

Furthermore, object detection algorithms have gained popularity in food recognition tasks due to their ability to detect and localize multiple objects within an image. The You Only Look Once (YOLO) algorithm family has shown promising results in object detection, and its latest iteration, YOLOv5, has achieved notable performance improvements. YOLOv5 has been successfully applied to various domains, including general object detection [7], pedestrian detection [8], and vehicle detection [9].

In the context of food recognition, YOLO-based approaches have also emerged. Wang et al. [10] proposed a YOLO-based method for food recognition and portion estimation, enabling calorie estimation for dietary monitoring. Their study focused on a general food dataset without specific cultural considerations. In contrast, our research extends the application of YOLOv5 to the domain of Egyptian cuisine, presenting a dataset dedicated to detecting Egyptian food and tailoring the training process accordingly.

Calorie estimation has been a topic of interest in the field of nutrition and health. Traditional approaches often rely on self-reported food diaries or simplified calculations based on fixed portion sizes. However, with the advent of computer vision and deep learning, there has been a shift toward automated methods for calorie estimation.

Image recognition techniques have been employed to recognize food items and estimate their caloric values. For instance, Kitamura et al. [11] developed a system that estimated caloric intake based on food images captured using a smartphone. Their approach utilized a combination of image recognition and a food database to estimate calorie intake accurately.

In this paper, we build upon these image recognition techniques and integrate them into the calorie estimation process. By utilizing the YOLOv5 algorithm for food recognition, we enhance the accuracy of calorie estimation by matching recognized nutrients to their respective caloric values. The creation of a comprehensive calorie database specific to Egyptian cuisine further refines the estimation process.

While previous studies have made significant contributions to food recognition and calorie estimation, our research stands out by focusing on the specific domain of Egyptian cuisine and leveraging the capabilities of the YOLOv5 algorithm. By tailoring the training process and dataset to the cultural context, we aim to achieve more accurate and contextually relevant results.

Therefore, the existing literature has explored various approaches for food recognition and calorie estimation. The application of transfer learning, CNN architectures, and object detection algorithms has significantly advanced the field. However, our research extends the state of the art by employing the YOLOv5 algorithm for food recognition in the domain of Egyptian cuisine, facilitating more precise calorie estimation and promoting cultural relevance in dietary assessment.

3 Research Methodology

The research methodology employed in this study involved a systematic approach to investigate food recognition and calorie estimation using the YOLOv5 algorithm with a specific focus on Egyptian cuisine. The methodology encompassed the following key steps:

The initial phase of the research involved clearly defining the problem statement and outlining the objectives. The primary goal was to develop an accurate food recognition and calorie estimation model specifically tailored to Egyptian food. This entailed addressing the challenges associated with recognizing and estimating the nutritional content of diverse Egyptian dishes.

A comprehensive literature review was conducted to gather relevant information on food recognition techniques, calorie estimation methodologies, and existing studies in the field. This review provided a foundation of knowledge, identified gaps in the current research, and informed the selection of appropriate methodologies and techniques for this study.

3.1 Data Collection

Acquiring a specialized dataset for Egyptian food proved challenging as it was not readily available on platforms like Kaggle or Roboflow. To overcome this obstacle, we took the initiative to create a dedicated dataset tailored specifically to the requirements of studying Egyptian cuisine.

To ensure the dataset's diversity and representativeness, we conducted extensive research across various sources. We scoured online recipe databases and food photography platforms and sought contributions from local food experts and enthusiasts.

Our primary objective was to curate a dataset that encompassed a wide range of Egyptian dishes, both traditional and modern. By including diverse meal types, ingredients, and culinary presentations, our aim was to present a comprehensive and accurate portrayal of the richness and variety found in Egyptian cuisine.

3.2 Data Annotation and Preprocessing

To facilitate effective training of the food recognition model, the collected dataset underwent a meticulous annotation process. Expert annotators were engaged to label each image in the dataset by drawing bounding boxes around individual food items and assigning appropriate class labels. The annotation process was conducted using the Makesense AI annotation platform, which streamlined the labeling process and ensured consistency in labeling conventions across the dataset.

The custom data labeling efforts on the Makesense Web site were instrumental in creating a richly annotated dataset specifically tailored to Egyptian food. This not only enhances the value and uniqueness of the dataset but also ensures that the subsequent food recognition and calorie estimation models are trained and evaluated on accurate and meticulously labeled data.

3.3 YOLOv5 Architecture and Training

The YOLOv5 algorithm was chosen as the core architecture for food recognition due to its real-time object detection capabilities and high accuracy. The YOLOv5

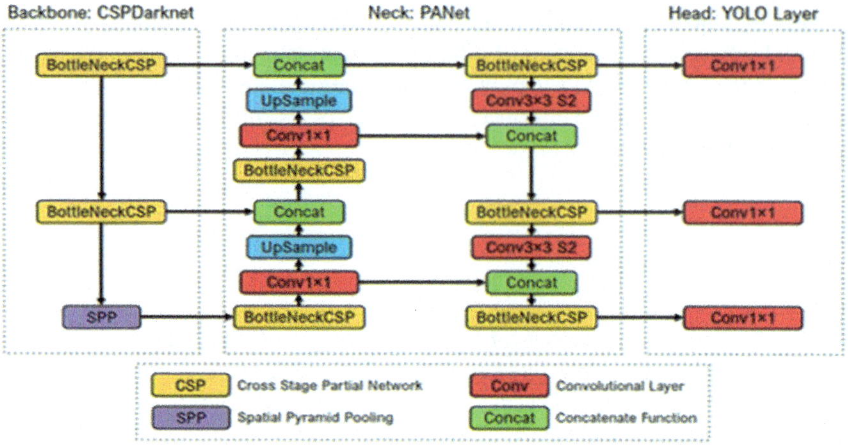

Fig. 1 The architecture of the YOLOv5 model

architecture comprises three main components: the backbone, neck, and head as shown in (Fig. 1).

3.3.1 Backbone

The backbone component is responsible for extracting high-level features from input images. In this research, the YOLOv5 architecture utilized a modified version of the EfficientNet as the backbone network. The EfficientNet is known for its efficiency and effectiveness in extracting informative features. It consists of multiple layers, including convolutional and pooling layers, which capture and encode the visual information present in the input images.

3.3.2 Neck

The neck component serves as a bridge between the backbone and the head. Its main function is to fuse and refine the features extracted by the backbone network. In this study, the YOLOv5 architecture employed a PAN (path aggregation network) neck. The PAN neck lead to improved performance in object detection tasks.

3.3.3 Head

The head component of the YOLOv5 architecture is responsible for predicting bounding box coordinates, class probabilities, and confidence scores for each detected object. In this research, the YOLOv5 architecture utilized a YOLO (You

Only Look Once) head, which consists of additional convolutional layers followed by a final detection layer. The YOLO head performs anchor-based object detection, where anchor boxes of different sizes and aspect ratios are used to predict the position and class of objects in the image.

During the training phase, the YOLOv5 model was optimized using the collected and annotated dataset specific to Egyptian food. Transfer learning techniques were employed to initialize the model's weights using pre-trained weights obtained from a large-scale dataset. By leveraging the pre-trained weights, the model was able to benefit from the general features learned from a diverse range of images. Fine-tuning was then performed on the Egyptian food dataset to adapt the model to the specific task of food recognition as shown in Fig. 2.

The training process involved optimizing various parameters as shown in Fig. 3, including the learning rate, batch size, and the number of training epochs. These parameters were fine-tuned to achieve optimal performance on the Egyptian food dataset. The model was trained using a suitable loss function, such as the YOLOv5-specific loss, which takes into account the predicted bounding box coordinates, class probabilities, and confidence scores.

By employing the YOLOv5 architecture with the customized backbone, neck, and head components, and fine-tuning the model on the Egyptian food dataset, this research aimed to develop a highly accurate and efficient food recognition model tailored to the unique characteristics of Egyptian cuisine.

Fig. 2 Predicted food after training

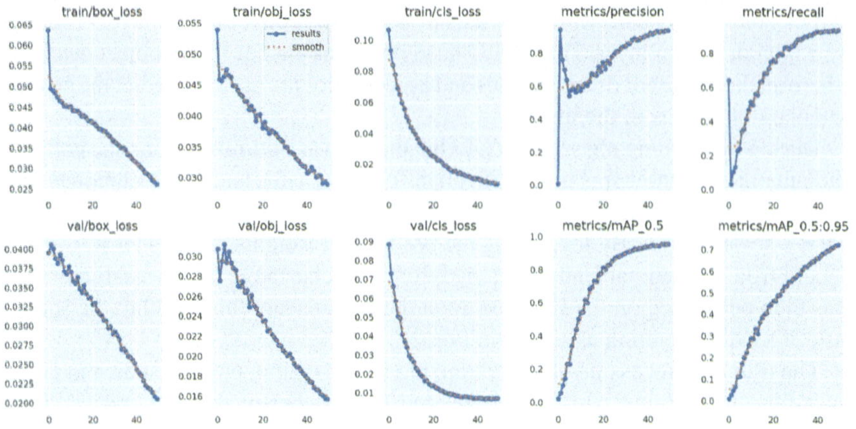

Fig. 3 Relationship between number of epochs, mAP matrices where it is clear that the higher the number of epochs, the lower the loss and the greater the accuracy

4 Evaluation and Performance Analysis

We thoroughly evaluated the performance of the trained YOLOv5 model using a range of metrics, including accuracy, recall, and average mean accuracy (mAP). To ensure a comprehensive assessment, we also employed the precision–recall (PR) curve and F1 curve.

We split the dataset into separate training and validation sets to prevent bias. This enabled us to assess the model's performance using hypothetical data. Metrics like precision, recall, and accuracy were calculated during the evaluation by comparing the model's predictions to the ground truth annotations.

The PR curve visually illustrates the trade-off between precision and recall at different classification thresholds, as depicted in Fig. 4. This curve provides valuable insights into the model's performance across various decision thresholds, aiding us in optimizing the model's performance for food item recognition.

We also employed the F1 curve, which combines precision and recall into a single metric, to evaluate the overall performance of the model. The F1 score offers a balanced measure of the model's capability to accurately identify positive samples while minimizing false positives and false negatives. A higher F1 score indicates superior performance in terms of both precision and recall, and you can see this as depicted in Fig. 5. This metric gives us valuable insights into how well the model strikes the right balance between accurate detections and avoiding misclassifications, enhancing our confidence in its effectiveness for food recognition.

Furthermore, a confusion matrix was employed to interpret the performance results in more detail. The confusion matrix provides insights into true positives, false positives, true negatives, and false negatives, allowing for a comprehensive analysis of the model's accuracy, recall, and other performance metrics specific to Egyptian food recognition.

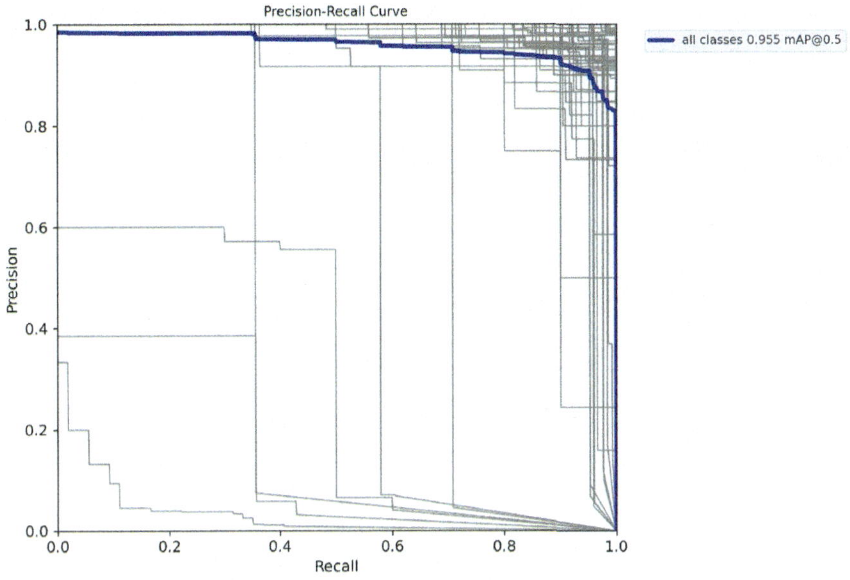

Fig. 4 PR curve where the curve shows the relationship between precision and recall for each phagosome and shows the high accuracy provided by the YOLO, where the blue curve is the mean

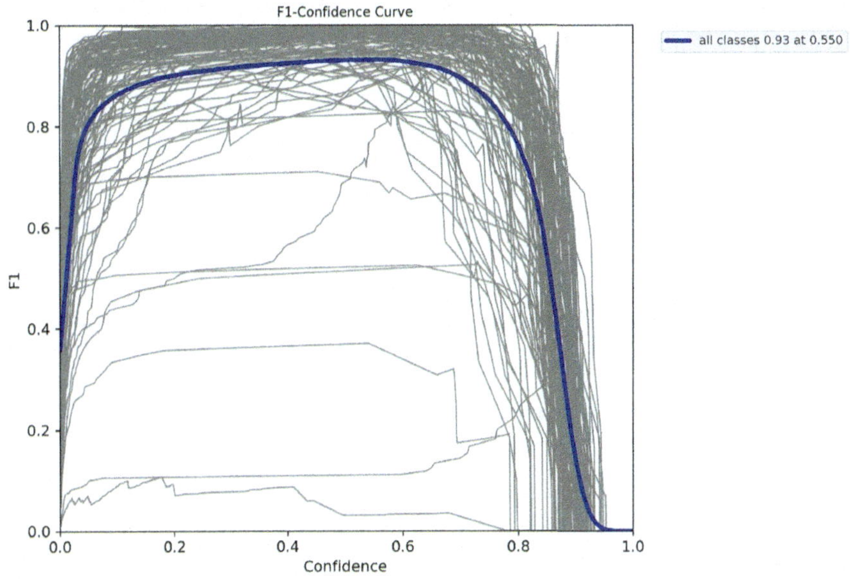

Fig. 5 F1 curve that we have shown has many categories, so we have many lines which show the high accuracy of most dishes

By evaluating the model's performance using various metrics, this research aimed to gain a comprehensive understanding of the trained model's capabilities in recognizing Egyptian food accurately and efficiently. These evaluation measures provided insights into the strengths and limitations of the model and guided further analysis and potential improvements in food recognition and calorie estimation within the context of Egyptian cuisine.

4.1 Loss Function

The loss function used in the YOLOv5 training process is typically a combination of multiple components. One common formulation is the sum of the following terms as in Eq. (1):

$$\text{Total Loss} = \text{Localization Loss} + \text{Confidence Loss} + \text{Class Loss} \tag{1}$$

Localization Loss: This term measures the discrepancy between the predicted bounding box coordinates and the ground truth bounding box coordinates. One commonly used localization loss is the mean squared error (MSE) loss.

Confidence Loss: The confidence loss quantifies the difference between the predicted confidence scores (abjectness scores) and the ground truth abjectness values. It can be calculated using a loss function such as binary cross-entropy.

Class Loss: The class loss evaluates the dissimilarity between the predicted class probabilities and the true class labels. It is typically computed using a suitable loss function, such as categorical cross-entropy.

4.2 Precision

Precision (Prec.) is a statistical function and is determined as the ratio of true positives (TP) to the sum of true positives and false positives (FP):

$$\text{Prec.} = \text{TP}/(\text{TP} + \text{FP}) \tag{2}$$

4.3 Recall

The model's ability to accurately identify all positive samples is measured by recall (Rec.), also known as sensitivity or true positive rate. It is a ratio of true positive divided by summation of true positives and false negatives.

$$\text{Rec.} = \text{TP}/(\text{TP} + \text{FN}) \tag{3}$$

4.4 F1 Score

In our paper, we employed the F1 score as a balanced measure of the model's performance, combining perfection and recall. The F1 score is calculated using the equation:

$$\text{F1 Score} = 2 * (\text{Precision} * \text{Recall})/(\text{Precision} + \text{Recall}) \tag{4}$$

These equations played a vital part in assessing the delicacy and effectiveness of our food recognition and calorie estimation system. We used them to quantify the loss during training and measure perfection and recall. The F1 score helped us understand how well the model rightly linked food particulars while minimizing misclassifications, furnishing a comprehensive evaluation of our system's performance.

5 Conclusion and Future Improvements:

Based on the data we collected, we divided it in 80:20 ratio for training: validation, respectively. The model was trained in 120 food classes, and we obtained some promising results. The accuracy of mAP_0.5:0.95 reached 0.7, which is quite good considering the number of food classes we worked with. Additionally, the accuracy of mAP_0.5 reached 0.98, showing the model's ability to make precise detections.

During the training process, we observed that increasing the number of epochs led to better accuracy. The loss in learning was minimized to 0.02, and in validation, it was even lower at 0.01. The F1 score, which represents the percentage of correct predictions for all food classes, achieved a commendable 0.93.

While we are pleased with the test results, our goal for the future is to further enhance the model's efficiency and reliability by increasing these percentages even more.

In the paper, we also discussed the findings and insights gained from our food recognition and calorie estimation experiments. We addressed the limitations and challenges encountered during data collection, annotation, and training. Additionally, we suggested possible improvements for future research, focusing on extending the application of YOLOv5 to other culinary traditions.

By following a systematic research methodology, our study aims to contribute to the field of food recognition and calorie estimation, with a specific focus on the unique context of Egyptian cuisine.

Authors' Declaration

Authors declare that there is no conflict of interest and no funding source. The researchers contributed in all parts of the research equally, and data will be made available on request.

References

1. Hassan E, Shams MY, Hikal NA, Elmougy S (2023) Covid-19 diagnosis-based deep learning approaches for covidx dataset: a preliminary survey. In: Artificial intelligence for disease diagnosis and prognosis in smart healthcare, p 107
2. Shams MY, Elzeki OM, Abouelmagd LM, Hassanien AE, Abd Elfattah M, Salem H (2021) HANA: a healthy artificial nutrition analysis model during COVID-19 pandemic. In: Computers in biology and medicine, vol 135, p 104606
3. Hendriks S, Soussana J-F, Cole M, Kambugu A, Zilberman D (2023) Ensuring access to safe and nutritious food for all through the transformation of food systems. In: Science and innovations for food systems transformation, p 31
4. Bossard L, Guillaumin M, Van Gool L (2014) Food-101–mining discriminative components with random forests. In: Computer Vision–ECCV 2014: 13th European Conference, Zurich, Switzerland, September 6-12, 2014, Proceedings, Part VI 13. Springer, Heidelberg, pp 446–461
5. Paymode AS, Malode VB (2022) Transfer learning for multi-crop leaf disease image classification using convolutional neural network VGG. Artif Intell Agric 6:23–33
6. Ung HT, Dang TX, Thai PV, Nguyen TT, Nguyen BT (2020) Vietnamese food recognition system using convolutional neural networks based features. In: Nguyen NT, Hoang BH, Huynh CP, Hwang D, Trawiński B, Vossen G (eds) Computational collective intelligence. Lecture Notes in Computer Science. Springer International Publishing, Cham, pp 479–490. https://doi.org/10.1007/978-3-030-63007-2_37
7. Bochkovskiy A, Wang C-Y, Liao HYM (2020) Yolov4: optimal speed and accuracy of object detection. arXiv preprint arXiv:2004.10934
8. Redmon J, Farhadi A (2018) Yolov3: an incremental improvement. arXiv preprint arXiv:1804.02767
9. Redmon J, Farhadi A (2017) YOLO9000: better, faster, stronger. In: Proceedings of the IEEE conference on computer vision and pattern recognition, pp 7263–7271
10. Wang Y, Chen J, Ngo C-W, Chua T-S, Zuo W, Ming Z (2019) Mixed dish recognition through multi-label learning. In: Proceedings of the 11th workshop on multimedia for cooking and eating activities, pp 1–8
11. Kitamura K, Yamasaki T, Aizawa K (2009) Foodlog: capture, analysis and retrieval of personal food images via web. In: Proceedings of the ACM multimedia 2009 workshop on Multimedia for cooking and eating activities, pp 23–30

Fuzzy Cognitive Security Techniques for Data Aggregation in Capillary Networks

Jayshree D. Mallapur, Kirankumar B. Balavalad, Kirankumar Y. Bendigeri, Santosh B. Kumbalavati, and Nandini Sidnal

Abstract IoT is a revolutionary area of computing technology that allows many different devices to work together over existing internet infrastructure. Capillary networks are the intermediate interface between the IoT devices and cellular e-nodes of the present IoT development, making sensor network nodes to connect and to make use of the capacity of cellular networks through the gateways. The global connectivity can also be enabled with short-range frequency and with the well-managed security and virtual services of the network. But biggest challenge they all face is security provisioning while data aggregation. Our paper has come with unique solution for security problem for data aggregation in capillary networks. We have proposed fuzzy cognitive security system, which considers few influential parameters for fuzzification to normalize the uncertainty and carrying out data aggregation. The proposed scheme shows that there is an increase in throughput and a decrease in the packet loss ratio.

Keywords Fuzzy · Capillary networks · Cognitive · Security

1 Introduction

The capillary network technology is very much different from a typical over-the-top solution in many ways. Alongside furnishing connectivity, the capillary network platform incorporates a multi-tenant PaaS (platform as a service) offering. This solution enables multiple users and operators to create and tailor services to their specific requirements. The system combines mobile network security and data management, delivering a comprehensive solution for managing security threats and connectivity

J. D. Mallapur · K. B. Balavalad (✉) · K. Y. Bendigeri · S. B. Kumbalavati
Department of Electronics and Communication Engineering, Basaveshwar Engineering College, Bagalkot, Karnataka 587103, India
e-mail: kiranb4004@gmail.com

N. Sidnal
Software Engineering and IT, Torrens University, Ultimo, Sydney, Australia

B. R. Devi et al. (eds.), *Proceedings of Fifth International Conference on Computer and Communication Technologies*, Lecture Notes in Networks and Systems 897,
https://doi.org/10.1007/978-981-99-9704-6_22

across the entire spectrum. The capillary network technology is a network aware platform compared to the IoT platforms. In capillary network applications with limited resources, data collection stands as a foundational task aimed at minimizing energy consumption in sensor nodes and enhancing network longevity. Nevertheless, any deviation can significantly impact the collected data's quality, particularly concerning data aggregation and its security. Data aggregation presents an effective solution for addressing challenges within capillary networks. This technique involves transmitting samples from multiple sensors, which possess a level of redundancy, instead of sending raw data. Within this approach, certain nodes function as aggregators, eliminating superfluous information received from sensor nodes and ensuring the precision of the collected data. The fuzzy cognitive security concept is applied to the capillary network. The mediator facilitates the collection of received data from terminals within the capillary network by the fuzzy cognitive engine. Fuzzy cognitive security involves conscious intellectual processes, such as understanding and perception, and could potentially be simplified to empirical factual information. Fuzzy cognitive security introduces both cognitive and fuzzy elements by leveraging technologies like fuzzy systems, machine learning, and knowledge representation.

2 Related Work

Currently, there is a focus on addressing privacy concerns within the realm of data mining [1]. The paper [2] introduces SPINS, a security protocol specifically designed for enhancing sensor network security. SPINS comprises two foundational elements: SNEP and μTESLA. SNEP encompasses aspects such as ensuring data confidentiality, data authentication, and providing proof of recent data. In highly resource-limited settings, μTESLA offers authenticated broadcast capabilities. In the paper [3], the timing-sync protocol for sensor networks (TPSN) is introduced, offering comprehensive network-wide time synchronization within a sensor network. In the article [4], it is explained that the capillary network functions as a nearby network utilizing short-range radio access technologies to establish connectivity among a group of devices. The paper [5] introduces a novel secure data aggregation approach called differentially private data aggregation with fault tolerance (DPAFT), which accomplishes both differential privacy and resilience to faults. Article [6] emphasizes the concept of the internet of things (IoT) and characterizes it as an empowering technology that seamlessly integrates various technologies and communication solutions. Paper [7] places significant emphasis on legislative aspects, including the right to access information, limitations on the utilization of internet of things mechanisms, and regulations pertaining to IT security. In this paper [8], attention is given to review of the research progress in IoT and security by analyzing the security architectures and features. The IoT architecture requires two technologies: data acquisition and networking technologies. Article [9] categorizes various technologies that are commonly used in IoT. Paper [10] addresses the changing world of security systems, and possible ways to improve them. In the context of secure data aggregation, paper

[11] employs fuzzy judgment to enhance the process. The SDAFPS protocol encompasses three stages: Initially, it governs topology by selecting proficient aggregator nodes; subsequently, it employs the concept of generating and utilizing category pattern codes to minimize data size and perform data aggregation; finally, the resulting aggregated data undergo encryption through partial equations of SHAP contribution and decryption using SHAP contribution equations. In the study presented in paper [12], a secure data aggregation technique with a foundation in fuzzy logic is introduced. The technique involves a series of steps, including clustering and the election of cluster heads. Within each cluster, calculations are made for factors like power consumption, distance, and trust values for each member. These parameters are subsequently utilized in a fuzzy logic approach to determine the selection of secure and reliable node members for the purpose of data aggregation. Paper [13] employs a secure data aggregation technique based on fuzzy logic for the selection of trustworthy and fault-free node members. In the context of paper [14], the application of the fuzzy c-means methodology involves clustering sensors according to data similarity, followed by the establishment of a robust support degree function for detecting outliers. Paper [15] utilizes fuzzy logic within a wireless sensor network (WSN), subsequently employing min–max normalization to obtain normalized weights for the graph's specific edges. The membership value signifies the degree to which an element belongs to a set, while the node identified with the least total of weights is assigned the role of the parent node.

3 Proposed Work

The block diagram Fig. 1 shows the proposed scheme for fuzzy cognitive security technique for data aggregation in capillary networks. In this scheme, the data from the sensor nodes are sent to the server, and the server sends to data storage system as well as to the data processor. The fuzzy controllers access the new data collected and compute the new data window for security window. The fuzzy controller has been fed with three parameters such as packet length, queue length, and time stamp difference, which are more influential in affecting data window and security window. A cognitive security in data aggregation algorithm should be able to adjust security levels to prevent potential participant functioning or malfunctioning nodes based on these criteria and comparisons with fuzzy data. The fuzzy agent examines the identity, or ID parameter, of the nodes that is suspected of being a possible disrupter. When a defuzzification parameter crosses the thresholds, the window sizes are reset for the new value. The agent adjusts the capillary network nodes' transmission characteristics to increase the bidirectional frames delivered. If the node is untrustworthy, the frame delivery is reset with rearrangement. The agent also informs the capillary network's management entity about the trustworthy nodes and malfunctioning nodes. Instead, if the suspicious node is reliable, the agent alerts the management entity with the node ID that the node has been compromised (Fig. 2).

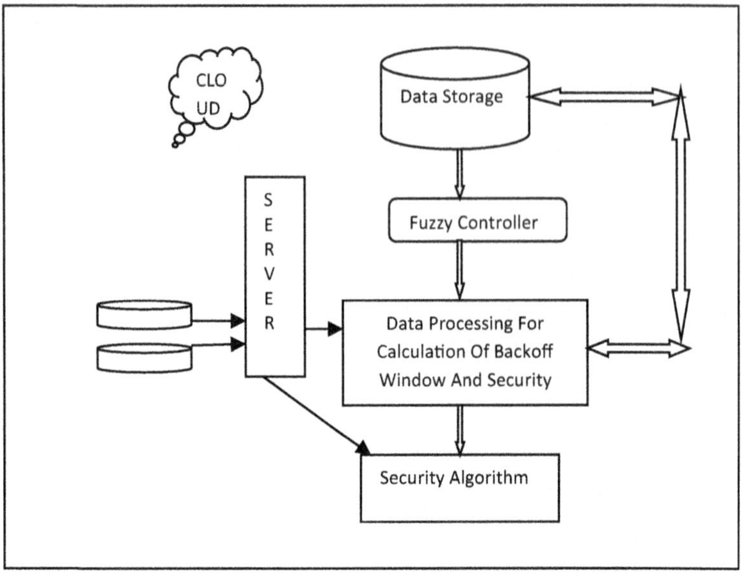

Fig. 1 Fuzzy cognitive security system

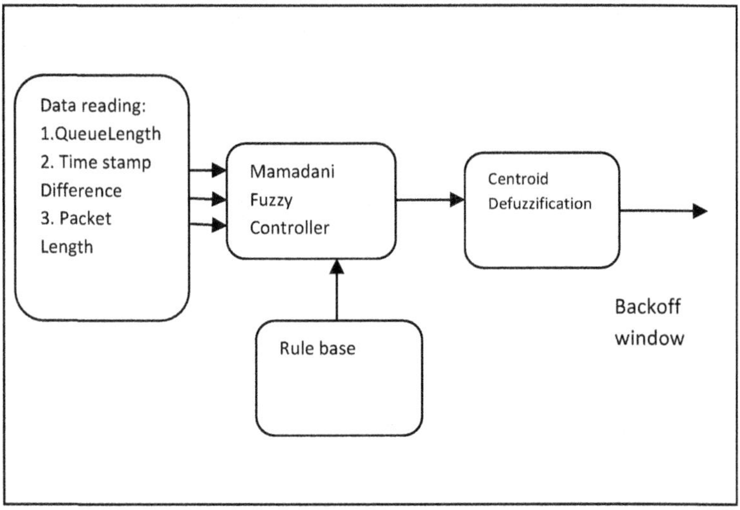

Fig. 2 Fuzzy controller

Algorithm 1:	Algorithm 2:
Fuzzy cognitive security in data aggregation Main: Nomenclature: N = set of sensor nodes F = the set of acceptance links Ql = the data queue length Ts = time difference of frames between the communication for each node Pl = packet lengths F = the transmission frequencies Begin: 1 Collect the data packets from the N number of sensor nodes 2 Collect the information about the queue length, time stamp difference, and packet length 3 CALL Algorithm 2 4 Set the new backoff window size 5 Reset the new security window size 6 Stop End	Backoff window size and security window size calculation Begin: 1. For $k = 1$ to j data set 2. Initialize fuzzy controller with queue length, packet length, and time difference of frames 3. Find the membership function for queue length packet length and time difference of frames 4. Compute the new backoff window size and security window size 5. Return the new window size 6. Stop End

Thus, possible actions are as follows:

- The modified parameters are accessed by agent to set nodes, based on the information attained through the application.
- The modified parameters due to fuzzy controller such as backoff window and security window are used for the next iterations.
- Every time the fuzzy controller sets new parameter, so that data aggregation done with valid nodes.
- The valid node data aggregation improves performance.
- The agent adjusts the time validity of the security keys to prevent replay attacks or address clock desynchronization.

4 Results

All the simulations are conducted in MATLB (fuzzy tool box) and NS2. The simulation parameters considered are %data rejection ratio versus packets per sec, secured data aggregation versus system latency and packet loss ratio. All the above-mentioned parameters are simulated for various securities like fuzzy, cognitive, and standard. Figure 3 describes the graph of percentage data rejection versus data packets per sec. It is evident from the graph that the %data rejection ratio for fuzzy security scheme is less compared to the other security schemes like cognitive and standard.

Figure 4 shows the plot of secured aggregation percentage against system latency for various security schemes. The graphs clearly illustrate that the fuzzy security scheme provides better secured data aggregation compared to other two schemes. Figure 5 is the plot of secured data aggregation percentage versus packet loss ratio, and the graphs obtained from the simulation results describe that for a given secured aggregation percentage, the packet loss ratio for fuzzy security scheme is minimum compared to the other schemes.

5 Conclusions

A solution addressing IoT security concerns has been introduced, centering on time-based mechanisms for generating and refreshing keys to ensure secure transactions involving both unidirectional and bidirectional non-IP devices. This approach utilizes the local transmitter's timestamp in the internet of things (IoT) security and privacy framework to establish the encryption key. The recipient then leverages this timestamp to autonomously select the appropriate decryption key, thereby eliminating the need for key exchanges or additional broadcast messages. This methodology notably reduces the likelihood of security breaches while also streamlining device functionalities, a critical factor in IoT contexts. Finally, the notion of fuzzy cognitive security is introduced and incorporated into this time-oriented security solution.

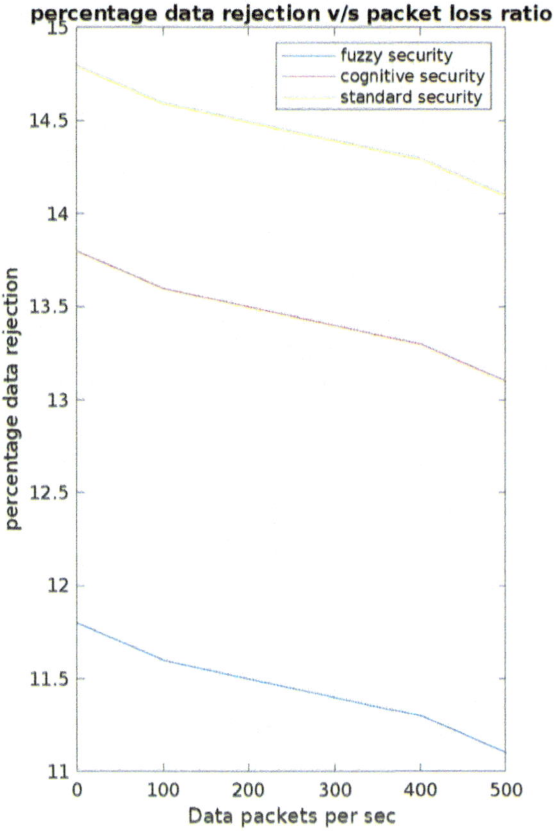

Fig. 3 Secured data aggregation percentage versus packet per second

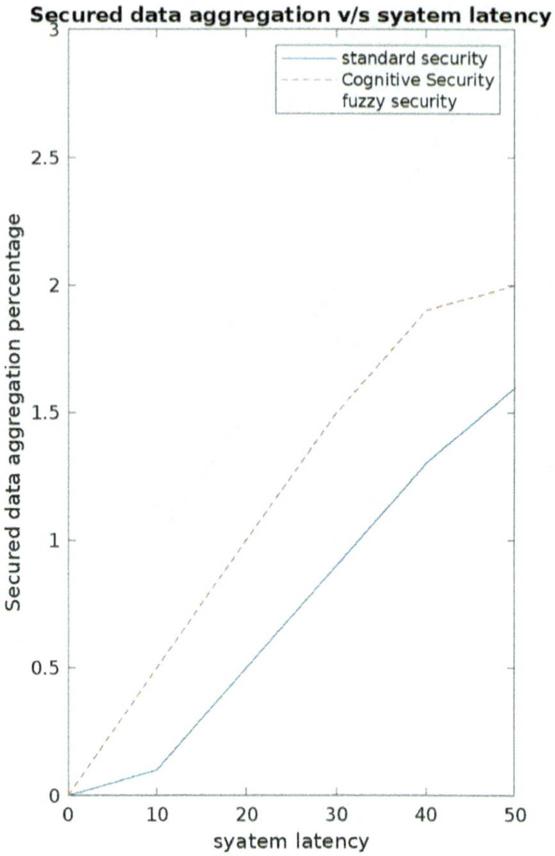

Fig. 4 Secured data aggregation percentage versus system latency

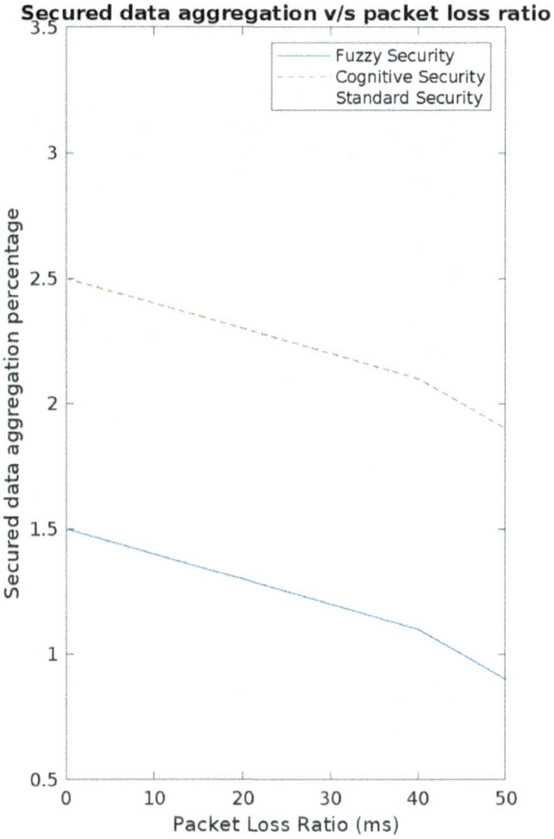

Fig. 5 Secured data aggregation percentage versus packet loss ratio

References

1. Chan H, Perrig A (2003) Security and privacy in sensor networks. Computer 36(10):103–105. https://doi.org/10.1109/MC.2003.1236475
2. Perrig A, Szewczyk R, Tygar D, Wen V, Culler D (2002) SPINS: security protocols for sensor networks. Wireless Netw 8(5):521–534
3. Ganeriwal S, Kumar R, Srivastava MB (2003) Timing-sync protocol for sensor networks. In: Proceedings of conference on embedded networked sensor systems (SenSys), pp 138–149
4. Sachs J, Beijar N, Elmdahl P, Melen J, Militano F, Salmela P (2014) Capillary network—a smart way to get things connected. In: Ericsson review. http://www.ericsson.com/res/thecompany/docs/publications/
5. Bao H, Lu R (2015) A new differentially private data aggregation with fault tolerance for smart grid communications. IEEE Internet Things J 2(3):248–258. https://doi.org/10.1109/JIOT.2015.2412552
6. Atzori L, Iera A, Morabito G (2010) The IoT a survey. Comput Netw 54:2787–2805. https://doi.org/10.1016/j.comnet.2010.05.010
7. Weber RH (2010) Internet of Things: new security and privacy challenges. Comput Law Security Rev 26(1):23–30. https://doi.org/10.1016/j.clsr.2009.11.008

8. Suo H, Wan J, Zou C, Liu J (2012) Security in the Internet of Things: a review. In: Proceedings of International Conference on Computer Science and Electronics Engineering (ICCSEE), vol 3, pp 648–651. https://doi.org/10.1109/ICCSEE.2012.373
9. Tan J, Koo SGM (2014) A survey of technologies in internet of things. In: 2014 IEEE international conference on distributed computing in sensor systems, Marina Del Rey, CA, USA, pp 269–274. https://doi.org/10.1109/DCOSS.2014.45
10. Witold K (2012) Towards cognitive security systems. In: 2012 IEEE 11th International Conference on Proceedings of Cognitive Informatics Cognitive Computing (ICCI*CC), pp 539–539. https://doi.org/10.1109/ICCI-CC.2012.6311207
11. Reshma S, Shaila K, Venugopal KR (2021) SDAFPS: secure data aggregation using fuzzy judgment, pattern category and SHAP contribution. SN Comput Sci 2:86. https://doi.org/10.1007/s42979-021-00475-1
12. Hevin Rajesh D, Paramasivan B (2012) Fuzzy based secure data aggregation technique in wireless sensor networks. J Comput Sci 8(6):899–907
13. Rajesh DH, Paramasivan B (2012) Fuzzy logic based performance optimization with data aggregation in wireless sensor networks. Proc Eng 38:3331–3336. ISSN 1877-7058. https://doi.org/10.1016/j.proeng.2012.06.385
14. Wan R, Xiong N, Hu Q et al (2019) Similarity-aware data aggregation using fuzzy c-means approach for wireless sensor networks. J Wireless Com Netw 2019:59. https://doi.org/10.1186/s13638-019-1374-8
15. Bhushan S, Kumar M, Kumar P et al (2021) FAJIT: a fuzzy-based data aggregation technique for energy efficiency in wireless sensor network. Complex Intell. Syst. 7:997–1007. https://doi.org/10.1007/s40747-020-00258-w

Performance Analysis on Optimal Data Integrity and Security-Constrained Cloud Storage Framework with Meta-heuristic-Based Mechanism

K. Brindha, K. Karuppasamy, and K. Kalaivani

Abstract The cloud computation applications need a novel and effective framework to balance the security of their data. In cloud computing systems, security issues are considered becomes complicated, where the data storage is situated in various locations in the universe. Cloud users worry about the main security constraints of cloud computing are data confidentiality and data integrity. Therefore, this problem mentioned above in cloud computing is motivated for developing an efficient and secured cloud storage system by utilizing the optimization algorithm of the Pathfinder Algorithm (PFA). Further, the data are gathered from the benchmark data set. Then, the collected data is fed into the optimal sensitive attribute selection stage along with the PFA application. Further, the optimal sensitive attributes are secured with the help of Optimal Minimum-Storage Regenerating Code (OM-SRC), where parameter optimization occurs with the PFA. The remaining non-sensitive attributes are secured with the utilization of the Shuffling Algorithm. Finally, the attributes like sensitive and non-sensitive are fused and uploaded into the cloud environment. The storage mechanism efficiency is tested and correlated with the other comparative algorithms. Finally, an effective examination of the suggested technique is done to show the effectiveness of changing the parameters.

Keywords Optimal Minimum-Storage Regenerating Code · Data Integrity and Security · Pathfinder Algorithm · Cloud Storage System · Shuffling Algorithm

K. Brindha (✉) · K. Karuppasamy
CSE, RVS College of Engineering and Technology, Coimbatore, India
e-mail: brindha8892@gmail.com

K. Karuppasamy
e-mail: karuppasamyrvs@gmail.com

K. Kalaivani
AIML, Coimbatore Institute of Technology, Coimbatore, India
e-mail: kalai.vani980@gmail.com

B. R. Devi et al. (eds.), *Proceedings of Fifth International Conference on Computer and Communication Technologies*, Lecture Notes in Networks and Systems 897,
https://doi.org/10.1007/978-981-99-9704-6_23

1 Introduction

In present years, utilizing the cloud platform in health care, education and other sectors is common practice to store and process business data for performing backup in personal data like contact details, storage space, mail archives, photographs and computing resources [1, 2]. Moreover, different outages and data loss incidents occur in the cloud services, and also the Cloud Service Provider (CSP) faces various failure issues occasionally that reject the hardly accessed files and don't display the data loss error [3]. Unless, the Data Owner (DO) and Third Party Auditor (TPA) identify the data integrity and data confidentiality level in the outsourced data. Hence, ensuring data integrity in the untrusted cloud server with cloud data became the main concern [4, 5]. The verifiers are widely utilized in public data auditing and private data auditing. Auditing private data DO attain the responsibility to identify the precise of the stored data [6]. Later, integrity verification is executed with data extractions that use more bandwidth in the network and increase the communication rate with overhead problems [7]. The utilization volume of data increased rapidly because cloud storage offers the outsourced data inevitable to the resource-constrained data owners associated with organizations [8–10]. Thus, it is required to design a new cloud storage framework by considering the limitations of the classical cloud storage systems.

Different contributions of the suggested framework are given as below.

- To adapt a novel cloud storage system with data integrity and security constraints for providing data protection rate and avoid data leakage in the cloud system.
- To select the optimal sensitive attributes with the help of PFA to optimize the threshold. Thus, it tends to maximize the variance, mutual information and correlation.
- To construct a new heuristic model named PFA to optimize the threshold rate in sensitive attributes and to tune computation time in the OM-SRC.
- To examine the efficacy of the recommended cloud storage framework over different performance measures with conventional techniques.

The major parts of the recommended cloud storage system are given below. The related works based on cloud data storage models are explored in Sect. 2. Attribute merging techniques utilized for privacy protections in cloud storage are offered in Sect. 3. In Sect. 4, a new privacy preservation model is given. Multiple experimental evaluations are evaluated using the developed model are provided in Sect. 5. The suggested technique is presented in Sect. 6.

2 Literature Survey

Related Works

In the recent days, cloud storage systems are generally used to offer remote storage services to data owners. Moreover, the cloud user can utilize the outsourced data, not including any local storage burden. In some cases, the cloud storage systems are not trustworthy and also their remote data are corrupted.

3 Dual Attribute Merging Scheme for Privacy Protection in Cloud Storage Framework

3.1 OM-SRC-Based Optimal Sensitive Attribute Protection

In this phase, OM-SRC techniques are utilized to choose the optimal sensitive attributes for enhancing performance in the cloud. Based on the input data, the sensitive attributes are selected. M-SRC [11] Code regeneration in the distributed storage system has attracted more professionals. Generally, the MSR code $[c, v, b] - (\alpha)$ is used over Fi files containing symbols like α, v over M_p in the node c, and the sorting is given as α. Here, the files are recovered while downloading the content of v symbols and c nodes. Then, the content attained from the failed node is reconstructed by accessing the balance $c - 1$ nodes and also downloading the symbols $\alpha / (b - v + 1)$ from every node. Moreover, the files Fi are presented in the non-coded format in the nodes c, and v that is termed as systematic nodes. In systematic nodes, the node-pair conditions are performed to obtain the optimal repair bandwidth, and these nodes are termed the systematic-repair MSR codes. The fitness function of the developed model is offered in Eq. (1).

$$Ft_1 = \underset{\{Th_h^{omsrc}\}}{\arg\min}\left(\frac{1}{CT}\right) \tag{1}$$

Here, the term Th_h^{omsrc} indicates the threshold of OM-SRC, which is tuned in the limit $[1, 5]$ using PFA, and the term CT denotes the computation time. The time length needed to execute the validation process is termed as computation time, provided in Eq. (2).

$$CT = R * j * Q \tag{2}$$

The variable R represents the instructions, j represents the average cycles per instruction and Q indicates the clock cycle time.

3.2 Non-sensitive Attribute Protection Based on Shuffling Algorithm

In input data, a non-sensitive attributes are utilized for the analysis. Shuffling [12] is termed the prehistoric cryptographic process. In the shuffling process, parties re-encrypt the data and shuffle the input data into a similar amount of outcome data and also publically achieve a better performance rate in different performance observations. Furthermore, it is performed in complicated in, large shuffling circuits and $F \times F$ matrices. Here, the term F denotes the input count, and also they utilize equality proof in the exponent products.

4 Focusing on Data Integrity and Security Using Privacy Preservation Technique

4.1 Proposed Cloud Storage Model

The architecture of the designed framework is provided in Fig. 1.

Further, the data are given into the selection of optimal sensitive attributes. Here, the attributes like sensitive and non-sensitive are chosen with the help of input data. Then, the attributes are chosen optimally with the help of PFA and offered to the

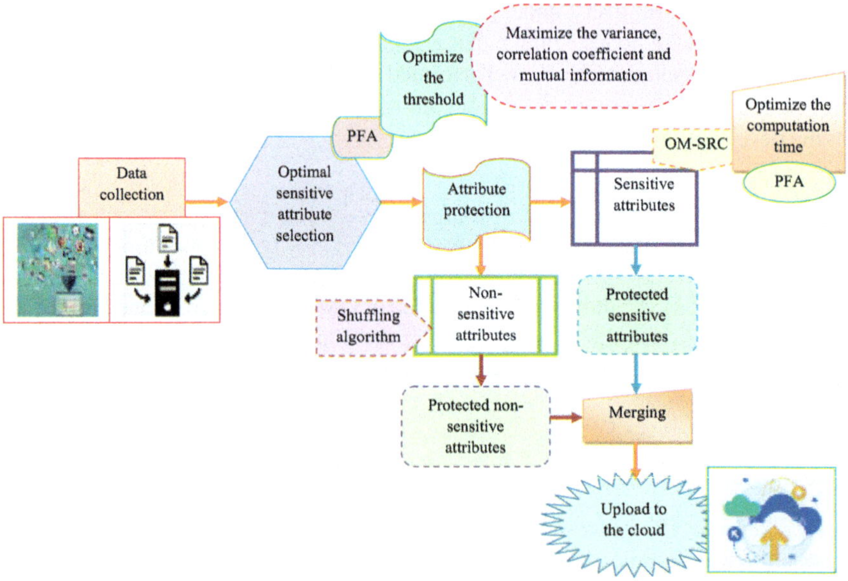

Fig. 1 Structural illustration of initiated cloud storage framework

attribute protection phase [13, 14]. Moreover, the sensitive attributes are secured using the developed OM-SRC technique. The parameter of the threshold is optimized by the PFA, which helps to improve variance, mutual information and correlation coefficient. With the help of shuffling techniques, the non-sensitive attributes are secured [15, 16]. In the phase of merging and uploading, the protected attributes of sensitive and non-sensitive attributes are given into the cloud storage system. Several experiments are done with the existing approaches to achieve an enhanced data security rate than the conventional techniques [17].

4.2 Data Collection

The data set used to validate cloud data storage is presented as Da_h^{Inp}, where $h = 1, 2, \ldots, H$ and the attained cloud data are given as H.

Data set 1: This data set is collected through the following link "https://github.com/IvanLetteri/MTA-KDD-19/find/master:Access date: 2023-07-01".

Data set 2: This data set is collected through the following link "https://datahub.io/machine-learning/lung-cancer:Access data: 2023-07-03".

4.3 Optimal Selection of Sensitive Attribute with PFA

The input data Da_h^{Inp} are utilized for choosing the optimal sensitive attributes by the PFA algorithm. In this phase, the sensitive attributes are chosen, and then, the remaining attributes are considered as the non-sensitive attributes. The objective function is suggested to maximize correlation, variance and mutual information, given in Eq. (3).

$$Ft = \underset{\left\{At_h^{optimal}\right\}}{\arg\min} \left(\frac{1}{\text{var} + cor + MI} \right) \tag{3}$$

Here, the term $At_h^{optimal}$ indicates the optimal sensitive attributes. The degree of spread in the data set over the data mean value is referred to as variance, and it is given in Eq. (4).

$$\text{var} = \frac{\sum \left(L_t - \overline{L}\right)^2}{j - 1} \tag{4}$$

The variable L_t specifies the value of a single observation. Here, the entire mean value of total observation represents as $\overline{L}$ and the total analysis is defined as j.

The relationship between the original value and the predicted value attained in the statistical observations is termed a correlation and is provided in Eq. (5).

$$cor = \frac{\sum\left(H_l - \overline{H}\right)\left(K_u - \overline{K}\right)}{\sqrt{\sum\left(H_l - \overline{H}\right)^2 \sum\left(K_u - \overline{K}\right)^2}} \tag{5}$$

Here, the values attained from h variable in the sample are offered as H_l, variable h mean values are presented as $\overline{H}$, values obtained from k variable are given as K_u and the mean values of k variable are provided as $\overline{K}$. The relationship validation among two random variables sampled similarly is termed mutual information.

PFA [18] is used to resolve the optimization issues in various architectures, and also it needs to use iteration count and population for the observation. The pathfinder method for prey detection is provided in Eq. (6).

$$Z(Q + \Delta Q) = Z^0(Q) \cdot w + j_a + j_b + \varepsilon \tag{6}$$

Here, time is indicated as Q, position vector is offered as Z, unit vector without any angle is presented as w, the pair-wise communication over the individuals y_m and y_n are given as j_a. The global force according to the global optimal location of the pathfinder is represented as j_b and the vibration vector is termed as ε. Further, the pathfinder position is renewed based on Eq. (7).

$$Z_b(Q + \Delta Q) = Z_b(Qi) + \Delta Z + B \tag{7}$$

The position vector of the pathfinder is given as Z_b, the span of the pathfinder that moves from one point to another is presented as ΔZ and the fluctuation rate of the vector is referred to as B. The main objective of the suggested model is to resolve the optimization issues and also to identify the optimum value. Modifications presented in the pathfinder model are offered in Eq. (8).

$$Z_a^{u+1} = Z_a^u + S_1 \cdot \left(Z_c^u - Z_a^u\right) + S_2 \cdot \left(Z_b^u - Z_a^u\right) + \varepsilon, \quad a \geq 2 \tag{8}$$

Here, the current iteration is offered as u, the position vector of the a-th member and c-th the member is given as Z_a and Z_c in the pathfinder model. The random vectors are presented as S_1 and S_2 in the above equation. The next modification attained in the framework is offered in Eq. (9).

$$Z_b^{u+1} = Z_b^u + 2e_3 \cdot \left(Z_b^u - Z_b^{u-1}\right) + B \tag{9}$$

The random vector produced uniformly in the bound of [0, 1] is presented as e_3.

$$\varepsilon = \left(1 - \frac{u}{u_{Mx}}\right) \cdot v_1 \cdot A_{ac} \tag{10}$$

$$A_{ac} = \|Z_a - Z_c\| \tag{11}$$

$$B = v_2 \cdot e^{\frac{-2Z}{ZMx}} \tag{12}$$

The random vectors attained in the bound $[-1, 1]$ are given as v_1 and v_2, the spans among the members are presented as A_{ac} and the numbers of iterations are referred as ZMx. Here, the term B and ε offers random movement for entire members in the limited range with the proper value. Furthermore, they modified quickly in the initial phase to offer better searching ability in the exploitation and exploration phases.

5 Experimentation of Result Analysis

5.1 Validation Set-Up

An inventive cloud storage system was designed using MATLAB 2020a. Here, a better cloud data storage rate was attained by utilizing the chromosome length as 21, the maximum iteration count as 25 and the population count as 10. Moreover, the efficacy of the developed model is computed with the conventional approaches. In the experimental observation, different optimization models like Salp Swarm Algorithm (SSA) [19], Whale Optimization Algorithm (WOA) [20], Butterfly Optimization Algorithm (BOA) [21], Grey Wolf Optimization (GWO) [22] and PFA [18] were employed.

5.2 State-of-the-Art Observation of the Designed Technique Over Data Sets

State-of-the-art validation executed using the suggested model is presented in Figs. 2 and 3.

5.3 Key Sensitive Analysis on the Implemented Framework

Key sensitive analysis executed in the recommended cloud data storage model is given in Table 1.

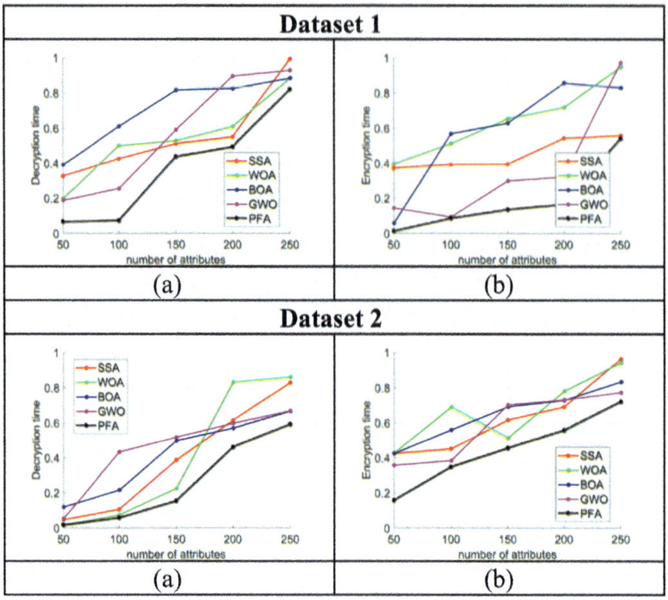

Fig. 2 State-of-the-art observation on the initiated framework over the attributes count in data set 1 and data set 2 with **a** decryption time and **b** encryption time

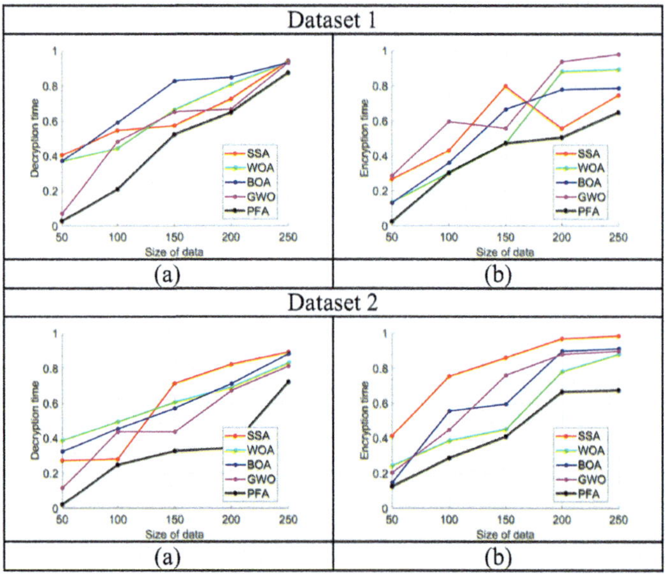

Fig. 3 State-of-the-art observation on the initiated framework over the size of data in data set 1 and data set 2 with **a** decryption time and **b** encryption time

Table 1 Key sensitive analysis on the offered mechanism in data set 1 and data set 2

Percentage (%)	SSA [19]	WOA [20]	BOA [21]	GWO [22]	PFA [18]
Data set 1					
10	0.22381	0.2551	0.1966	0.24352	0.13862
20	0.34039	0.50596	0.25108	0.25428	0.14929
30%	0.49836	0.69908	0.34998	0.81428	0.25751
50	0.58527	0.75127	0.54722	0.84072	0.47329
70	0.95974	0.8909	0.95929	0.92926	0.61604
Data set 2					
10	0.26297	0.083821	0.078176	0.084436	0.004634
20	0.60198	0.15238	0.44268	0.25987	0.10665
30	0.65408	0.39978	0.53834	0.77491	0.22898
50	0.68921	0.80007	0.82582	0.8173	0.45054
70	0.86869	0.91334	0.99613	0.9619	0.74815

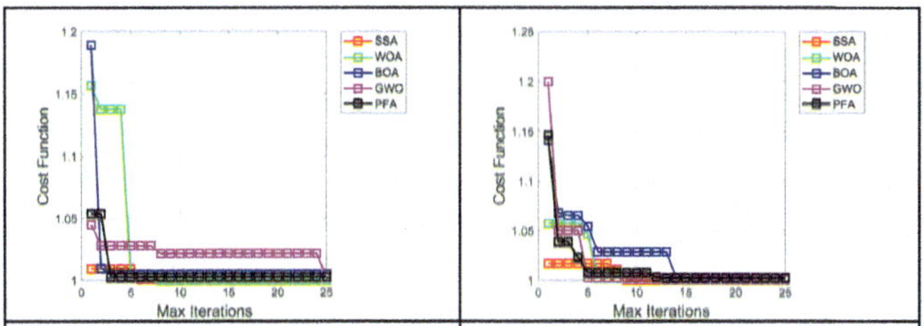

(a) (b)

Fig. 4 Analysis of convergence in the implemented framework over Heuristic models in **a** data set 1 and **b** data set 2

5.4 Analysis of Convergence in the Suggested Model

Convergence observation carried out in the recommended cloud storage mechanism over the traditional heuristic methods is represented in Fig. 4.

5.5 Statistical Validation of the Implemented Method

The experimental validation of the recommended cloud data storage model for data sets 1 and 2 is computed using the statistical analysis, which is presented in Table 2.

Table 2 Statistical examination on the recommended Cloud Data Storage system

Performance metrics	SSA [19]	WOA [20]	BOA [21]	GWO [22]	PFA [18]
Data set 1					
Standard deviation	0.003505	0.053275	0.036803	0.00671	0.014203
Worst	1.0099	1.1576	1.1896	1.0452	1.0535
Best	1.0009	1.0007	1.0055	1.0025	1.0022
Median	1.0014	1.0007	1.0055	1.0221	1.0022
Mean	1.003	1.024	1.013	1.0238	1.0063
Data set 2					
Standard deviation	0.007555	0.021872	0.032399	0.041327	0.029568
Worst	1.0178	1.0585	1.1415	1.2005	1.1454
Best	1.0013	1.003	1.004	1.0034	1.0016
Median	1.0013	1.003	1.0294	1.0034	1.0016
Mean	1.0064	1.0137	1.0273	1.017	1.0126

6 Conclusion

A novel, effective cloud data storage approach was constructed with privacy preservation techniques to improve the data preservation rate of the sensitive data in the cloud platform. Initially, the data from the benchmark resources were collected and fed to the optimal sensitive attribute selection region. Here, 20 attributes were optimally selected by the PFA technique and then fed to the attribute protection region. While considering the OM-SRC technique, the sensitive attributes are protected. Moreover, the non-sensitive attributes were protected by the shuffling technique. Further, both data were integrated and uploaded to the cloud data storage system. The analysis of decryption time is carried out using the implemented cloud storage system achieved a better performance rate as 17.3% better than SSA, 23.21% good than WOA, 47.5% more efficient than BOA and 29.50% superior to GWO over the number of attributes. Thus, the offered cloud data storage method secured a better data protection rate in the cloud storage system than the conventional mechanism.

References

1. Shen W, Qin J, Yu J, Hao R, Hu J, Ma J (2021) Data Integrity auditing without private key storage for secure cloud storage. IEEE Trans Cloud Comput 9(4):1408–1421
2. Yan H, Gui W (2021) Efficient identity-based public integrity auditing of shared data in cloud storage with user privacy preserving. IEEE Access 9:45822–45831
3. Li T, Chu J, Hu L (2023) CIA: a collaborative integrity auditing scheme for cloud data with multi-replica on multi-cloud storage providers. IEEE Trans Parallel Distrib Syst 34(1):154–162
4. Wang Q, Wang C, Ren K, Lou W, Li J (2011) Enabling public auditability and data dynamics for storage security in cloud computing. IEEE Trans Parallel Distrib Syst 22(5):847–859

5. Dhakad N, Kar J (2022) EPPDP: an efficient privacy-preserving data possession with provable security in cloud storage. IEEE Syst J 16(4):6658–6668
6. Zhang X, Zhao J, Xu C, Wang H, Zhang Y (2022) DOPIV: post-quantum secure identity-based data outsourcing with public integrity verification in cloud storage. IEEE Trans Serv Comput 15(1):334–345
7. Shao B, Bian G, Wang Y, Su S, Guo C (2018) Dynamic data integrity auditing method supporting privacy protection in vehicular cloud environment. IEEE Access 6:43785–43797
8. Yang A, Xu J, Weng J, Zhou J, Wong DS (2021) Lightweight and privacy-preserving delegatable proofs of storage with data dynamics in cloud storage. IEEE Trans Cloud Comput 9(1):212–225
9. Bian G, Fu Y, Shao B, Zhang F (2022) Data integrity audit based on data blinding for cloud and fog environment. IEEE Access 10:39743–39751
10. Liao Y, Liang Y, Oyewole AW, Nie X (2019) Security analysis of a certificateless provable data possession scheme in cloud. IEEE Access 7:93259–93263
11. Wang Z, Tamo I, Bruck J (2016) Explicit minimum storage regenerating codes. IEEE Trans Inf Theory 62(8):4466–4480
12. Peng K, Boyd C, Dawson E (2005) Simple and efficient shuffling with provable correctness and ZK privacy. In: CRYPTO 2005: advances in cryptology, pp 188–204
13. Thangavel GM, Varalakshmi P (2020) Enabling ternary hash tree based integrity verification for secure cloud data storage. IEEE Trans Knowl Data Eng 32(12):2351–2362
14. Huang P, Fan K, Yang H, Zhang K, Li H, Yang Y (2020) A collaborative auditing blockchain for trustworthy data integrity in cloud storage system. IEEE Access 8:94780–94794
15. Yue D, Li R, Zhang Y, Tian W, Huang Y (2020) Blockchain-based verification framework for data integrity in edge-cloud storage. J Parallel Distrib Comput 146:1–14
16. Narayanan E, Muthukumar B (2022) A machine learning framework for providing data integrity and confidentiality for sensitive data cloud applications. Int J Syst Assur Eng Manage
17. Yu Y, Au MH, Ateniese G, Huang X, Sus W (2017) Identity-based remote data integrity checking with perfect data privacy preserving for cloud storage. IEEE Trans Inf Forensics Secur 12(4):767–778
18. Yapici H, Cetinkaya N (2019) A new meta-heuristic optimizer: pathfinder algorithm. Appl Soft Comput 78:545–568
19. Mirjalili S, Gandomi AH, Mirjalili SZ, Saremi S, Faris H, Mirjalili SM (2017) Salp swarm algorithm: a bio-inspired optimizer for engineering design problems. Adv Eng Softw 114:163–191
20. Mirjalili S, Lewis A (2016) The whale optimization algorithm. Adv Eng Softw 95:51–67
21. Arora S, Singh S (2019) Butterfly optimization algorithm: a novel approach for global optimization. Soft Comput 23:715–734
22. Rezaei H, Bozorg-Haddad O, Chu X (2017) Grey Wolf Optimization (GWO) Algorithm

K. Brindha AP/CSE working in RVS COLLEGE OF ENGINEERING AND TECHNOLOGY, COIMBATORE. I have teaching experience of 2 years. She completed her UG AND PG Degree in Info institute of engineering and Dr. N.G.P institute of engineering and technology area if interests is cloud computing.

Computer Vision-Based Model for Classification of Citrus Fruits Diseases with Pertinent Image Preprocessing Method

Ashok Kumar Saini, Roheet Bhatnagar, and Devesh Kumar Srivastava

Abstract In the agricultural industry, diseases that harm plants are mostly responsible for the reduced profit that results in monetary loss. Nowadays, approaches based on computer vision and image processing are often used for identifying and categorizing plant diseases. The major goal of this paper is to offer an automated system for correct disease prediction at an early stage so that appropriate action can be taken at the right time and loss can be minimized because manual prediction is time-consuming. As a result, we proposed a classifier based on deep convolutional neural networks (DCNN). It has four primary steps, i.e., 1. preprocess captured images, 2. segment infected images, 3. feature extraction, and 4. classification. Firstly, preprocessing is ended to lift the image's eminence by histogram equalization. To reduce overfitting, a data augmentation approach is used to increase the dataset. As a feature extractor, the AlexNet model is then used. The numerous types of citrus illnesses are finally classified using the proposed model and is validated with baseline classifiers like Support Vector Machine and Random Forest. The proposed model gives 98.26% test accuracy, which is better than other classifiers and is very useful for farmers.

Keywords Citrus fruit · Diseases · Computer vision · Deep learning · Histogram equalization · Accuracy

A. K. Saini (✉) · R. Bhatnagar
Department of CSE, School of Computer Science & Engineering, Manipal University Jaipur, Rajasthan, India
e-mail: shksaini@gmail.com

R. Bhatnagar
e-mail: roheetbhatnagar@yahoo.com

D. K. Srivastava
Department of IT, School of Information Technology, Manipal University Jaipur, Rajasthan, India
e-mail: devesh988@yahoo.com

B. R. Devi et al. (eds.), *Proceedings of Fifth International Conference on Computer and Communication Technologies*, Lecture Notes in Networks and Systems 897,
https://doi.org/10.1007/978-981-99-9704-6_24

1 Introduction

One of the fruits that is most often grown around the world in about 137 countries is citrus. It is useful for a healthy lifestyle for human beings with its nutrients and vitamin C. It is also utilized as a raw material in many agro-industries. But for the past few years, citrus production has been highly impacted due to various diseases, e.g., melanoses, canker, black spot, scab, greasy spot, and greening[1]. Every year, a sizable fraction of high-quality exported fruit is rejected because of illnesses. So, early disease detection is very important to cut down on losses and improve the quantity and quality of production.

Currently, laboratory biochemical analysis and field analysis are the two basic diagnostic techniques for citrus diseases. Field analysis is the quickest way to diagnose citrus diseases. It is a straightforward and uncomplicated procedure that doesn't need any equipment help [2]. However, this approach is subjective, necessitates a high level of expertise, and has madly mutable accuracy. Detecting diseases using laboratories is time-consuming as labs are situated remotely from farms and transporting diseased fruits from farm to lab takes a long time. During this time, transporting infected fruits may change their characteristics. It may lead to wrong disease prediction and wrong treatment. That is harmful for consumers, nature, and financial losses as well. These techniques require considerable time and economic expenditures for big citrus orchards, which does not help promote actual agricultural productivity [3]. With the advancement of computer vision, ML, and AI technologies, and powerful computing machines and processors, it is expected to be possible to create automated models that can quickly and effectively diagnose plant leaf diseases [4]. This idea is crucial in creating, managing, sustaining, and increasing agricultural production efforts. The smart farming approach, which is recognized for incorporating new technology, algorithms, and equipment into agriculture, is at its foundation [5]. Based on how well CNN methods work at extracting features and classifying them [6, 7], we present a classifier that is based on the DCNN model.

Liu et al.'s [8] classification of citrus illnesses from photographic images made use of a deep learning model. Relevant parameters and layers in CNN play an important role in the performance of the model [9]. Our model consists of pertinent image preprocessing methods called histogram equalization, a different set of convolutional layers, and a parameter set that improves the model's accuracy. We ran the model through several iterations for testing and comparison with existing research. Run tests with multiple CNN model iterations and compare the results to yardstick studies. The proposed model is an effective method for correctly classifying citrus diseases. The paper has the following sections: The first section is constituted by an introduction. The second and third sections have an overview of similar works and the suggested CNN Model along with the dataset and data preparation, respectively. Section 4 conducts all the tests; discusses the suggested model; and further research. Section 5 concludes the paper.

2 Related Work

By combining several layers, the proposed CNN model captures complementing discriminative characteristics [9]. The dataset consisting of citrus images, both manually collected from different sources and PlantVillage datasets, was compared to other cutting-edge deep learning algorithms. With a test accuracy of 94.55%, the CNN model is a useful tool for farmers who want to categorize citrus fruit and leaf illnesses. To reduce the computation cost and parameter count in this study, we moved from regular convolution to depth-separable convolution [10]. To analyze the model's efficacy, other factors like size of the batches, dropouts, and variations in epochs were considered. Various diseases were categorized in cucumbers by Ma et al. [11] Real-time images were used for classification using DCNN. An R-CNN framework that is speedier than a region-based convolutional neural network was developed for leaf spot detection in sugar by varying the model's parameters getting accuracy of 95.48% when trained on 155 images [12]. An augmentation technique-based stochastic pooling method was used to improve the results [13]. Multiple classifiers classified the data after being used to extract features using a deep convolutional neural network (DCNN). Overall, 72% of classifications were correctly made [14].

2.1 *Identified Research Gap*

Considering the reality that basic ML and DL algorithms have been demonstrated given in Table 1 to be effective and widely utilized in crop disease prediction, but most of them had difficulty in raising classification accuracy to some extent. Furthermore, there is some performance deterioration in the neural network model owing to inadequate parameter and layer selection. In the proposed DCNN model for diagnosing citrus diseases, we employ a varied number of layers and parameter values. We also investigated other ML/DL model versions and compared the outcomes to the baseline investigations. We present a DCNN model with many layers for reliably diagnosing citrus diseases from images.

2.2 *Our Contribution*

The following contributions are made to the proposed DCNN model: (i) It employs a cutting-edge DCNN model to categorize citrus diseases into distinct classifications, including melanoses, canker, black spot, scab, greasy spot, and greening. (ii) The proposed deep learning model has an adequate number of layers, kernels, ReLU, maxpooling, batch size, epochs, learning rate = 0.01, dropout optimizer, and momentum. (iii) The proposed approach is validated using machine learning and

Table 1 Summary of the literature work

References	Year	Network Used	Dataset	Plant	Diseases	Accuracy (%)	Parameters
Khattak et al. [9]	2021	CNN	PlantVillage	Citrus	Several	95.65	Batch size, dropout, layers, filters, batch
Hassan et al. [10]	2021	InceptionV3, InceptionResNetV2, MobileNetV2,	PlantVillage	Multiple	Several	98.42, 99.11, 97.02,	Batch size, dropout, epochs
Ma et al. [11]	2018	DCNN	PlantVillage	Cucumber	Several	93.4	SGDM, learning rate, drop factor, momentum, batch, epoch
Ozguven et al. [12]	2019	R-CNN	Sugar beet leaf images	Sugar beet	Several	95.48	Batch size, dropout, epochs, learning rate
Jahanbakhshi et al. [13]	2020	CNN	341 images	Lemons	Several	100	Stochastic pooling based on sparsity
Gandhi et al. [14]	2018	CNN	PlantVillage	Multiple	Several	92	Batch size, dropout, epochs, learning rate

deep learning model that is already in existence. This model aids in the development of increasingly complex practical applications for identifying plant diseases based on visual indicators.

3 Proposed Work

The flow of our projected model is exposed in Fig. 1. In this paper, a DCNN-based classifier is projected for the classification of citrus fruits exaggerated by canker, black spot, greening, anthracnose, scab, and healthy images. This network is inspired by the AlexNet design. The projected technique entails image acquisition from the benchmarked repository and real-time images, followed by preprocessing of images. As images are in different sizes and shapes, some images are of lower quality, which impacts the classification accuracy. This step involves resizing images, contrast enhancement, and data augmentation. This class labeling and splitting of the dataset into training (80%) and testing (20%) is done.

Then the model is trained for learning. The model has five convolutional layers for extracting deep-level features. It must flatten to a single dimension vector followed by classification. This trained model is tested for appraising the concert of the projected model.

3.1 Materials and Methods

Images from both real-time captured images and the PlantVillage repository are included in the dataset. The total number of images is 3498, out of which, black spot, canker, greening, scab, anthracnose diseases, and healthy images are 1308, 1080, 160, 220, 508, and 150, respectively, were utilized. The proximity of the camera to the illness allowed for the preservation of more details. These photographs are tagged with the appropriate classes based on the category.

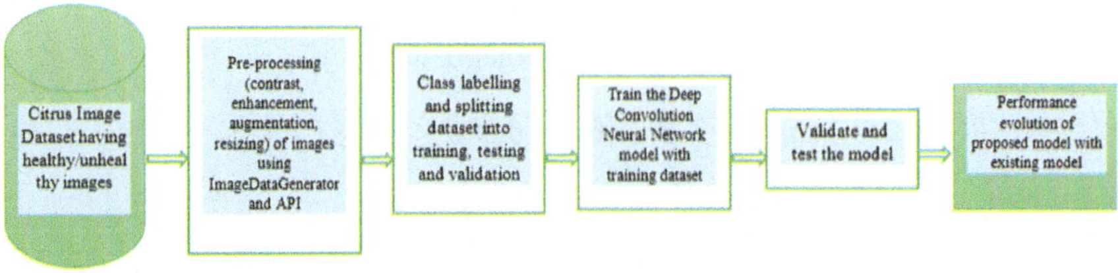

Fig. 1 Flow of our projected model

3.2 Preprocessing of Images

The picture is subjected to a median filter to reduce noise and reflections. We use the ImageDataGenerator class and API to carry out Keras image preprocessing. Firstly, enhance the contrast and rescaling by 128 × 128 of training and testing images. For this goal, two separate techniques, the central square crop technique and the histogram equalization approach for contrast enhancement, are used for the full set of the dataset. Let f's image be represented by a matrix m_r x m_c and the pixel intensities range is 0 to L-1, where L varies by 256. For each possible intensity, let p stands for the normalized histogram of f.

$$p_n = \frac{\text{the quantity of pixels with } n \text{ intensity}}{\text{total amount of pixels}} \quad \text{where } n = 0 \text{ to L - 1}$$

The equalized picture with a defined histogram is

$$g_{i,j} = \text{floor}(\text{L - 1}) \sum_{n=0}^{f_{i,j}} (p_n) \tag{1}$$

In this case, the floor () function returns the lowest possible integer. It is the same as enlarging the function by intensities of pixel k of the picture f. $T(k) = \text{floor}\Big((\text{L - 1})\Big(\sum_{n=0}^{f_{i,j}} p_n\Big)\Big)$, The idea that the strengths of f and g can be seen as random variables that change over time, with *X* and *Y* on [0, L-1] and *Y* standing for

$$\text{Y} = \text{T(X)} = (\text{L} - 1) \int_0^x p_X(x)dx, \tag{2}$$

where $p_{X,}$ represents the probability of density function of f and *T* is governed by dividing *X* times (L-1) by *X*'s cumulative distributive function. For the purpose of simplicity, let's presume that *T* is both invertible and differentiable shown in Eqs. (3, 4). This may be shown that the *Y* specified by *T(X)* is evenly dispersed across the range [0, L-1] when $pY(y) = \frac{1}{\text{L - 1}}$,

$$\begin{aligned} \int_0^y p_Y(z)\text{d}(z) &= \text{possibility that } 0 \le Y \le y \\ &= \text{possibility that } 0 \le X \le T^{-1}(y) \\ &= \int_0^{T^{-1}(y)} p_X(w)\text{d}(w) \end{aligned} \tag{3}$$

$\frac{\text{d}}{\text{dy}}\left(\int_0^y p_Y(z)\text{d}(z)\right) = p_Y(y) = p_X\ T^{-1}(\text{y})) \frac{\text{d}}{\text{dy}}\ T^{-1}(\text{y})$. Recall that $\frac{\text{d}}{\text{dy}}\ T(T^{-1}(\text{y})) = \frac{\text{d}}{\text{dy}}\ y = 1$, so

$$\left.\frac{dT}{dx}\right| x = T^{-1}(y)\frac{d}{dy}\left(T^{-1}(y)\right)\Big| = (L - 1)p_X((T^{-1}(y))\frac{d}{dy}(T^{-1}(y)) = 1, \quad (4)$$

this implies $p_y\ (y) = \frac{1}{L-1}$

Equation 1 transformation is like Eq. 2, and our discrete histogram approximates (x). Although the histograms produced by the discrete version won't be completely flat, they will be flattened, which will enhance the contrast of the picture [15]. The pictures are also scaled by Eq. (5) using the center square crop technique.

def centered_Crop(image; new_height; new_width) (5)

3.3 Deep Convolutional Neural Network

Due to high-ranking computational machines like the Graphical Processing Unit (GPU), DL-based applications that are inspired by neural networks have an exponential growth in performance [16]. The proposed architecture shown in Fig. 3 has five convolutional layers. In the first layer, the image matrix is convoluted by a filter of (3 × 3). Then the ReLU activates all neurons at the same time. After ReLU, batch normalization happened to increase the learning rate and make learning simpler. Then, maxpooling2D(3,3) is required for extracting glossy and precise features and also for reducing computations and variance. It extracts modest features like points, edges, etc. Then dropout (0.25) precludes overfitting in training data. If it is not used, then the first batch of training will influence the learning in a disproportionately positive manner.

Similarly, all rest 4 layers have their own roles and structure shown in Fig. 2. It is followed flatten, which converts a 2D array of pooled feature maps into a single extended linear vector. Finally, a dense network with a multi-class classification function called softmax is responsible for providing the class of a particular disease.

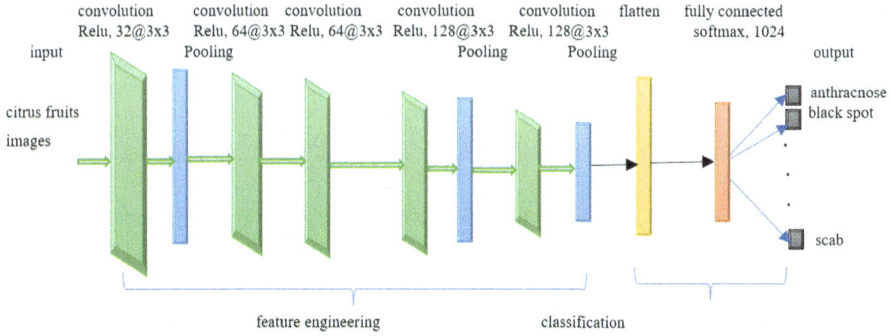

Fig. 2 Proposed deep convolutional neural network layered architecture

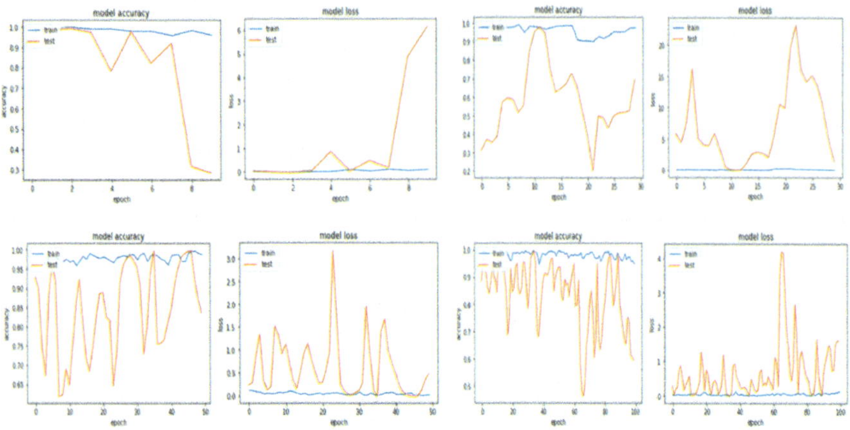

Fig. 3 Accuracy vs epochs and loss versus epochs at epoch = 10, 30, 50, and 100

The convolutional layer, which is CNN's first layer, that is responsible for feature extraction. The learnable filters and image matrices are convolved to generate the feature matrix in the convolutional layer. K and I are filters and image matrices for convolution and the formation of a function map F. The created feature map F is: $F_{\mathrm{r,c}} = R(K_{\mathrm{w,b}}\ \mathrm{o}\ I_{\mathrm{r+w-1}}, c + h - 1 + b)$, where r and c represent matrix rows and columns, h and w are filter height and width, o is convolution function, R is the activation function ReLU, I is the filter's maximum width and height, and b is the bias value[8]. Classification: The model estimates the probability of citrus fruit diseases class by changing the dense layer with various neurons and employing the softmax function as "add (Activation("softmax"))". Consequently, $u_j = \sum_i^{l} w_i x_i + b$, where x, b, and w are the input vector, bias, and weight vector, respectively.

4 Results and Discussion

The TensorFlow framework and Python were used to execute the proposed model's training and testing methods. The training process was conducted using the CUDA framework with an NVIDIA GTX1080 card's GPU. In this test, Windows 10, 8 GB of RAM, a 500 GB hard drive, and an Intel i5 7-700 CPU with 3.60 GHz were used.

Evaluating the proposed network's accuracy during both its training and testing phases. Evaluating outcomes in comparison to other cutting-edge techniques. Figure 3 demonstrates the correctness and damage of the proposed model. Accuracy and loss vary according to the number of epochs (Table 2, Fig. 4). The model is trained at a different number of epochs, as shown in Fig. 7. For example, epoch = 10, 30, 50, 70, 100 the accuracy = 97.33%, 96.21%, 97.96%, 97.86%, 98.26%, and loss = 5.9% 10.80%, 5.93%, 5.21%, 5.21%, respectively. At epoch = 100, the

Table 2 Analysis of accuracy vs loss at epoch = 10, 30, 50, 70, and 100 font

No of epoch	Accuracy	Loss
10	0.9733	0.05900
30	0.9621	0.10802
50	0.9796	0.05936
70	0.9786	0.05210
100	0.9826	0.05210

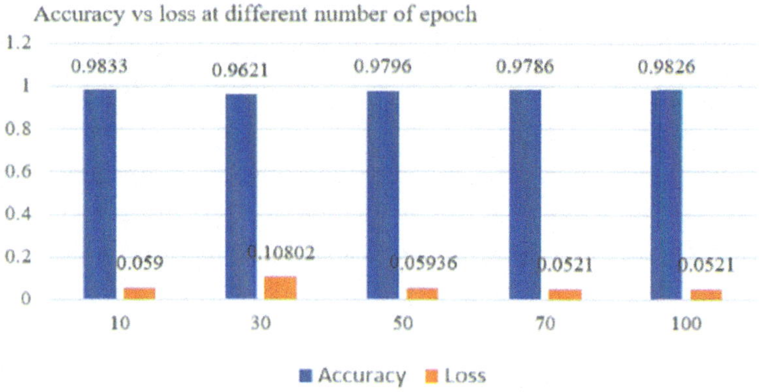

Fig. 4 Accuracy versus loss (%)

Table 3 Performance measure of SVM, RF and proposed model

Model	SVM (%)	RF(%)	Proposed (%)
Accuracy	92.8	94.8	98.26
Loss	6.49	6.49	5.21
Missing report rate	7.3	5.2	1.74
False report rate	2.45	2.45	0

accuracy is maximum, and loss is minimum. SVM's highest accuracy is 92.8% and RF's accuracy is 94.8% (Table 3, Fig. 5).

5 Conclusion and Future Work

The proposed DCNN model classifies various citrus diseases and healthy images. It has the following steps: data acquisition, data preprocessing, feature extraction, and classification. The proposed model used histogram equalization for enhancing the image's contrast, data augmentation for increasing the dataset by transforming images in various ways, and used multiple convolutional layers for extracting features for improving classification accuracy. The first layer extracts low-level features, and

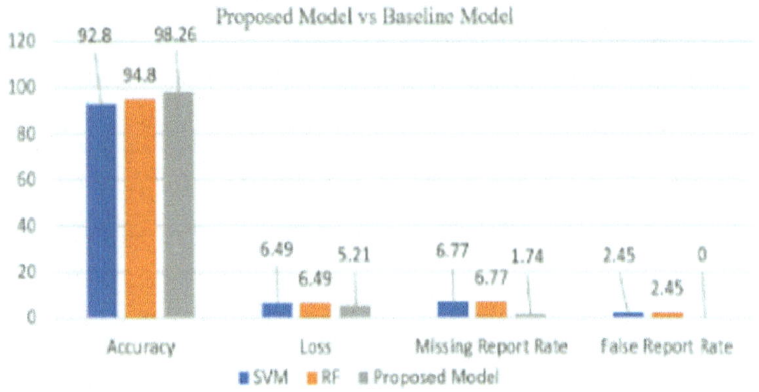

Fig. 5 Comparison of performance (%) of SVM, RF, and proposed model

the last four layers gather the high-level features from the images. We investigated SVM, RF, and proposed models for classification at different epochs. The best accuracy of the proposed model among various sets of epochs is 98.26% at epoch = 100, which is the highest among other models. The proposed model is compared with the SVM and RF classifiers, having accuracies of 92.8% and 94.8%, respectively. In the future, we would like to work on large datasets with specific diseases and incorporate new technologies like pruning and parameter quantization with our framework.

References

1. Dobbert JP (2011) Food and agriculture organization of the United Nations. Chem Environ Heal 309–324. https://doi.org/10.1201/b11064-23
2. Hall DG, Shatters RG, Carpenter JE, Shapiro JP (2010) Research toward an artificial diet for adult Asian citrus psyllid. Ann Entomol Soc Am 103(4):611–617. https://doi.org/10.1603/AN10004
3. Graham LL et al (2017) Abstracts of presentations at the 5th international research conference on Huanglongbing. J Citrus Pathol 4(1). https://doi.org/10.5070/c441034714
4. Ferentinos KP (2018) Deep learning models for plant disease detection and diagnosis. Comput Electron Agric 145:311–318. https://doi.org/10.1016/j.compag.2018.01.009
5. Singh UP et al (2019) Multilayer convolution neural network for the classification of mango leaves infected by anthracnose disease. IEEE Access 7:43721–43729. https://doi.org/10.1109/ACCESS.2019.2907383
6. Luaibi AR et al (2021) Detection of citrus leaf diseases using a deep learning technique. Int J Electr Comput Eng 11(2):1719–1727. https://doi.org/10.11591/ijece.v11i2.pp1719-1727
7. Doh B et al (2019) Automatic citrus fruit disease detection by phenotyping using machine learning. In; ICAC 2019—2019 25th IEEE international conference automation computing. https://doi.org/10.23919/IConAC.2019.8895102
8. Liu Z et al (2021) Image recognition of citrus diseases based on deep learning. Comput Mater Contin 66(1). https://doi.org/10.32604/cmc.2020.012165
9. Khattak A et al (2021) Automatic detection of citrus fruit and leaves diseases using deep neural network model. IEEE Access 9:112942–112954. https://doi.org/10.1109/ACCESS.2021.3096895

10. Hassan S et al (2021) Identification of plant-leaf diseases using CNN and transfer-learning approach. Electronics 10(12). https://doi.org/10.3390/electronics10121388
11. Ma J et al (2018) A recognition method for cucumber diseases using leaf symptom images based on deep convolutional neural network. Comput Electron Agric 154:18–24. https://doi.org/10.1016/j.compag.2018.08.048
12. Ozguven MM et al (2019) Automatic detection and classification of leaf spot disease in sugar beet using deep learning algorithms. Phys A Stat Mech Appl 535. https://doi.org/10.1016/j.physa.2019.122537
13. Jahanbakhshi A et al (2020) Classification of sour lemons based on apparent defects using stochastic pooling mechanism in deep convolutional neural networks. Sci Hortic (Amsterdam) 263. https://doi.org/10.1016/j.scienta.2019.109133
14. Gandhi R et al (2018) Plant disease detection using CNNs and GANs as an augmentative approach. In: Proceedings of IEEE international conference on innovative research development (ICIRD), Bangkok, Thailand, pp 1–5. https://doi.org/10.1109/ICIRD.2018.8376321
15. Gajanand S et al (2022)Cognitive framework and learning paradigms of plant leaf classification using artificial neural network and support vector machine. J Experimental Theor Artif Intell 1–26
16. Muhammad W et al (2020) An introduction to deep convolutional neural networks with keras. Mach Learn Deep Learn Real-Time Appl 231–272. https://doi.org/10.4018/978-1-7998-3095-5.ch011

FLAML-HDPS Model: An Efficient and Intelligent AutoML Approach for Heart Disease Prediction

P. Deepan, R. Vidhya, B. Rajalingam, R. Santhoshkumar, and N. Arul

Abstract Heart health is vital for the survival of all living things. Heart-related illnesses require more precision, accuracy, and correctness in diagnosis and prognosis since even a small mistake might have fatal consequences. Each year, numerous people die from heart-related causes, and this figure is sharply rising. As a first step toward solving the problem, a disease prediction system and more public awareness are required. An increasing number of people are suffering from heart disease every minute, and it has now become a major public health issue. A complicated mix of medical and pathological data is frequently required for the diagnosis of cardiac disease which results in significant medical costs, which have an impact on the standard of care. To make predictions about heart disease based on medical information, this study was undertaken. Using Fast and Lightweight AutoML (FLAML), we analyzed a large dataset with 14 different categories, such as age, gender, medical measures, and test results. Machine learning algorithms, hyperparameters, and model ensembles were all selected, regulated, and optimized automatically by the FLAML framework. High levels of accuracy, precision, recall, and ROC-AUC in our heart disease prediction models were generated by using FLAML. FLAML's streamlined process saves doctors and nurses precious minutes every day, paving the way for more accurate risk assessments and more effective treatment of heart disease thanks to the power of predictive analytics.

Keywords AutoML · FLAML · Optimization · Predictive techniques · Data mining · Clinical data and machine learning

P. Deepan (✉) · N. Arul
St. Peter's Engineering College, Hyderabad 500043, India
e-mail: deepanp87@gmail.com

B. Rajalingam · R. Santhoshkumar
St. Martin's Engineering College, Hyderabad 500100, India

R. Vidhya
B.S. Abdur Rahman Crescent Institute of Science and Technology, Chennai 600048, India

B. R. Devi et al. (eds.), *Proceedings of Fifth International Conference on Computer and Communication Technologies*, Lecture Notes in Networks and Systems 897,
https://doi.org/10.1007/978-981-99-9704-6_25

1 Introduction

Heart diseases are a significant challenge to the human community worldwide leading to high mortality rates. An accurate and timely diagnosis of heart disease is crucial for taking earlier treatments and reducing risks. In the recent times, Machine Learning (ML) techniques have been used in predicting heart diseases, but the challenges in complexity of model selection, feature engineering, and hyperparameter tuning make it difficult for non-experts of the data science field to opt for this powerful technology. In the recent years, predictive modeling using machine learning algorithms has emerged as a promising approach to assess an individual's risk of developing heart disease. By analyzing large datasets encompassing various risk factors, lifestyle habits, and medical history, these models can uncover hidden patterns and provide insights into factors that contribute to heart disease development. This predictive capability holds tremendous potential for enhancing clinical decision-making and guiding personalized prevention strategies.

To overcome these issues and develop an accessible heart disease prediction system, an AutoML comes as a promising solution. This AutoML allows the non-experts to apply ML effectively and at ease. AutoML automates and streamlines the entire process of prediction starting from data preprocessing, hyperparameter tuning, model evaluation, model selection, and result evaluation. These complex and time-consuming tasks are automated and benefit a wider range of stakeholders. AutoML results in efficient and intelligent predictions, promoting enhanced disease prediction outcomes. In this research paper, we implement a FLAML-based heart disease prediction technique which leads to an efficient and robust prediction model.

The following sections in this work are organized as follows: Important efforts on heart disease prediction systems are summarized in Sect. 2. The developed approach is described in Sect. 3, which is based on the Fast and Lightweight AutoML methods. In Sect. 4, we detail the methodology that was used and the findings from the experiments. The final conclusion of the research work is discussed in Sect. 5.

2 Literature Works

The ML models' hyperparameter selection and optimization are automated in the study using AutoML (AML) techniques [1]. With AML, manual adjustment and professional assistance are less necessary, making it easier for non-experts to use. The report does not go into detail on the exact AML methods applied in the study. The importance of ML for predicting heart disease in the medical field is highlighted in this research [2]. The accuracy of patient diagnosis and treatment may be improved by using machine learning algorithms to analyze massive amounts of medical data, identify patterns, and derive meaningful information. Given the considerable impact of cardiovascular illnesses on public health, the research underlines the need for precise heart disease prediction. The authors emphasize the need for reliable and effective forecasting techniques that can take use of healthcare data's ability to improve diagnostic precision [3]. The article [4] probably highlights the significance of early identification in reducing the severity and effect of heart disease, which is a complicated global source of illness leading to death. Early intervention can greatly enhance patient outcomes while lowering medical expenses.

In order to effectively forecast the likelihood of heart illness based on input data, the study in [5] outlines the construction of a system for predicting heart diseases employing convolutional neural networks. The article [6] makes recommendations on how to address these issues and achieve minimal latency while delivering a significant number of AutoML models. It probably includes methods for parallelizing, load distribution, and other performance-improving approaches, as well as strategies for improving the model serving infrastructure. The proposed [7] domain-customized AutoML for networking, along with potential use cases and experimental findings to support their claims, are likely to be covered in more detail in the paper. The methodology and findings pertaining to interpretable feedback for AutoML are also likely to be covered in greater detail. This strategy [8] enables the system to sustain strong performance throughout time, especially in surroundings that change quickly. The system includes techniques for automatically developing features, choosing models, and tweaking hyperparameters, which are updated in response to newly observed data.

In an effort to make machine learning more accessible by increasing the process' automation and efficiency, the authors [9] examine the ideas of meta-learning for transfer learning and AutoML. The authors [10] perform tests utilizing standard datasets from OpenML, a system that offers the use of an extensive variety of ML datasets, to assess the capability of AutoML. On these datasets, they contrast the results achieved by people and AutoML models in terms of performance. The usefulness of AutoML in contrast to models created by humans is assessed using a variety of metrics and performance measurements.

3 Proposed Methodologies

We propose a novel FLAML-based heart disease prediction system to predict the heart-related diseases in advance. The generic workflow of the model is depicted in the following sections.

3.1 Dataset Description

Heart_disease.csv is a publicly available benchmark dataset which is cited from the kaggle repository. This dataset comprises information regarding age, blood pressure, cholesterol, and whether or not heart disease exists, as indicated in Table 1. The variables in the dataset allow us to forecast an individual's probability of developing heart disease and identify risk factors. The 'target' variable represents heart disease of a patient denoted by a binary variable. The integer variable '0' denotes 'No Heart disease' and '1' denotes 'Heart Disease'.

Table 1 Dataset description representing heart disease

S. No	Variable name	Type	Explanation	Measure/Range
1	Age	Numerical	Age of the individual at data collection	Age-29 to Age-65
2	Sex	Nominal	Gender of the individual	Male is 1 Female is 0
3	Cp	Nominal	Description of chest pain experienced	Typical angina is 1, atypical angina is 2, non-anginal pain is 3, asymptomatic is 4
4	trestbpd	Numerical	Resting blood pressure in mmHg	92–200
5	serumCho	Numerical	Cholesterol level in mg/dL	126–564
6	fbs	Nominal	Fasting blood sugar level in mg/dL	Yes is 1, No is 0
7	restecg	Nominal	Results of the resting electrocardiogram	Normal is 0, having ST-T wave abnormality is 1, showing probable or definite left ventricular hypertrophy is 2
8	thalach	Numerical	Highest heart rate achieved during exercise testing	82–185
9	exang	Nominal	Presence of angina during exercise	Yes is 1, No is 0
10	oldpeak	Numerical	ST segment changes during exercise	71–202
11	peakSlope	Numerical	Slope of the ST segment during exercise	1–3
12	numVessels	Numerical	Count of major blood vessels with fluoroscopy abnormalities	0–3
13	thal	Nominal	Results of the thallium stress test	3 is normal, 6 is fixed defect, 7 is reversible defect
14	Disease	Nominal	Existence (1), Absence (0)	Yes is 1, No is 0

3.2 Workflow of Fast and Lightweight AutoML

Fast and Lightweight AutoML (FLAML) is an automated machine learning framework that eliminates the need for considerable human configuration in order to facilitate the construction of predictive models in a quick and effective manner. The following activities are commonly included in the workflow of FLAML for the prediction of cardiovascular disease:

Data Preprocessing: In the process of data analysis and machine learning, data preprocessing plays a crucial role. Data preprocessing is the process of converting unprocessed data into a format that may be used for further analyses and model training. There are several preprocessing methods, including data gathering, data cleaning, feature engineering, feature selection, and many more. Data cleaning means the process of handling missing values, eliminating duplicates, and addressing any data inconsistencies.

Splitting the Dataset: The heart_disease.csv dataset is contains 376 records of patient information. From the dataset, 75% of the dataset is used for the training (282 records) and the remaining 25% is used to predict heart disease (94 records). The 282 records are trained by using Fast and Lightweight AutoML optimization algorithms.

Fast and Lightweight AutoML Configuration: The architecture block diagram based on Fast and Lightweight AutoML for the proposed work is illustrated in Fig. 1. FLAML is an automated ML framework including several optimization algorithms and efficient heuristic search techniques for tuning the ML model. The working sequential flow of FLAML begins with specifying the objective function $f(x)$. x is the set of hyperparameters for the proposed machine learning model. Each x belongs to the hyperparameter space (Θ).

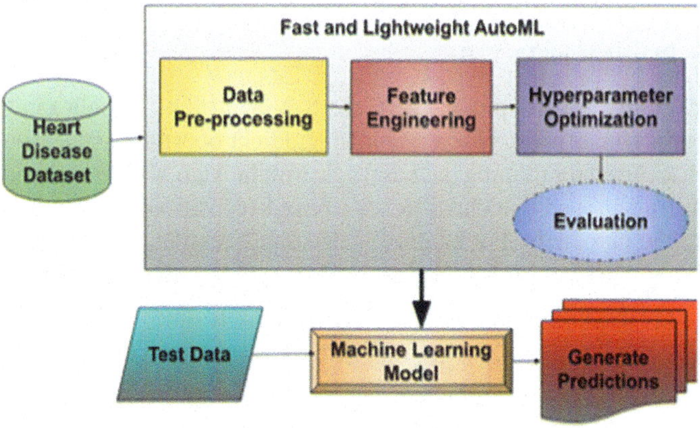

Fig. 1 FLAML architecture block diagram

Therefore, it can be represented as $x \in \Theta$. Then, the FLAML model selects the best model and the hyperparameter configuration based on the proposed objective. Hence, we denote it as $f(x^*)$. The optimal hyperparameters are found from the hyperparameter space through various optimization algorithms and heuristics approaches. FLAML combines the Bayesian optimization with Tree-based search (TPE) and results in enhanced optimization. The objective function is modeled using Bayesian optimization, representing the hyperparameters x, using Gaussian processes. Using Bayesian optimization with techniques like Expected Improvement (EI) acquisition function, it explores the optimal set of hyperparameters as represented in Eq. (1).

$$\mathrm{EI}(X) = \max(0, \mu(x) - \eta) * \phi(\mathrm{z})\sigma + (x) * \varphi(z) \tag{1}$$

where $\mu(x)$ represents posterior distribution, η represents best found performance metrics value, and $\phi(z)$ and $\varphi(z)$ represent Cumulative Distribution Function (CDF) and Probability Distribution Function (PDF). The standard normalized difference between mean $\mu(x)$ and best performance is denoted as shown in Eq. (2).

$$Z = ((x) - \eta)/\sigma(x) \tag{2}$$

TPE explores the hyperparameter space and suggests the new set of hyperparameters for evaluation. There are two different probability density functions one for Good (A) and other for Bad (B) regions in the hyperparameter space. The probability of improvements (PI) for each set of hyperparameters in the dataset (D) is calculated as shown in Eqs. (3) and (4).

$$\mathrm{PI}(x\_\mathrm{A}) = \mathrm{P\_A}(x\_\mathrm{A}|\mathrm{D})/(\mathrm{P\_A}(x\_\mathrm{A}|\mathrm{D}) + \mathrm{P\_B}\ (x\_\mathrm{A}|\mathrm{D})) \tag{3}$$

$$\mathrm{PI}(x\_\mathrm{B}) = \mathrm{P\_B}(x\_\mathrm{B}|\mathrm{D})/(\mathrm{P\_A}(x\_\mathrm{B}|\mathrm{D}) + \mathrm{P\_B}\ (x\_\mathrm{B}|\mathrm{D})) \tag{4}$$

The hyperparameters with the higher PI value are chosen to maximize the PI by comparing PI(x_A) and PI(x_B).

Train the Model: In this phase, there are five classifiers, namely random forest, logistic regression, support vector machine, gradient boosting, and k-nearest neighbors which are used for training the Fast Lightweight Auto Machine Learning model. In the above-mentioned five classifiers, we have used various parameter settings for fine-tuning the classifiers. For example, logistic regression, gradient boosting, and k-nearest neighbor classifiers used various parameters settings as mentioned below Table 2.

Table 2 Parameter fine-tuning for classifiers

S. No	Classifiers	Parameter	Parameter_Values
1	Logistic regression	'C'	{0.1, 1.1, 0.5}
2	Logistic regression	'Penalty'	{'l1', 'l2'}
3	K-nearest neighbors	'weights'	{'uniform', 'distance'}
4	K-nearest neighbors	'Algorithm'	{'auto', 'ball_tree', 'kd_tree', 'brute'}
5	Gradient boosting	'Learning rate'	{'0.1', '1.0', '1.1'}
6	Gradient boosting	'C'	{0.1, 1.1, 0.5}

4 Experimental Results and Analysis

Traditional ML-based heart disease prediction systems have a hard time finding the most accurate and generalizable predictive model because of the manual and time-consuming processes of feature engineering, hyperparameter tuning, and algorithm selection. These processes require domain expertise and extensive experimentation to achieve optimal model performance. In order to handle this issue, we have to use the Fast and Lightweight AutoML (FLAML) prototype for predicting heart disease. The prototype model is an automated machine learning tool designed to optimize the machine learning pipeline efficiently. The prototype automates the process of hyperparameter tuning, model selection, feature selection and reduces the human effort as well as reduces the time consumption. The workflow of automated FLAML-based heart disease prediction systems is given in Table 3 (Fig. 2).

This FLAML method has selected five classifiers, namely random forest, logistic regression, support vector machine, gradient boosting, and k-nearest neighbors. As shown in Fig. 3 and Table 4, when comparing the accuracy, precision, recall, and F1-score of several FLAML-optimized models for heart disease prediction, k-nearest

Table 3 Algorithm for heart disease prediction using FLAML

Algorithm for heart disease prediction using FLAML
Input: Patient Information (Like age, Bp, sex, trestbpd) **Output**: Heart disease prediction (Yes or No) 1. Load the Dataset 2. Perform the pre-processing operation 3. Define the model 4. If the model is good then 5. Goto Tool 6. Else 7. Try with another model selection 8. Tool: compile the model and predict the test result 9. If the test results are not good then fine tune the parameters of the model. (Repeat until all the possibilities) 10. Deploy the model 11. Final prediction results

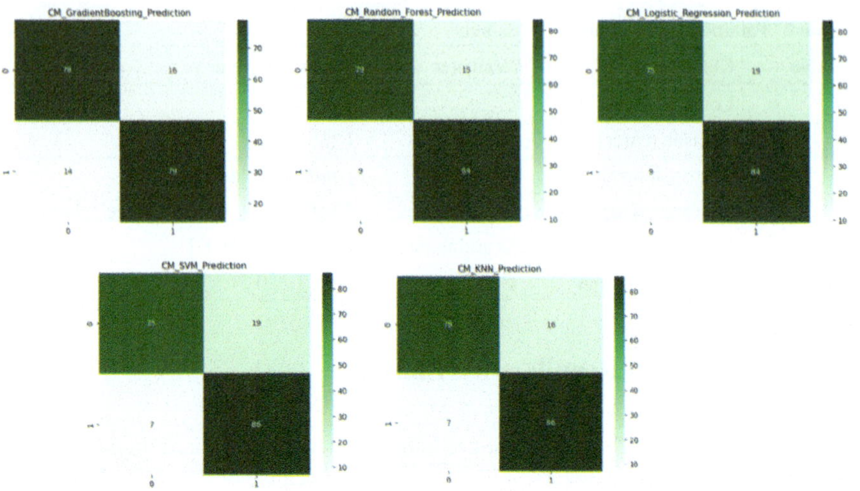

Fig. 2 FLAML confusion matrix of various classifiers (4a) gradient boosting, (4b) random forest, (4c) logistic regression, (4d) support vector machine, and (4e) k-nearest neighbors

neighbors came out on top with an accuracy of 87.70%, precision of 84.31%, recall of 92.4%, and an F1-score of 88.21%. Random forest came in second with an accuracy of 86.63%, precision of 85.42%, recall of 88.17%, and an F1-score of 86.77%. The accuracy of heart disease prediction achieved by using gradient boosting classifiers was just 83.96%, which was the lowest achieved by all of the five classifiers.

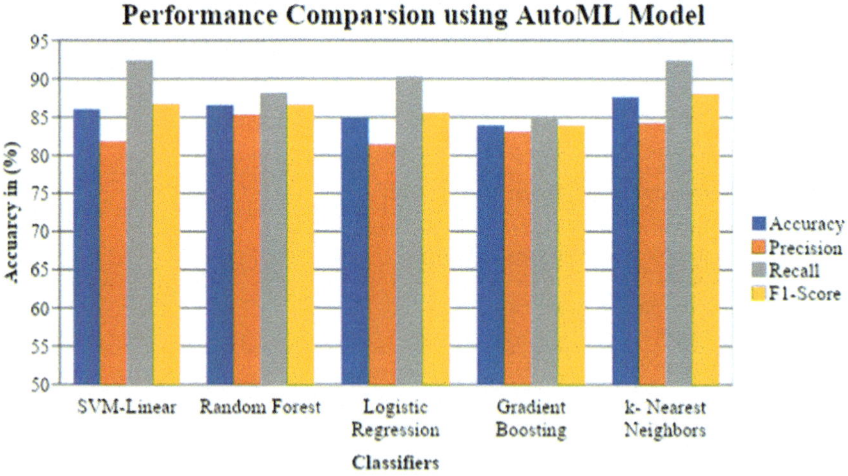

Fig. 3 Performance comparison of all the FLAML classifiers

Table 4 Performance metrics of AutoML model

S. No	Classifiers	Accuracy	Pr	Re	F1-Sc
1	SVM-linear	86.10	81.90	92.47	86.87
2	Random forest	86.63	85.42	88.17	86.77
3	Logistic regression	85.03	81.55	90.32	85.71
4	Gradient boosting	83.96	83.16	84.95	84.04
5	K-nearest neighbors	87.70	84.31	92.47	88.21

5 Conclusions

In conclusion, the heart disease prediction system using Fast and Lightweight AutoML (FLAML) demonstrates its efficacy as an automated machine learning tool for this critical medical task. The FLAML model automates the machine learning process, which allows for the optimization of hyperparameters, selection of the models that perform the best, and identification of the features that are most relevant. This automated process not only saves valuable time and computational resources but also ensures that the predictive model is well-tuned and capable of making accurate predictions. Experimental results and analysis showed that FLAML can outperform conventional manual machine learning methods for predicting heart disease. When compared to the manual traditional ML models, the FLAML-optimized model performed better in terms of computational time as well as accuracy.

References

1. Chatterjee D, Chandran S (2019) Prediction and classification of heart disease using AML and power BI. In: Proceedings of the 11th international joint conference on computational intelligence. https://doi.org/10.5220/0008381505080515
2. Nikhar S, Karandikar AM (2023) Prediction of heart disease using machine learning with data mining. Phys Sci Biophys J 7(1):1–6
3. Pati A, Parhi M, Pattanayak BK (2022) IHDPM: an integrated heart disease prediction model for heart disease prediction. Int J Med Eng Inf 14(6):1–12
4. Shuvo S (2023) An early detection of heart disease using machine Learning (recurrent neural network). https://doi.org/10.14293/s2199-1006.1.sor-.ppgv4id.v1
5. Varkala K (2022) Heart disease prediction system using convolutional neural networks. https://doi.org/10.21203/rs.3.rs-2009078/v2
6. Agarwal M (2020) Serving very large numbers of low latency AutoML models. In: 2020 IEEE infrastructure conference. https://doi.org/10.1109/ieeeconf47748.2020.9377614
7. Arzani B, Hsieh K, Chen H (2021) Interpretable feedback for AutoML and a proposal for domain-customized AutoML for networking. In: Proceedings of the twentieth ACM workshop on hot topics in networks. https://doi.org/10.1145/3484266.3487373
8. Celik B, Singh P, Vanschoren J (2022) Online AutoML: an adaptive AutoML framework for online learning. Mach Learn 112(6):1897–1921. https://doi.org/10.1007/s10994-022-06262-0
9. Dai W (n.d.) Lower the barrier of machine learning: meta learning for transfer learning and automl. https://doi.org/10.14711/thesis-991012010569503412

10. Hanussek M, Blohm M, Kintz M (2020) Can AutoML outperform humans? An evaluation on popular OpenML datasets using AutoML benchmark. In: 2020 2nd international conference on artificial intelligence, robotics and control. https://doi.org/10.1145/3448326.3448353

An Efficient Selective Features Approach to Detect Hypothyroid Using Machine Learning

N. Subhash Chandra, Srinivasa Rao Dhanikonda, Dhanamma Jagli, and Nalla Siddhartha

Abstract The thyroid gland is a hormone-producing gland located in the upper part of the neck. It helps to regulate the metabolism of the body. Its malfunction might lead to either inadequate or excessive thyroid hormone production. Thyroid issues are classified as hyperthyroidism or hypothyroidism. Predicting thyroid illness is now more crucial than ever. Machine learning algorithms and data mining techniques are critical for dealing with today's massive amounts of data and information, particularly in the health-care system. A specific dataset is created that contains patient-related medical data. Based on the attributes considered, it is possible to categorize thyroid disease into three categories: primary hypothyroid, compensated hypothyroid, and routine. To identify earlier, various conventional machine learning methods are utilized, including Random Forest, Decision Tree, and Artificial Neural Network (ANN). These algorithms carry out the classification and determine the status but fall short of the desired results, particularly in the medical sector. In order to accomplish the goal, this research developed an effective classification strategy to improve the performance of hypothyroid prediction and it is divided into two stages: data preprocessing and classification methods. During the preparation phase, the dataset was preprocessed using two methods: PCA and an Extra Tree classifier, and the data was enriched with the Synthetic Minority Oversampling Technique (SMOTE). The proposed approach's outcomes were compared to traditional methods.

N. S. Chandra
Department of Computer Science and Engineering, CVR College of Engineering, Hyderabad, India
e-mail: subhashchandra@cvr.ac.in

S. R. Dhanikonda (✉)
Department of Information Technology, BVRIT HYDERABAD College of Engineering for Women, Hyderabad, India
e-mail: srinivasarao.dhanikonda@gmail.com

D. Jagli
V.E.S.Institute of Technology, University of Mumbai, Mumbai, India

N. Siddhartha
Indian Institute of Technology Tirupati, Tirupati, India

B. R. Devi et al. (eds.), *Proceedings of Fifth International Conference on Computer and Communication Technologies*, Lecture Notes in Networks and Systems 897,
https://doi.org/10.1007/978-981-99-9704-6_26

Keywords Decision Tree · Random Forest · PCA · ANN · SMOTE

1 Introduction

Thyroid hormones, primarily triiodothyronine (T3) and thyroxine (T4), are secreted by the thyroid gland. These hormones are in charge of metabolism, sleep, growth, sexual function, and mood. They play an important role in the body's energy levels, influencing children's growth and development, metabolic rate, and protein synthesis. Hypothyroidism can result in health issues such as high cholesterol and heart problems. Cold sensitivity, constipation, dry skin, and weight gain are all common symptoms of hypothyroidism.

Thyroid conditions are becoming more prevalent in India, according to statistics. One in ten adults in India experience thyroid issues. Thyroid illness is thought to affect 42 million people worldwide. Thyroid Stimulating Hormone (TSH), T3 and T4 hormones, is used by doctors to assess whether a patient has thyroid issues. A person may have hypothyroidism if their TSH levels are elevated. An individual may have hyperthyroidism if their TSH level is low. Depending on the type of thyroid condition a person has, their T3 and T4 levels may also change. Higher TSH levels do not, however, guarantee that a person has hypothyroidism. TSH levels in our bodies might rise due to factors such as iodine shortage, radiation therapy, certain poisons, medicines, or supplements such as arsenic, opioids, dopamine inhibitors, and so on. As a result, determining whether a person has thyroid issues becomes extremely difficult.

Machine learning can help doctors diagnose ailments and minimize their workload. Machine learning is critical in determining whether or not a person has the disorders. It can assist us forecast whether or not the person is hypothyroid. As a result, machine learning can help clinicians diagnose ailments and minimize their workload.

The remainder of this paper is divided into sections. Section 2 discusses cutting-edge research on detecting and classifying thyroid disorders. Section 3 discusses the recommended methodology for addressing the problem of thyroid illness prediction. Section 4 describes and compares the experimental results to traditional methods. The contributions are summarized in Sect. 5.

2 Literature Review

In their work, Chaganti et al. [1] introduced a method that explores feature engineering techniques for both machine learning and deep learning models. The adoption of forward feature selection and backward feature deletion. The suggested method foresaw several thyroid conditions. They developed models using deep learning techniques like CNN as well as a variety of machine learning methods like Random Forest,

Logistic Regression, SVM, ADA, and GBM. K-fold cross-validation was used for evaluation. The study is limited by the small feature set that remains after the key features have been chosen. As a result, there were fewer training occurrences, which in turn decreased the deep learning model's dependability.

Based on patient medical data, Aversano et al. [2] estimated the treatment trend of thyroid disease. Patients' hormonal markers, such as LT4, were considered, and the therapy trend for hypothyroid patients was projected. For classification, they used a variety of techniques, including ensemble algorithms, boosting algorithms, and lazy learners. The K-cross-validation technique was used for validation, and models were evaluated based on metrics like as accuracy, $F1$-score, recall, and precision. The key constraint of this study is the dataset's quality. The required target classes were not equally represented since the original dataset was excessively unbalanced. As a result, minority classes had extremely few cases for training, resulting in discrepancies during prediction.

Alyas et al. [3] suggested a system for determining whether or not a person has thyroid issues. A dataset is utilized as input, and models are built using the KNN, ANN, Naive Bayes, and Random Forest algorithms. There were only two target classes for positive and negative classifications, 1 and 0, respectively, which was a significant limitation. Another problem of this methodology is the little amount of data available for training deep learning models. As a result, deep learning models are not very trustworthy.

Rajulislam et al. [4] suggested a technique for predicting hypothyroidism in its early stages. Three feature selection strategies are used, as well as various categorization techniques. RFE, UFS, and PCA were used to select features. Following feature selection, those features were utilized to assess the performance of each algorithm independently. SVM, Decision Tree, Random Forest, Logistic Regression, and Naive Bayes were the methods employed. The primary restriction of this study is the magnitude of the dataset. The dataset was extremely small, with only 519 cases. As a result, the models created are untrustworthy.

Chandan et al. used multiple supervised machine learning classifier approaches to analyze the blood test dataset. The hyperthyroidism dataset was used, with the labels hyper and hypo. To prepare the data for classification, it was preprocessed. To sanitize the data, null data and superfluous data were removed. Cleaned data was utilized as training and test data for the algorithm, which was provided as input. Models were created using Decision Tree, KNN, ANN, SVM, and Logistic Regression techniques. The size of the data sample was a major drawback in this case. It had 5 features and 215 instances. This makes it impossible to train a deep learning model. To adequately train the model for performance evaluation, a sizable dataset is needed [5].

With this literature review, it is concluded that, there is a need of research to address these limitations.

3 Research Methodology

The proposed strategy is divided into two phases, with two distinct strategies used in the first phase. The data's features are chosen using a classifier called an Extra Tree and Principal Component Analysis. Use the Borderline SMOTE approach for data augmentation to balance the dataset. Apply several machine learning algorithms for categorization in the final phase.

3.1 Principal Component Analysis

Principal Component Analysis (PCA) is a statistical method used to transform correlated features into linearly uncorrelated Principal Components (PCs). By reducing the dimensionality of the data while retaining the most important variance, PCA helps to find essential patterns in the dataset. It seeks a lower-dimensional subspace that preserves the maximum variance when projecting high-dimensional data. The PCs are orthogonal, meaning they have no correlation and represent independent patterns. Figure 1 gives the representation of PCA and it serves as a feature extraction technique, retaining only the most relevant variables and simplifying the data representation. Overall, PCA is a powerful tool for data analysis, visualization, and dimensionality reduction in various machine learning and statistical applications.

PCA is widely used for data visualization, pattern recognition, and noise reduction. When projecting high-dimensional data onto lower-dimensional spaces, PCA seeks out the subspace with the lowest dimensionality that preserves the maximum variance

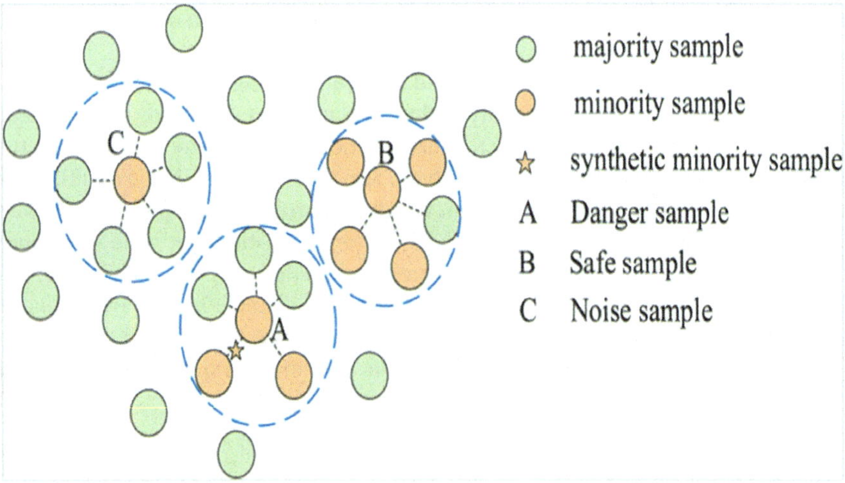

Fig. 1 SMOTE borderline

in the data. It allows for capturing the essential structure of the data while discarding the less informative or noisy components.

Each of the newly generated principal components, which are orthogonal to one another and do not correlate, captures a distinct and independent pattern in the data. Since the components are uncorrelated, they can be considered as independent axes in a new coordinate system.

By retaining only, the most important principal components, PCA serves as a feature selection technique, where it identifies the essential variables driving the variation in the data. This results in a simplified representation of the data, making it easier to analyze and interpret the underlying patterns.

3.2 Extra Trees Classifier

The Extra Trees Classifier is an ensemble learning technique used for classification tasks. It combines multiple de-correlated Decision Trees into a "forest" to make accurate predictions. One of its main applications is feature selection, where it randomly selects subsets of features from the dataset during tree construction. This randomness and diversity among the trees improve the model's performance and reduce overfitting. Unlike other ensemble methods, it trains each Decision Tree on the entire dataset without using bootstrap replicas. Overall, the Extra Trees classifier is a powerful tool for improving classification accuracy and selecting the most relevant features in high-dimensional datasets [6].

3.3 Borderline SMOTE

Synthetic Minority Oversampling Technique (SMOTE) is a data augmentation algorithm used to address class imbalance in machine learning datasets. By extrapolating between current samples, it creates fictitious data points for the minority class, balancing the distribution of classes. The synthetic data helps the model learn from both classes effectively and reduces the risk of overfitting. However, traditional SMOTE may encounter challenges when minority class observations are scattered among majority class regions. To overcome this, Borderline SMOTE was introduced, which focuses on generating synthetic samples near the class boundaries, improving the minority class representation and reducing noise. Figure 2 shows the basic architecture of SMOTE Borderline. Both SMOTE and Borderline SMOTE are valuable techniques for handling class imbalance and enhancing the performance in imbalanced classification tasks [7].

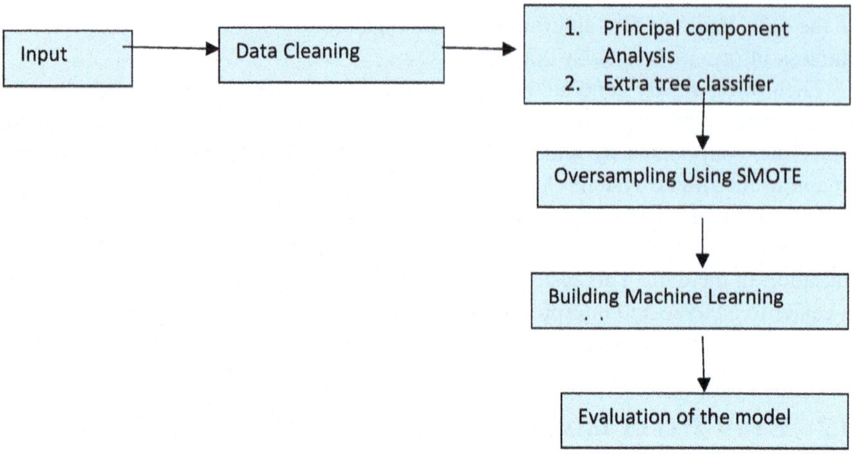

Fig. 2 Flowchart diagram

3.4 Decision Trees

Decision Trees are powerful and interpretable supervised learning models used for classification and regression tasks. They create judgments via a series of feature-based splits within a tree-like structure, where internal nodes stand in for features, branches for decision-making steps, and leaf nodes for the outcomes or forecasts. Decision Trees are graphical representations of possible solutions to a problem given specific parameters. They are versatile, handling both categorical and numerical data, and their interpretability makes them valuable for understanding the decision-making process. Decision Trees are often used in ensemble methods to enhance their performance. Figure 3 shows basic structure of Decision Tree.

3.5 Random Forest

Random Forest is an ensemble learning technique that combines multiple uncorrelated Decision Trees to solve complex problems and improve predictive accuracy. It aggregates the predictions of these trees through averaging or majority voting, reducing overfitting and enhancing robustness. The low correlation between individual trees allows them to collectively produce more accurate predictions. Random Forest (Fig. 4) is efficient in handling high-dimensional data, making it widely used in various machine learning tasks. Its power lies in leveraging diverse models to achieve superior performance and is known for its accuracy, scalability, and ability to handle complex datasets.

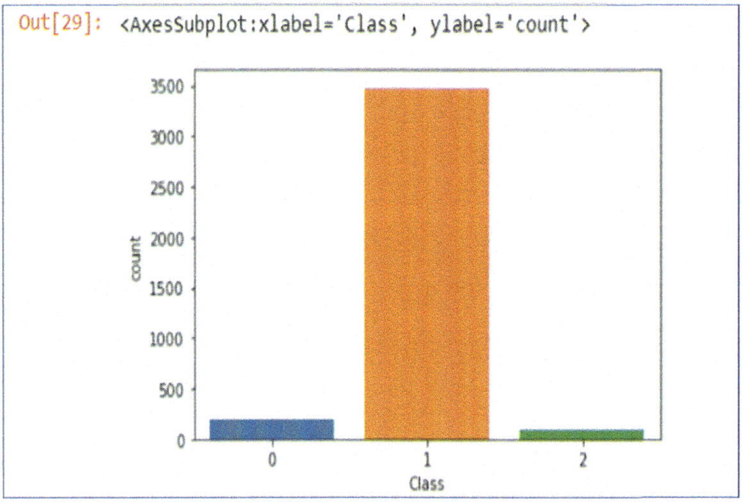

Fig. 3 Count plot

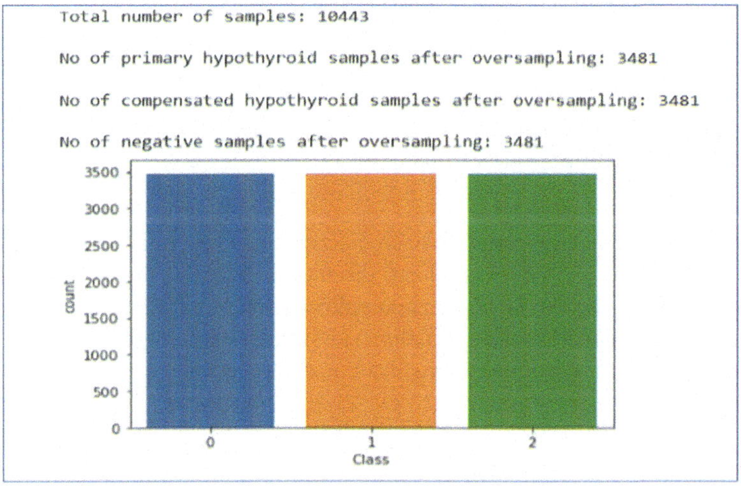

Fig. 4 Count plot balanced

3.6 *Artificial Neural Networks (ANN)*

Artificial Neural Networks (Fig. 5) comprise three layers: Input Layer, Hidden Layer, and Output Layer. The Input Layer accepts various forms of input data from the outside world. The Hidden Layer performs mathematical computations to uncover hidden patterns and characteristics. It transforms the input into a variety of outputs communicated through this layer. The computed weighted total is passed through

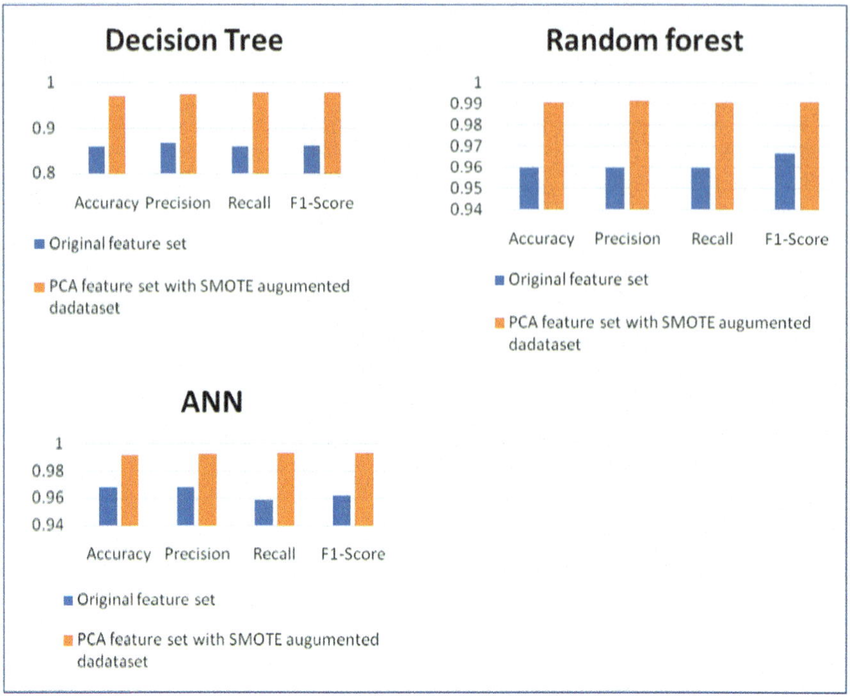

Fig. 5 Comparative results of machine learning models using the original feature set and PCA feature set with SMOTE-augmented dataset

an activation function, determining neuron activation. Only activated neurons make it to the Output Layer, which produces the final predictions or results. ANNs are powerful models capable of learning complex patterns and are commonly used in various machine learning tasks, including classification and regression.

A transfer function is used to visualize this computation.

$$\sum_{i=1}^{n} Wi * Xi + b$$

4 Proposed Methodology

To implement the hypothyroid detection system in Fig. 6, the steps are given and it is essential to install all the necessary dependencies in the working directory. Once that was confirmed, the dataset for hypothyroid diseases was downloaded which was

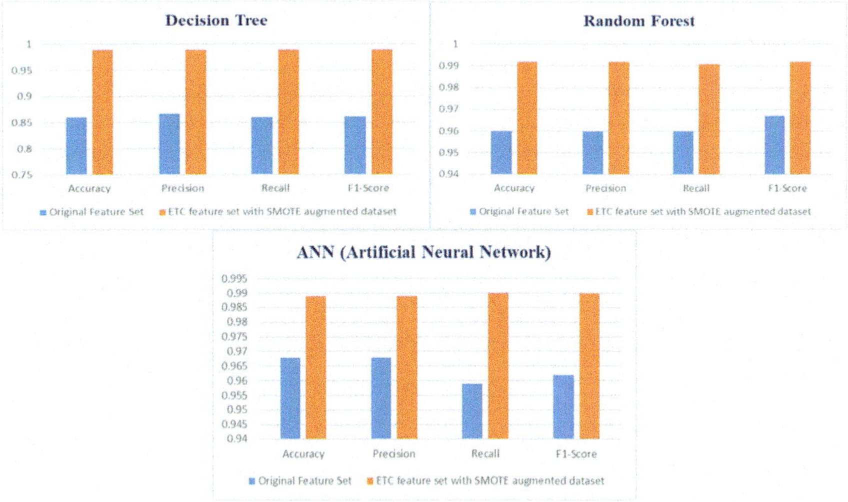

Fig. 6 Comparative results of machine learning models using the original feature set and the extra-tree classifier feature set with SMOTE-augmented dataset

to be used to extract interesting patterns about the disorder. The dataset was acquired from Kaggle and had 3770 rows and 30 columns at first [8–10].

4.1 Data Cleaning

Cleaning involved locating and addressing null and garbage values that were present in the dataset. NaN values were used in place of the garbage values to make handling them simpler. The median values of the corresponding columns were used to fill up the gaps left by the missing values. Nominal categorical values include characteristics like gender, which lack both numerical values and a clear preference hierarchy. To address this issue, label encoding was applied to these characteristics, which will give each label a numerical value, such as male and female being mapped to 0 and 1.

4.2 Feature Selection

Principal Component Analysis (PCA) and feature importance utilizing Extra Trees classifier were the two feature selection methods employed. Data of higher dimensions are converted into data of fewer dimensions using the dimensionality reduction approach known as PCA. The entropy values of each feature were determined using the Extra Trees Classifier, and the best features were then chosen from the group.

4.3 Oversampling

After performing feature selection, a count plot was used to analyze the class distribution of all target classes in the dataset. It was then observed that the instances for primary and secondary hypothyroid were very few when compared to the negative instances. This caused the dataset to be highly imbalanced and that is not good for building the models.

Borderline Synthetic Minority Oversampling Technique (SMOTE) was used to balance the dataset when it was discovered that it was very unbalanced and did not equally represent all of the target classes. Borderline SMOTE takes into account border points and generates fictitious observations in their vicinity. All of the noise points are ignored.

4.4 Building Models

After performing oversampling, the dataset was split into training and testing datasets. 70% of the data was used for training and 30% of the data was used for testing. Classification algorithms were applied on training datasets to build models. The algorithms that were used are: Decision Trees, Random Forest, Artificial Neural Networks (ANNs). These algorithms were applied on the features selected by PCA and Extra Trees Classifier.

4.5 Evaluating Models

Accuracy, precision, recall, $f1$-score, and confusion matrix were calculated as performance metrics. For the confusion matrix of each model, heat maps were made and visualized. To determine the best model, K-cross-validation was used on all of the models, using a k value of 5. The web application was constructed using the cross-validation model with the maximum accuracy. The serialized model is kept in a pickle file.

5 Results and Discussion

Based on the evaluation criteria, such as Accuracy, Precision, Recall, and $F1$-score, the performance of our proposed thyroid detection system is assessed. The outcomes of the web application are also made available.

In Table1, original feature set and the respective machine learning models shown. Decision Tree, Random Forest, and ANN models all perform well in terms of all evaluation criteria, with accuracy ratings of 0.86, 0.96, and 0.968, respectively. Even with a short feature set and a little dataset, the PCA-based feature set for enhanced data can still perform better.

Table 2 presents the outcomes of machine learning models utilizing the initial feature set. With accuracy scores of 0.86, 0.96, and 0.968, respectively, Decision Tree, Random Forest, and ANN models perform well across all evaluation criteria. Even with a short feature set and a small dataset, the tree-based ensemble feature set for augmented data can still outperform other feature sets.

Table 1 Comparative results of machine learning models using original feature set proposed approach

	Original feature set				PCA feature set with SMOTE augmented dataset			
ML model	Accuracy	Precision	Recall	$F1$-score	Accuracy	Precision	Recall	$F1$-score
Decision tree	0.86	0.867	0.86	0.862	0.978	0.978	0.978	0.978
Random forest	0.96	0.96	0.96	0.967	0.991	0.991	0.991	0.991
Artificial neural network (ANN)	0.968	0.968	0.959	0.962	0.99.23	0.993	0.993	0.993

Table 2 Comparative results of machine learning models using original feature set with proposed approach

	Original feature set				Extra tree classifier feature set with SMOTE augmented dataset			
ML model	Accuracy	Precision	Recall	$F1$-score	Accuracy	Precision	Recall	$F1$-score
Decision tree	0.86	0.867	0.86	0.862	0.989	0.989	0.989	0.989
Random forest	0.96	0.96	0.96	0.967	0.992	0.992	0.991	0.992
Artificial neural network (ANN)	0.968	0.968	0.959	0.962	0.989	0.989	0.990	0.99

6 Conclusion

In the recent years, the identification of thyroid illnesses has emerged as a major medical concern, necessitating the development of effective computerized prediction algorithms. This paper proposed an approach to accurately predict the thyroid disease that the patient has. After performing appropriate data preprocessing and applying feature selection techniques, namely PCA and feature importance using Extra Trees Classifier, three machine learning algorithms were applied to build the models. Those algorithms were Decision Tree, Random Forest, Artificial Neural Network (ANN). These models were evaluated based on proper evaluation metrics. The results obtained demonstrated that Random Forest model built on the features selected using PCA had the highest cross-validation accuracy of 98.7648%. This model was used to build the application to allow users to predict their thyroid diseases. In the future, the designed system can be enhanced by applying algorithms on bigger datasets and on real-time data to obtain better results. The application can also have more components that can help the user to take preventive measures against thyroid diseases.

References

1. Chaganti R, Rustam F, De La Torre Díez I, Mazón JLV, Rodríguez CL, Ashraf I (2022) Thyroid disease prediction using selective features and machine learning techniques. Cancers 14:3914
2. Aversanoa L, Bernardia ML, Cimitileb M, Iammarinoa M, Macchiac PE, Nettore IC, Verdonea C (2021) Thyroid disease treatment prediction with machine learning approaches. ScienceDirect Proc Comput Sci 192:1031–1040
3. Alya T, Hamid M, Alissa K, Faiz T, Tabassum N, Ahmad A (2022) Empirical method for thyroid disease classification using a machine learning approach. BioMed Res Int 2022. Article ID 9809932
4. Riajuliislam M, Rahim KZ, Mahmud A (2021) Prediction of thyroid disease (hypothyroid) in early stage using feature selection and classification techniques. In: 2021 international conference on information and communication technology for sustainable development (ICICT4SD). 978-1-6654-1460-9/21
5. Vasan CRC, Chethan MS, Devikarani HS (2021) thyroid detection using machine learning. Int J Eng Appl Sci Technol 5(9):173–177, ISSN No 2455-2143
6. Begum A, Parkavi A (2019) Prediction of thyroid disease using data mining techniques. In: International conference on advanced computing & communication systems (ICACCS). Coimbatore, India, pp 342–345
7. Tyagi A, Mehra R (2018) Interactive thyroid disease prediction system using machine learning techniques. published on ResearchGate
8. Pavya K, Srinivasan B (2017) Feature selection algorithms to improve thyroid disease diagnosis. IEEE international conference on innovations in green energy and healthcare technologies (ICIGEHT'17), pp 1–5, 02 November, 2017
9. Aswathi AK, Antony A (2018) An intelligent system for thyroid disease classification and diagnosis. In: 2nd international conference on inventive communication and computational technologies (ICICCT 2018), pp 1261–1264, 27 Sept 2018
10. Idarraga AJ, Luong G, Hsiao V, Schneider DF (2021) False negative rates in benign thyroid nodule diagnosis: machine learning for detecting malignancy. J Surg Res 268:562–569

Kullback–Leibler Divergence-Based Feature Selection Method for Image Texture Classification

M. Subba Rao, Guntoju Kalpana Devi, Suraya Mubeen, Badam Prashanth, Tazzeen Fatima, K. Reddy Madhavi, Voruganti Naresh Kumar, and Charan Yadav Chintalacheri

Abstract Nowadays, with the Big Data phenomenon, the need of the usage of feature selection (FS) methods is critical. FS can assist to address photograph analysis, and in reality it has been gaining importance within a rapid few years ago, however also to relieve the computational burden required for extracting records from the snapshots. An exhaustive assessment and evaluation of the latest contributions of FS to the field of image analysis is necessary. By eliminating unimportant features from the original set, selection of features significantly enhances the efficiency of texture classification, which is a key factor in machine learning performance. By selecting a characteristic, one can achieve accuracy in database type while also accelerating the class rate. The primary goal of the endeavor is to choose the most substantial skills in the function set to carry out a particular undertaking. Kullback–Leibler (KL) divergence approach showed that accuracy in classification is proven with the use of the present methods when feature selection is applied to Feature extraction.

M. S. Rao
Department of Computer Science and Engineering, Annamacharya Institute of Technology and Sciences (Autonomous), Rajampet, Andhra Pradesh 516126, India

G. K. Devi · V. N. Kumar
Department of CSE, CMR Technical Campus, Hyderabad, Telangana, India

S. Mubeen
Department of ECE, CMR Technical Campus, Hyderabad, Telangana, India

B. Prashanth
Department of CSE (AI&ML), CMR Technical Campus, Hyderabad, Telangana, India

T. Fatima
Department of ECE, Shadan Women's College of Engineering and Technology, Hyderabad, Telangana, India

K. R. Madhavi (✉)
AI&ML, School of Computing, Mohan Babu University, Tirupati, India
e-mail: kreddymadhavi@gmail.com

C. Y. Chintalacheri
Masters in Computer Science, University of Kansas, Lawrence, USA

B. R. Devi et al. (eds.), *Proceedings of Fifth International Conference on Computer and Communication Technologies*, Lecture Notes in Networks and Systems 897,
https://doi.org/10.1007/978-981-99-9704-6_27

Keywords Image texture classification · Feature selection · Extraction · KL divergence · Performance measures

1 Introduction

Pattern recognition [1] is concerned with the analysis or understanding of a pattern in an image. It focuses on extracting information of the image in order to recognize its components. Input data are already in the form of computerized two-dimensional input image or textures including the pattern to be identified. The study and identification of patterns, such as images and textures, is becoming increasingly complex and multi-faceted due to the movement of patterns from general to specific, significant fluctuations and intense noise in patterns. Texture class is one of the special varieties of strategies for pattern recognition. Selection of capabilities plays a key position in [2, 3] enhancing system learning performance because it significantly enhances the overall performance of texture category by eliminating irrelevant capabilities from the original collection. Accuracy in the classification of databases can be accomplished by the selection of features. At the same time, the classification rate can be enhanced.

1.1 Texture Classification

Using a non-uniform spatial distribution, texture [4, 5] is defined as depth values. Texture class is the system by way of which the particular textures of the listed photos are classified. Although the explanation of the textures itself frequently appears meaningless in its own manner, it's far feasible to answer a large wide variety of real-existence issues associated with the precise textures of various materials. The goal of texture classification [6] is to categorize a amid a variety of considered texture types, an example image. Approaches of classifying textures are utilized in a number of real-world applications, along with content-based photograph restoration, face recognition, rock class and timber species reputation, material class and geographic panorama segmentation.

Structural pattern recognition [7] also referred to as syntactic pattern recognition due to its roots in formal language theory and syntactic grammar to distinguish between datasets from various classes on the basis of the morphological interrelationships within the content. The utility of structural pattern recognition systems is limited as a result of the fundamental complexities [8, 9] associated during the execution of the definition and categorization tasks. Because of their capacity to learn from training datasets, neural networks (NN) have attracted the attention of pattern recognition researchers.

1.2 Feature Extraction and Feature Selection

The emergence of Big Data troubles, particularly those with an excessive variety of attributes, has generated an vital necessity for techniques that can effectively lessen dimensionality. To solve this problem, there are two major approaches such as feature extraction and feature selection. Feature selection (FS) [10] is one way of dimensional reduction method and its goal is to remove redundant or irrelevant data. Feature extraction is another way of dimensional reduction method and has the electricity to generate a new set of functions, which are greater compact and of stronger discriminating electricity. The major difference among them is that FS strategies paintings by way of selecting a fixed of the original features, while characteristic extraction techniques are capable of lessen the dimensionality by way of combining the prevailing features and growing new ones.

2 Existing System

2.1 Feature Selection

Filters work by using the use of most effective the overall traits of the datasets, without an impact of any classification method. Filter methods select a subset of features completely independent [11] of the classifier used. Their advantage is speed and versatility, but they provide worse results. Function discovery correlation-based filter (FDCF) is a basic filter algorithm that ranks subsets of features and discovers the value of a function or subset of features. Fast correlation-based filter (FCBF) is a method that separately evaluates the relevance and redundancy of symptoms. Wrapper strategies build the subset of relevant features based on a gaining knowledge of set of rules. Wrapper methods are narrowly categorized into sequential sorting algorithms and heuristic search algorithms. The remaining functions are added independently to the current subset from the second step and the new subset is tested. If it provides optimal classification [12, 13] specificity, the individual attribute is permanently included in the subset. Wrapper methods additionally construct the subset of applicable capabilities based on a gaining knowledge of algorithm methods that are directly part of the learner and are collectively referred to as embedded algorithm. Embedded approaches are unique to the style of learning. For the learning model, the selection is made during the training process. Different types of decision tree algorithms include popular built-in methods including CART, C4.5, and Random forest.

3 Proposed System

The suggested approach is designed to achieve a set of features with a minimal number of features that could provide an effective classification. The classification efficiency of the new system for the collection of features is compared to that of all the original methods. The accuracy of the classification shall be obtained using the classification algorithm for SVM and *K*-Nearest Neighbors.

Pre-processing is a method for enhancing the appearance of photographs and is an imperative phase in the selection of features for texture classification, since imprecise pre-processing has a direct impact on the outcome of the selection of features and the accuracy of the classification. Denoise of an image is a technique used in the pre-processing phase of fuzzy inclusion and fuzzy entropy measures.

3.1 KL Divergence-Based Feature Selection Method

The second phase introduces a new feature selection method, Kullback–Leibler (KL) divergence-based feature selection for texture classification. The significance of the choice of features is to lessen fees and enhance the efficiency of the type of the mastering set of rules. An efficient technique is that the distance measurements between two class-conditional distributions and probability distributions are used to evaluate probability distributions and are also evaluated based totally on both dependent and independent and multivariate features is proposed in this work.

Entropy is the measure of uncertainties and information, and entropy of Shannon is still a measure of the variation due to the random variables. In many cases, the nature of raw data is vague and imprecise. It is therefore of great importance to have adequate steps to evaluate the results. Fuzzy sets have the membership value of its members, who play a vital role in data accuracy. To evaluate and check the consistency of the feature subsets which are hesitant at some point, the design of a new model, which can describe the data as it is, is essential and useful. Intuitionistic fuzzy sets give the details of both membership and non-membership along with hesitancy margin of its elements. The measure with these qualities can be used to create new models needed including one with a dynamic property. An intuitionistic fuzzy entropy measure (IFEM) and temporal intuitionistic fuzzy entropy measure (TIFEM) are used to pick appropriate features for the classification purpose.

Divergence is the fundamental measure of the inconsistency between different regions of probability that is used in various applications. The KL divergence is a knowledge-theoretic metric that a real surprise from image compression, communication, probability theory, statistics, optimization techniques, machine learning and several other fields of study. The estimator production for the KL divergence denoted by using $D_{KL}(P||Q)$ is simple.

Specifically, KL divergence is given by

$$D_{\mathrm{KL}}(P \| Q) = -H(P) - \sum_{i=1}^{k} P_i \log q_i \tag{1}$$

The negated sum of the entropy of Shannon

$$H(P) = \sum_{i=1}^{k} -p_i \log p_i \tag{2}$$

and the cross-entropy

$$H(P) = \sum_{i=1}^{k} p_i \log q_i \tag{3}$$

$$\begin{aligned} D_{\mathrm{KL}}(PQ) &= -H(P) - \sum_{i=1}^{k} p_i \log q_i \\ &= \sum_{i=1}^{k} -p_i \log p_i + \sum_{i=1}^{k} p_i \log q_i \end{aligned} \tag{4}$$

The detailed construction is as follows:

Evaluate the non-smooth portion of a cross-entropy
Evaluate the non-smooth component of the Shannon entropy
Evaluate the smooth component of the cross-entropy
Evaluate the smooth component of the Shannon entropy
Calculate the total entropy by combing the cross- and Shannon entropies of smooth and non-smooth components.

3.2 Proposed Feature Selection Algorithm Based on KL Divergence

A new feature selection approach based on KL divergence is presented for non-temporal and temporal data sets. Information content of all the features are available by the computation of entropies. Since a feature with less fuzzy entropy possess less fuzziness [12, 13], the feature is chosen to obtain a better classification. To remove the features, a threshold value is fixed. The aim is to retain the features with less entropies as they could give a higher contribution to the classification results. The steps involved inside the processing of the proposed algorithm based on KL divergence are given.

3.3 Algorithm: KL Divergence-Based Feature Selection Method

Select a feature from the feature set

Construct the KL divergence or relative entropies for each interval
Determine the cross-entropy for smooth and non-smooth components based on Eq. 3
Determine the Shannon entropy for smooth and non-smooth components based on Eq. 2
Compute the total entropy of the selected feature using Eq. 4
Repeat the procedure for all the features
Select the features with minimum total entropy by setting a threshold.

4 Results of KL Divergence

An experimental results of the KNN classifier on the Iris flower data set are displayed in Table 1. A threefold cross-validation is performed using different distance measurements without and with a feature selected from the KL divergence approach. Two of the four features are chosen from the Iris data set with better outcomes.

Table 2 shows the experimental results of the Wisconsin breast cancer using KNN. Different distance measurements are used in a threefold and fivefold cross-validation, both with and without a feature picked from the KL divergence approach. With improved values, 20 of the 30 features are chosen for the Wisconsin breast cancer data collection.

Table 3 displays the experimental results of the KNN classifier on the Texas weather data set. A threefold, fivefold, and 10-fold cross-validation using different distance measurements is performed without feature selection and with feature selection from the KL divergence approach. In the Texas weather data set, 42 features with higher values were chosen out of 72. The best features are determined using the KL divergence approach with four distance measurements and may be validated

Table 1 Impact of feature selection with KNN distance measures on Iris dataset using threefold cross-validation

Impact of feature selection	KNN with distance measures			
	Chebychev	City block	Euclidean	Minkowski
Threefold cross-validation with 4 features (without feature selection)	96.69	94.68	95.38	96.03
Threefold cross-validation with 2 features (with KL divergence feature selection)	93.38	92.68	93.38	93.8

Table 2 Wisconsin breast cancer dataset's impact of feature selection using KNN distance measurements utilizing threefold and fivefold cross-validation

Impact of feature selection	KNN with distance measures			
	Chebychev	City block	Euclidean	Minkowski
Threefold cross-validation with 30 features (without feature selection)	92.48	94.73	95.61	95.44
Threefold cross-validation with 20 features (with KL divergence feature selection)	92.45	93.25	94.26	93.68
Threefold cross-validation with 30 features (without feature selection)	93.16	95.26	95.27	95.28
Threefold cross-validation with 20 features (with KL divergence feature selection)	92.87	94.96	94.8	94.11

using k-fold cross-validation with $k = 3$, 5, and 10. Tables 4a, b and c illustrate that the proposed strategy produced better outcomes with certain attributes.

Table 3 Impact of feature selection with KNN distance measures on Texas weather dataset using threefold, fivefold and 10-fold cross-validation

Impact of feature selection	KNN with distance measures			
	Chebychev	City block	Euclidean	Minkowski
Threefold cross-validation with 72 features (without feature selection)	100	98.26	99.94	100
Threefold cross-validation with 42 features (with KL divergence feature selection)	89.4	91.1	90.6	90.8
Fivefold cross-validation with 72 features (without feature selection)	100	98.27	99.95	100
Fivefold cross-validation with 42 features (with KL divergence feature selection)	88.9	90.53	90	89.64
10-fold cross-validation with 72 features (without feature selection)	100	98.32	99.95	100
10-fold cross-validation with 42 features (with KL divergence feature selection)	89.11	90.48	89.89	90.16

Table 4 Accuracy comparison of feature selection methods through KNN and SVM on Iris dataset using threefold cross-validation

Type of validation	Feature selection method	KNN	SVM
Threefold cross-validation	Fuzzy entropy method	91.33	94.67
	Intuitionistic fuzzy entropy method	93.38	95.34
	KL divergence method	94.88	96.56

Classification Accuracy

Feature selection is made for different data sets to reduce total the number of attributes used in fuzzy entropy method (FEM), the intuitionist fuzzy [14] entropy method (IFEM), and the KL divergence method (KLDM) methods. KNN and SVM classification algorithms are used for classification purposes. The experimental results of the [15] Iris, Wisconsin and Texas weather data sets are used to compare existing [16, 17] methods of fuzzy entropy and intuitive fuzzy entropy. Out of 4 features, 2 are selected from the Iris data, 20 out of 30 features selected from Wisconsin, and 42 out of 72 are selected from the Texas weather data set. In this experiment, we use a held-out testing set to calculate the proposed model's accuracy, to verify its performance [18] on texture datasets as True Positive (TP), True Negative (TN), False Positive (FP) and False Negative (FN), and to compute the KL divergence model performance with respect to accuracy as follows:

$$\text{Accuracy} = \text{TP} + \text{TN} / \text{TP} + \text{TN} + \text{FP} + \text{FN} \tag{5}$$

Table 4 shows the performance of the KNN and SVM classifiers on the Iris data set with features chosen from FEM, IFEM and KLDM using threefold CV. Table 5 shows the performance of the *K*-Nearest Neighbor and SVM classifiers on the Wisconsin data set using features from FEM, IFEM and KLDM three and fivefold cross-validation (CV).

Table 6 represents the performance of *K*-Nearest Neighbor and SVM classifiers on Texas weather data set with features selected from FEM, IFEM and KLDM using threefold, fivefold and 10-fold cross-validation (CV). It is observed that maximum accuracy is achieved using the proposed KL divergence-based feature selection method on three datasets. SVM achieves better results than KNN on three datasets.

Table 5 Accuracy comparison of feature selection methods through KNN and SVM on Wisconsin breast cancer using threefold and fivefold cross-validation

Type of validation	Feature selection method	KNN	SVM
Threefold cross-validation	Fuzzy entropy method	93.67	94.9
	Intuitionistic fuzzy entropy method	93.68	96.49
	KL divergence method	95.98	97.56
Fivefold cross-validation	Fuzzy entropy method	94.21	95.25
	Intuitionistic fuzzy entropy method	95.8	96.32
	KL divergence Method	97.34	98.56

Table 6 Accuracy of feature selection approaches employing KNN and SVM on the Texas weather dataset using threefold, fivefold and 10-fold cross-validation

Type of validation	Method	KNN	SVM
Threefold cross-validation	Fuzzy entropy method	91.07	94.9
	Intuitionistic fuzzy entropy method	93.10	95.12
	KL divergence method	94.23	95.89
Fivefold cross-validation	Fuzzy entropy method	92.78	95.34
	Intuitionistic fuzzy entropy method	93.30	96.23
	KL divergence method	94.67	96.65
10-fold cross-validation	Fuzzy entropy method	94.12	96.23
	Intuitionistic fuzzy entropy method	95.48	97.76
	KL divergence method	97.43	98.89

5 Conclusion

Feature selection has been made using a new approach to extract the most appropriate features from a feature space. This, in effect, eliminates redundant and irrelevant features and increases the accuracy of classification with minimal computational effort. The primary goal of this work is to identify the optimal features for categorization of a large and diverse data set with multiple attributes. Fuzzy entropy and intuitional fuzzy entropy methods are used in the selection of features. The experimental results are compared to the existing method based on KL divergence. It can be noted that the maximum accuracy is achieved by using the proposed method using KL divergence. This method therefore shows better performance on the basis of the accuracy of the classification.

References

1. Tuceryan M, Jain AK (1998) Texture Analysis, The Handbook of Pattern Recognition and Computer Vision, 2nd edn. World Scientific Publishing, Singapore, pp 207–248
2. Manjunathi BS, Ma WY (1996) Texture features for browsing and retrieval of image data IEEE transaction on pattern analysis machine. Intelligence 18(8):837–842
3. Chen Y, Dougherty ER (1994) Gray-scale morphological granulometric texture classification. Opt Eng 33(8):2713–2723
4. Fu KS (1982) Syntactic pattern recognition and applications. Prentice Hall, Englewood Cliffs, New Jersey
5. Aparna Y, Somasekhar G, Bhaskar N, Divya G, Reddy Madhavi K (2023) Analytical approach for soil and land classification using image processing with deep learning. In: 2023 2nd international conference for innovation in technology (INOCON). Bangalore, India. Mar 3–5, 2023. IEEE. https://doi.org/10.1109/INOCON57975.2023.10101169
6. Kulkarni A (1994) Artificial Neural Networks for Image Understanding. Van Nostrand Reinhold, New York
7. Kohavi R, John GH (1997) Wrappers for feature subset selection. Artif Intell 97:273–324

8. Blum AL, Langley P (1997) Selection of relevant features and examples in machine learning. Artif Intell 245–270
9. Sunitha G, Rani BS, Bhukya SN, Mohammad H, Vittal RHS (2023) political optimizer-based automated machine learning for skin lesion data. In: Proceedings of third international conference on advances in computer engineering and communication systems: ICACECS 2022. Singapore: Springer Nature Singapore, pp 487–495
10. Avanija J, Sunitha G (2017) Semantic similarity based clustering of web documents using hybrid swarm intelligence and fuzzy C-means. Helix Int J 7(5):2007–2012
11. Sunitha G, Pushpalatha MN, Parkavi A, Boyapati P, Walia R, Kohar R, Qureshi K (2022) Modeling of chaotic political optimizer for crop yield prediction. Intell Autom Soft Comput 34(1):423–437
12. Zadeh LA (1968) Probability measures of fuzzy events. J Math Anal Appl 23(2):421–427
13. Atanassov KT (1986) Intuitionistic fuzzy sets. Fuzzy Sets Syst 20(1):87–96
14. Avanija J, Konduru S, Kura V, NagaJyothi G, Dudi BP (2022) Designing a fuzzy Q-learning power energy system using reinforcement learning. Int J Fuzzy Syst Appl (IJFSA) 11(3):1–12
15. Bogiatzis A, Papadopoulos B (2018) Binarization of texts with varying lighting conditions using fuzzy inclusion and entropy measures. Int Conf Num Anal Appl Math 1978(1):290006
16. Bogiatzis A, Papadopoulos B Producing fuzzy inclusion and entropy measures and their application on global image thresholding. Evolving Syst
17. Zheng H et al (2016) Deep learning for surface material classification using haptic and visual information. IEEE Trans Multimedia 18:2407–2416
18. Naseeba B, Sai PHR, Karthik BVP, Chitteti C, Sai K, Avanija J (2023) Toxic comment classification. In: Abraham A, Hong TP, Kotecha K, Ma K, Manghirmalani Mishra P, Gandhi N (eds) Hybrid intelligent systems. HIS 2022. Lecture Notes in Networks and Systems, vol 647. Springer, Cham. https://doi.org/10.1007/978-3-031-27409-1_80

Generative Adversarial Networks: Overview

Shivani Pachika, A. Brahmananda Reddy, Bhavishya Pachika, and Akhil Karnam

Abstract Deep neural networks (DNNs) have attained boundless accomplishments in dealing with high-dimensional data, particularly images. However, generating naturalistic images for tasks such as classification, detection, segmentation, reconstruction, and so on remains a difficult task. As a result, researchers are compelled to propose new algorithms and methods to overcome these challenges. These learning algorithms, on the other hand, frequently need a large volume of training data to be efficient. Also, its associated real-world challenges have limited data accessibility, and gathering sufficient amounts of data may be expensive. Additionally struggling with the issues of class imbalance, data shortage, and label propagation has turned the attention toward generative approaches, i.e., using modern procedures for producing new sample data with similar features as real sample data to enhance the efficiency of learning algorithms. Generative adversarial networks (GANs), a class of DNNs offer a novel way to model and generate data in an unsupervised manner. The deep learning community is increasingly pursuing an interest in GANs. The paper gives a basic introduction to GANs and discusses their variants in the study.

Keywords Neural networks (NN) · Deep neural networks · Generative modeling · Unsupervised · Minimax · Generative adversarial networks · Generative · Adversarial · Networks · Generator · Discriminator

S. Pachika
Output Media, Windsor, Canada

A. B. Reddy · A. Karnam (✉)
VNR Vignana Jyothi Institute of Engineering and Technology, Hyderabad, India
e-mail: akhilresearch18@gmail.com

B. Pachika
NIT Kurukshetra, Kurukshetra, India

B. R. Devi et al. (eds.), *Proceedings of Fifth International Conference on Computer and Communication Technologies*, Lecture Notes in Networks and Systems 897,
https://doi.org/10.1007/978-981-99-9704-6_29

Table 1 Acronym of GAN

1	Generative	Refers to producing new data built on training sets
2	Adversarial	Denotes two components of GANs, i.e., the generator and the discriminator, competing contrary to one another
3	Networks	Shows that the model comprises of two networks. The networks might be convolution neural networks (CNN), fully connected neural networks (FCNNs), recurrent neural networks (RNNs), autoencoders, long short-term memory (LSTM) neural networks, or any fusion of the above

1 Introduction

Generative modeling has the capacity for unsupervised learning of any type of data distribution. The popular generative modeling methods that produce data distributions include variational autoencoder (VAE), autoregressive (AR) models, and generative adversarial network (GAN). The ability of GANs to produce high-quality images and enhance data has created substantial attention from the researchers in recent years. Medical imaging, super-resolution, semantic segmentation, image deblurring, image captioning, synthesizing, text-to-image generation or scene generation, steganography speech recognition, object detection, industrial risk analysis, fault diagnosis, and natural language processing applications are just a few areas where GANs have been used successfully [1]. GANs eliminate the need for human supervision while resolving the issue of creating data when there is not enough of it, to begin with. GANs can offer an effective technique to close gaps in the training data's discrete distribution [2]. In 2014, Ian Goodfellow and his colleagues introduced GANs framework [3]. The acronym GANs is built on three words [4] as shown in Table 1.

The generator may be seen as a counterfeiter attempting to create some high-quality forged goods, while the discriminator might be seen as the cop attempting to find the fake goods. In this game, rivalry forces both teams to refine their strategies to the point where the fake goods cannot be distinguished from the real goods [5].

The rest of this paper is organized as follows. Section 2 presents GAN framework. A list of survey papers related to an overview of GANs and their various designs and frameworks are provided in Sect. 3. Next, in Sect. 4, a few GAN variants are explained briefly. Finally, presents future works and conclusions in Sect. 5.

2 GAN Framework

In the traditional GAN paradigm, two models are learned concurrently in an adversarial process that estimates generative models. A generative model called the Generator (G) captures the data distribution and creates new data. The Discriminator (D), a classifier in the second model, computes the possibility that a sample is from training data instead from G. The aim of the G's training procedure is to maximize the possibility that the D would misclassify the output. D's job is to evaluate the created data's quality and give G feedback. G is in charge of data producing. Under game-theory principles, these NN are optimized: The D is optimized to identify the input source, i.e., the real data set or the G, while G is optimized to produce data to deceive the D.

2.1 Discriminator

The D classifies things. It attempts to differentiate between real and synthetic data (from G) that it receives. Based on the data type being classified, it might use a variety of different network designs. Two loss functions, utilized in various training phases, are connected to the D network. Following the categorization of synthetic or real data, the D loss penalizes it for incorrect classifications, and its weights are revised through the network via backpropagation from its loss.

2.2 Generator

The G learns to produce synthetic data that closely look like the original data in significant ways using feedback from the D. Its objective is for the D to identify the generated data as real. The G is penalized for not tricking the D by the network when it generates some output from an arbitrary input (noise). The D then evaluates this output, which results in a G loss. GAN may be able to produce a varied range of outputs by noise adding and sampling from several regions throughout the target distribution (Fig. 1).

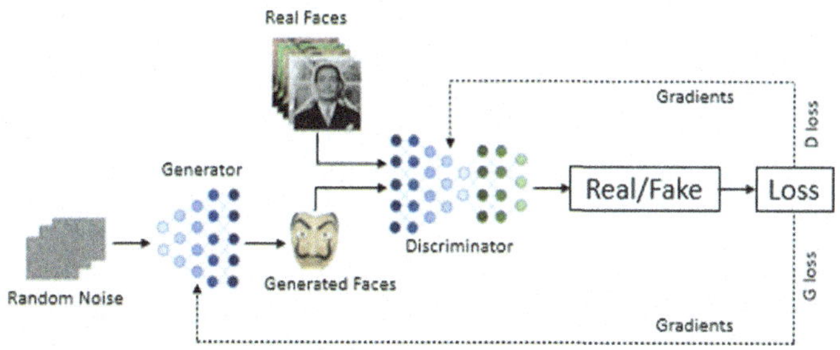

Fig. 1 Architecture of a GAN

2.3 Loss Function

The similarity between the real and generated data distributions is evaluated using loss functions during the GAN training process. To address this issue, numerous solutions have been put up. A "minimax loss" was added to the original "vanilla" GAN. The G and D losses in the original design are caused by scale of the distance between probability distributions, and the motive of training is to minimize the G loss and maximize the D loss. According to the training network, the two terms are updated alternately.

$$\text{Min G max D (D, G)} = E_{x\sim p(x)}\left[\log(s(x))\right] + E_{z\sim p(x)}\left[\log(1 - \text{D}(\text{G}(z))\right]$$

2.4 Optimizers

The model is updated by optimizers in response to the loss function's output. In essence, they can influence how a NN learns by determining the parameters that would result in the lowest possible loss function. The rate of learning is an important hyperparameter that adjusts the gradient and controls how quickly the model is updated. A gradient descent-based optimizer is used by most of the models. Network weights are iteratively updated using the optimization algorithm adaptive moment estimation (Adam).

2.5 Training

The networks D and G are separately trained, by different alternative procedures.

G: Sample from arbitrary noise = > new sample is generated in G = > D classifies it as real/fake = > loss is computed from D classification = > backpropagate through D and G to fetch gradients = > gradients utilized to alter weights of G.

D: D classifies G fake and real data = > D loss penalizes G for misclassifying fake and real instances = > D updates its weights through backpropagation from D loss through the D. The D classifies real images, x as $\text{D}(x) = 1$ and the fake images as $\text{D}(x) = 0$.

Training: Both models are engaged in a zero-sum game in which they are competing against one another. This implies that when the D successfully recognizes a sample, it is rewarded or model parameters are left as is, while the G is penalized with large changes to model parameters. In another way, when the G deceives the D, it is rewarded or its parameters are not updated, but the D is punished and its model parameters are altered. The networks are learned in an adversarial fashion to attain the Nash equilibrium, a state of optimum at which $D(x) = 0.5$ for each image x, infers thereby D is unable to distinguish between real and fake samples. Weights are changed during the training of a NN to lessen error or output loss. As the G is not directly related to its loss function, GANs are more complicated. The D is the output source that has an impact on the G (G loss). Each weight is adjusted by backpropagation by computing the impact of weight on the output. G's weight's impact is inclined by D weights it supplied into. Backpropagation starts at the output and flows back through D into G [6].

3 Related Works

In recent years, extensive research has been performed on GANs. Most of the survey publications [7–13] talk about GANs in general. In late 2017 [14], one of the first reviews of general GANs took place. At about the same time, Hitawala et al. [15]'s study of various GANs designs included a summary of the major advancements made to the GANs frameworks as well as metrics for evaluating them. The number of studies that examine GANs, in general, has increased in 2019 and 2020 [7–10, 12, 13, 16–18]. It is important to note that a longer and more in-depth general evaluation is available [19]. The most popular GAN models were introduced by Cao et al. [20], and their functionalities were contrasted considering various use cases.

4 GAN Variants

Deep Convolutional Generative Adversarial Networks (DCGANs)

For up-sampling, G is composed of fractionally strided convolutional layers, while D is composed of convolutional layers for down-sampling. Its main design eliminates max-pooling layers, fully connected layers, batch normalization except for G's output layer and D's input layer, ReLU (Rectified Linear Unit) as G's activation function (except at output), LeakyReLU as D's activation function, transposed convolution for up-sampling, and convolutional stride for down-sampling [21] (Fig. 2).

Conditional Generative Adversarial Networks (cGANs)

By adding some additional information (class labels or data from other modalities) as an additional input layer, GANs could be expanded to conditional model if conditioned on both G and D. The G creates a combined hidden representation out of the prior noise and added information [22] to produce output images with the desired attributes. cGANs falls under supervised learning.

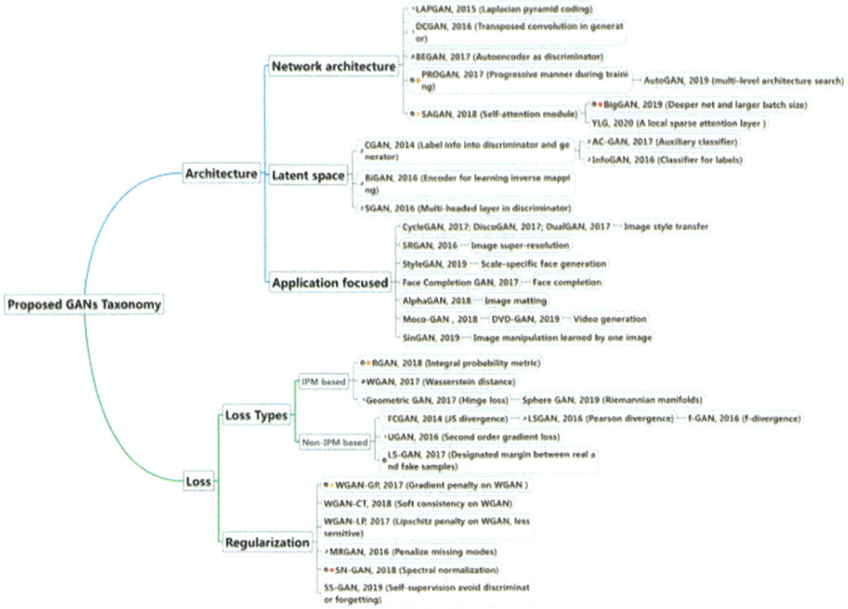

Fig. 2 The taxonomy of the recent GANs [16]

Bidirectional Generative Adversarial Network (BiGAN)

By adding an encoder that learns G'inverse, BiGAN expands the GAN framework. During the BiGAN training process, a mapping from latent space to data and vice versa can be learned simultaneously. The encoder is a nonlinear parametric function that may be learned via gradient descent [23].

Semi-supervised Generative Adversarial Network (SGAN)

It's semi-supervised learning that trains a G, supervised D, and unsupervised D all at once. When N datasets and classes are summed to match G's output, D is revised by predicting $N + 1$ classes [24].

Wasserstein GAN (WGAN) and Wasserstein GAN-Gradient Penalty (WGAN-GP)

WGAN is trained for lessening the Wasserstein distance (WD) between real and fake data distribution [25]. WGAN enforces an additional Lipchitz constraint to utilize WD as D's loss, which installs weight clipping to impose D's weights after each training batch to satisfy the Lipchitz constraint. WGAN's D's ability to learn is significantly reduced by weight clipping, and it frequently leads to converging failure. In place of weight clipping, WGAN-GP utilizes gradient penalty [2].

Least Squares Generative Adversarial Network (LSGAN)

The D in LSGANs uses a least square loss, which reduces several issues and enhances the generated sample's quality. By reducing the mean square error on a D's middle layer that addresses the vanishing gradient problem, the goal is to have the generated samples' statistics match with real data [26].

Cycle Generative Adversarial Networks (CycleGANs)

Two Gs (G_1 and G_2) and their associated two Ds (Di and Dj) make up cycleGANs. The G_1 attempts to transform data from domain I to domain J which is supplied to Dj to determine whether they are fake or real according to the J domain and likewise, and the G_2 attempts to transform data from domain J to domain I which is supplied to Di to determine whether they are fake or real according to I domain. The Gs are being taught to invert one another's tasks concurrently. The network is learned using a cycle-consistency loss along with the minimax loss [27].

Variational Autoencoder Generative Adversarial Network (VAEGAN)

By combining the VAE decoder with the GAN framework, VAEGAN jointly trains VAE and GAN. To enhance the quality of images produced by VAE, VAEGAN uses the GAN's feature-wise adversarial loss rather than VAE's element-wise reconstructive loss [28]. It uses content loss, sometimes referred to as perceptual loss, to compare two images based on pre-trained VGG network's high-level feature representation, along with adversarial loss and latent loss [29].

StyleGAN

Rather than supplying the latent code directly to G's input, it uses a style-based G architecture that first transforms the code into an intermediate latent space before using it to scale and shift the feature responses of normalized image calculated from every convolution layer [30].

Boundary Equilibrium Generative Adversarial Network (BEGAN)

An autoencoder serves as D in BEGAN [31]. Without explicitly matching the data distributions, it uses the equilibrium technique and adds a hyperparameter during training to balance G and D [7]. The distributions of reconstructive losses of real and the generated images are matched using WD.

Super-Resolution Generative Adversarial Network (SRGAN)

A low-resolution image is used as input by the SRGAN model, and it outputs a high-resolution image by upscaling the image's spatial resolution. The model produced high-quality photos with more realistic and natural-looking textures [32].

Spectrally Normalized Generative Adversarial Network (SN-GAN)

Weight normalization is the method suggested by SN-GAN to train D more steadily. It can be used for GANs that already exists and requires little computational effort [33] (Fig. 3).

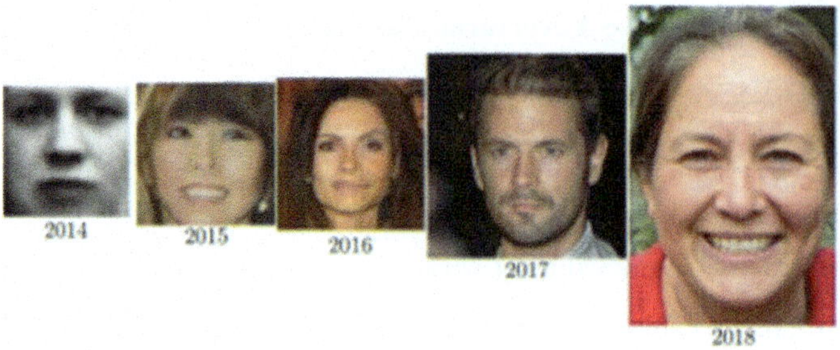

Fig. 3 GAN progress on face generation [34]

5 Future Works and Conclusion

Compared to other DNNs, the GAN's generator and discriminator can reach equilibrium slowly. Trying out new models or adapting the ones that already exist for diverse applications is challenging. To choose the best research directions to address them, one needs a thorough knowledge that places equal focus on both their theoretical and practical perspectives. The query of which GAN is the best has no conclusive answer. The application determines which GAN type should be used. GANs have been used in a variety of fields, including computer vision, semantic segmentation, time series synthesis, natural language processing, etc. We conclude that GANs have a bright future and that there are several potential for additional study which will increase the NN's potential and applications across a wide range of industries.

References

1. Pavan Kumar MR, Jayagopal P (2021) Generative adversarial networks: a survey on applications and challenges. Int J Multimedia Inf Retrieval 10(1):1–24
2. Sampath V, Maurtua I, Aguilar Martín JJ, Gutierrez A (2021) A survey on generative adversarial networks for imbalance problems in computer vision tasks. J Big Data 8(1):1–59
3. Goodfellow I, Pouget-Abadie J, Mirza M, Xu B, Warde-Farley D, Ozair S, Bengio Y (2014) Generative adversarial nets. Adv Neural Inf Process Syst 27
4. Aldausari N, Sowmya A, Marcus N, Mohammadi G (2022) Video generative adversarial networks: a review. ACM Comput Surv (CSUR) 55(2):1–25
5. Yi X, Walia E, Babyn P (2019) Generative adversarial network in medical imaging: a review. Med Image Anal 58:101552
6. Eckerli F, Osterrieder J (2021) Generative adversarial networks in finance: an overview. arXiv preprint arXiv:2106.06364
7. Hong Y, Hwang U, Yoo J, Yoon S (2019) How generative adversarial networks and their variants work: an overview. ACM Comput Surv (CSUR) 52(1):1–43
8. Jabbar A, Li X, Omar B (2021) A survey on generative adversarial networks: variants, applications, and training. ACM Comput Surv (CSUR) 54(8):1–49
9. Alqahtani H, Kavakli-Thorne M, Kumar G (2021) Applications of generative adversarial networks (gans): an updated review. Arch Comput Methods Eng 28(2):525–552

10. Gui J, Sun Z, Wen Y, Tao D, Ye J (2021) A review on generative adversarial networks: algorithms, theory, and applications. IEEE Trans Knowl Data Eng
11. Creswell A, White T, Dumoulin V, Arulkumaran K, Sengupta B, Bharath AA (2018) Generative adversarial networks: an overview. IEEE Signal Process Mag 35(1):53–65
12. Pan Z, Yu W, Yi X, Khan A, Yuan F, Zheng Y (2019) Recent progress on generative adversarial networks (GANs): a survey. IEEE Access 7:36322–36333
13. Zamorski M, Zdobylak A, Zięba M, Świątek J (2019, June) Generative adversarial networks: recent developments. In: International conference on artificial intelligence and soft computing. Springer, Cham, pp 248–258
14. Wang K, Gou C, Duan Y, Lin Y, Zheng X, Wang FY (2017) Generative adversarial networks: introduction and outlook. IEEE/CAA J Autom Sinica 4(4):588–598
15. Hitawala S (2018) Comparative study on generative adversarial networks. arXiv preprint arXiv: 1801.04271
16. Wang Z, She Q, Ward TE (2021) Generative adversarial networks in computer vision: a survey and taxonomy. ACM Comput Surv (CSUR) 54(2):1–38
17. Cheng K, Tahir R, Eric LK, Li M (2020) An analysis of generative adversarial networks and variants for image synthesis on MNIST dataset. Multimedia Tools Appl 79(19):13725–13752
18. Saxena D, Cao J (2021) Generative adversarial networks (GANs) challenges, solutions, and future directions. ACM Comput Surv (CSUR) 54(3):1–42
19. Zhang H, Goodfellow I, Metaxas D, Odena A (2019, May) Self-attention generative adversarial networks. In: International conference on machine learning. PMLR, pp 7354–7363
20. Cao YJ, Jia LL, Chen YX, Lin N, Yang C, Zhang B, Dai HH (2018) Recent advances of generative adversarial networks in computer vision. IEEE Access 7:14985–15006
21. Arora A, Shantanu (2022) A review on application of GANs in cybersecurity domain. IETE Tech Rev 39(2):433–441
22. Mirza M, Osindero S (2014) Conditional generative adversarial nets. arXiv preprint arXiv: 1411.1784
23. Di Mattia F, Galeone P, De Simoni M, Ghelfi E (2019) A survey on gans for anomaly detection. arXiv preprint arXiv:1906.11632
24. Odena A (2016) Semi-supervised learning with generative adversarial networks. arXiv preprint arXiv:1606.01583
25. Cai Z, Xiong Z, Xu H, Wang P, Li W, Pan Y (2021) Generative adversarial networks: a survey toward private and secure applications. ACM Comput Surv (CSUR) 54(6):1–38
26. Mao X, Li Q, Xie H, Lau RY, Wang Z, Paul Smolley S (2017) Least squares generative adversarial networks. In: Proceedings of the IEEE international conference on computer vision, pp 2794–2802
27. Zhu JY, Park T, Isola P, Efros AA (2017) Unpaired image-to-image translation using cycle-consistent adversarial networks. In: Proceedings of the IEEE international conference on computer vision, pp 2223–2232
28. Larsen ABL, Sønderby SK, Larochelle H, Winther O (2016, June) Autoencoding beyond pixels using a learned similarity metric. In: International conference on machine learning. PMLR, pp 1558–1566
29. Simonyan K, Zisserman A (2014) Very deep convolutional networks for large-scale image recognition. arXiv preprint arXiv:1409.1556
30. Karras T, Laine S, Aila T (2019) A style-based generator architecture for generative adversarial networks. In: Proceedings of the IEEE/CVF conference on computer vision and pattern recognition, pp 4401–4410
31. Berthelot D, Schumm T, Metz L (2017) Began: boundary equilibrium generative adversarial networks. arXiv preprint arXiv:1703.10717
32. Ledig C, Theis L, Huszár F, Caballero J, Cunningham A, Acosta A, Shi W (2017) Photo-realistic single image super-resolution using a generative adversarial network. In: Proceedings of the IEEE conference on computer vision and pattern recognition, pp 4681–4690
33. Miyato T, Kataoka T, Koyama M, Yoshida Y (2018) Spectral normalization for generative adversarial networks. arXiv preprint arXiv:1802.05957

34. Salehi P, Chalechale A, Taghizadeh M (2020) Generative adversarial networks (GANs): an overview of theoretical model, evaluation metrics, and recent developments. arXiv preprint arXiv:2005.13178

Horse Race Prediction Using Machine Learning Algorithms

Puja Chavan, Pankaj Kunekar, Prajakta Musale, Mrunal Shinde, Atharva Shinde, and Nikhil Shinde

Abstract Horse racing has been one of the most enjoyable forms of gambling long since. Winning streaks are difficult to achieve without the right knowledge and tools which makes the betting aspect of this game intimidating. This system predicts the horse with the highest chance of winning under certain circumstances using machine learning technology. For this model, 6348 races and 79,447 horse records were considered where a race had 14 horses each. This prevents the better from losing money to the immoral traps set by the bookie or the organization or due to his/her lack of knowledge and placing informed bets.

Keywords Winning horse · Machine learning · Python · Scikit learn · Classification

---

P. Chavan · P. Kunekar · P. Musale · M. Shinde · A. Shinde · N. Shinde (✉)
Vishwakarma Institute of Technology, Pune, India
e-mail: nikhil.shinde21@vit.edu

P. Chavan
e-mail: puja.cholke@vit.edu

P. Kunekar
e-mail: pankaj.kunekar@vit.edu

P. Musale
e-mail: prajakta.musale@vit.edu

M. Shinde
e-mail: mrunal.shinde211@vit.edu

A. Shinde
e-mail: atharva.shinde21@vit.edu

B. R. Devi et al. (eds.), *Proceedings of Fifth International Conference on Computer and Communication Technologies*, Lecture Notes in Networks and Systems 897,
https://doi.org/10.1007/978-981-99-9704-6_30

1 Introduction

Nobody has ever bet enough on a winning horse. For many years, horse race betting has aroused many emotions among the audience. Regal horses, excellent sportsmanship, competition and the possibility of winning bets, and making fast money attract the public effectively. The beauty of this activity is that anyone can enjoy it without having to learn complicated terms and mechanisms.

To predict a winning horse, we can use machine learning. Thus, to make betting easier for the users, our system takes into account different parameters that predict the highest chances of a horse winning a race. For the machine learning code with Python, we used Google Colab as it gives free access to GPUs (graphics processing units). Using the previous data, better suggestions could be provided to the users to place informed bets [1]. Horse race prediction is a solution for the prediction of the winning horse by taking into consideration different parameters such as the horse weight, jockey weight, horse's winning stats, etc. We are using machine learning, Python, and Scikit Learn. This system helps betters have fruitful returns from their betting pursuits.

As believed by our research and previous study, the relationship between the variation in odds and the winning horse was first done during this study [2]. The following paper discusses the various factors that affect the result of a horse race.

A few of them are field conditions, jockey winning stats, previous records, the draw number, etc. There have been various researches conducted. Zanjan et al., to predict the winner of the New York Aqueduct Race used a machine learning technique [3].

2 Literature Review

2.1 Structuring Your Paper

Chung, et al. [1] proposed a method for predicting horse racing results in Hong Kong. They got their data set from Hong Kong Jockey Club and using it for training SVM-based committee machine. Also, the accuracy of the prediction and the return rate were considered. The correct prediction ratio was found to be 7/10 for the horse to win the race. As a result, this method proved to be better than random forest and SVM in correctly predicting the outcome of the Race. For Hong Kong horse racing, the accuracy was high with 70.86%. Nemdharry et al. [2] predict the horse race winner at the Champ de Mars, which is organized by The Mauritius Turf Club (MTC). Collected around eight odds at different intervals. The module was trained using 232 races and tested on 27 races. This was the first study on the connection between odds and the rank of horses. There are many magazines and portals in Mauritius which provide plenty of information about horses. They tell us how we can also make a module or artificial network to guess the winner of horse races using the parameters such as jockey's weight and experience, stability, weighing capacity of the horse, and age of the horse. But the result was around 7.4% so they came to know the outcome of horse races at Champs de Mars does not depend on horse racing odds.

Chen et al. [4] highlight the importance of machine learning which involves using a computer algorithm to capture knowledge from data. A comparison is being made between three different human tracks using the machine learning technique. The author investigates a different problem-solving situation game-planning variable from eight competing dogs in the race. Approximately, 112 races were conducted every week. It was assumed that the behavioral capabilities of the participant were nearly stable which may improve or humble down. Further, a fair game is expected to be free from all the other external human factors. Jain et al. [5] proposed an approach for accurate RUL (Remaining Useful Life) prediction of milling cutters that are used in the industry such as CNC and VMC milling machines based on an artificial neural network (ANN). For this model, they are considering various statistical features such as average force, standard deviation, skewness, amplitude of force, etc. for its selection they are using stepwise regression. Statistics for including and excluding parameters in the model. This is a very effective technique as suggested by the authors. And also, for the training of the ANN model, a MATLAB neural network toolbox is used. As the cutter reaches the end of its life, the model successfully predicts the failures in impeding the milling cutters and allows us to take precautionary actions at the appropriate time.

Verma et al. [6] designed a model that accurately predicts the prices of houses. According to customers' needs and requirements, they first analyze various parameters and used them to find an ideal price for the house. Also, their data set contains all the required parameters that one needs to predict the house price. Using machine learning for prediction and analysis of the result they got, for this, they used linear regression, forest regression, and boosted regression for prediction. Further accuracy of the model can be increased by implementing neural network. Meena et al. [7] are predicting traffic flow information based on the factors affecting smooth traffic flow such as signals, accidents, repairing of roads or rallies and give better way or path to follow for saving time, which also reduces major problems such as accidents or traffic jams. In this project, they used machine learning, soft computing, and some deep learning algorithms for analyzing the data of transportation systems and also used image processing algorithms for the recognition of traffic signals for more accuracy. In future scope, they further add that they are focusing on deep learning and genetic algorithm which are important in data analysis and give higher accuracy than the existing project and also plans to make web server or an application. Some paper used the combination of machine learning and deep learning classifier to predict the seizure [8, 9].

Srinivas et al. [3] paper predicts the age, race, and gender of an individual using the CNN. The new and unique data set used was Wild East Asian Face Dataset (WEAFD). East Asian population's statistics were predicted using the same data set. The data set consisted of images captured in different environments. Gender prediction was a binary classification problem, whereas age and race was multi-class classification problem. CNN was used in order to predict statistics about the population using a

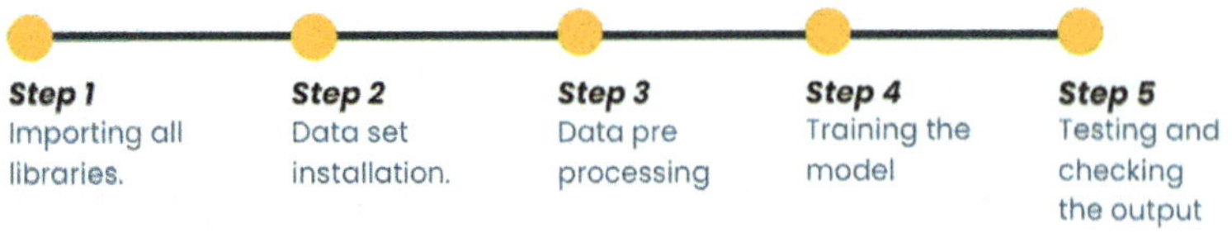

Fig.1 Flow of the proposed system

data set. Predicting the race of one individual was the most challenging part, followed by the prediction of age and gender. Using different network designs and different methods of data augmentation, the performance of the model can be significantly increased. Ganai et al. [10] puts forward an effective data race prediction method that has used lock recording-based incremental search on time-stamped locked histories. Record locking is the technique to prevent simultaneous access to data in a database, to prevent inconsistent results. This method can be easily understood with the 2 Bank clerk's example. Using this technique, the data is covered properly and the prediction accuracy also increases. A specially designed algorithm is used that can store and reuse past search results. Thus, the cost of reasoning useful data and overlapping searching is reduced. This algorithm can handle hundreds and thousands of different events and predict known–unknown results.

3 Methodology

3.1 Flow of Proposed System

3.2 Method

Figure 1 shows the flow of the proposed system.

Step 1: Importing all libraries

The required libraries such as NumPy (easy working with arrays/multidimensional arrays, mathematical computations, open-source, and easy to use), Pandas (fast and efficient for manipulating and analyzing data, easy handling of missing data), Matplotlib (data visualization and graphical plotting), Seaborn (it will be used to visualize random distributions), Scikit Learn (efficient tool for machine learning and includes models for classification, regression, clustering, etc.) are to be imported.

Step 2: Data set installation

https://www.kaggle.com/datasets/gdaley/hkracing

The horse racing data set from Hong Kong used for this project is downloaded from Kaggle. This data set has two files, the first being races.csv and the latter as runs.csv. The races.csv file has parameters such as race id, venue, the distance of the race, winning prize for the race also the type of race track surface and its condition, etc. The runs.csv file contains horse age, declared weight of the horse as well as the jockey, win odds, draw, won, etc.

Step 3: Data pre-processing

The two data set files, races.csv and runs.csv, were merged based on the race-id column so the variables make meaningful data. The parameters chosen for this study did not have any missing values. The venue and going columns in the data set had string inputs which were converted into numerical data by using an encoding feature provided by the Scikit Learn library. A label encoder was used for the race venue and an ordinal encoder was used for going.

Step 4: Training the model

ML algorithm was trained by feeding the data set in this step. The data set selected will be divided into 80% training and the 20% testing data set randomly, and the training data set will be used to train the model. The accuracy of different classification models has been tested including dummy classifier, logistic regression, decision tree, random forest, and support vector machine (SVM). Figure 2 shows the architecture of proposed model.

Step 5: Testing and checking the output

Fig. 2 Architecture

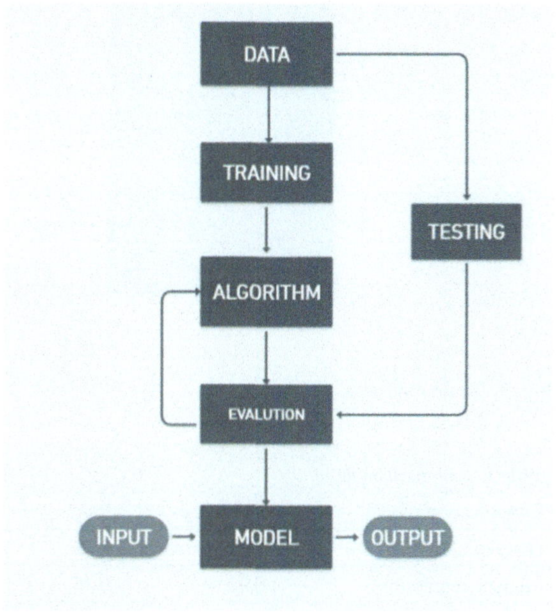

The testing part of the data set was used to test the ML model by supplying the testing data set as an input to the model. After testing, the accuracy of the classification model was compared.

4 Result and Discussions

The ML model was able to predict the winning horse from a race based on the data set that was used. From the predicted results, we also came to know which factors were the most important for winning and should be considered before placing any bets. Figure 3 shows the count plot of Won.

As the data set had more "not winning" horses than the "winning" horses, the "Dummy Classifier" always predicted "not winning" horses which gave an accuracy of 92% without actually having trained the model. Table 1 shows the comparison.

For "Logistic Regression" which is the trained classifier, the accuracy was found to be the same as that of the dummy classifier. Thus, for comparison of this classification

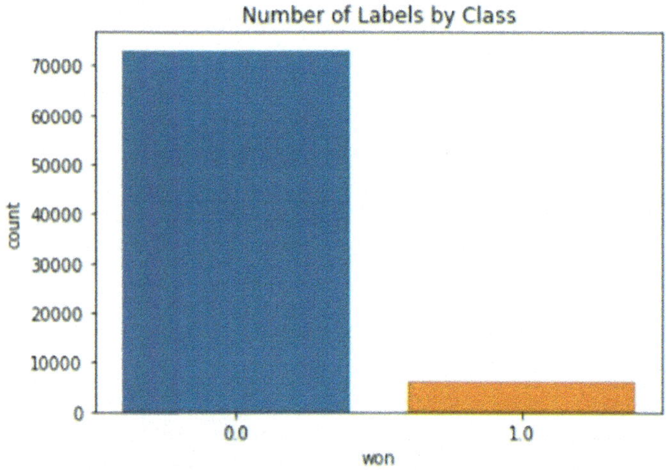

Fig. 3 Count plot: won

Table 1 Comparison table

Classification model	Accuracy	Precision	Recall	$F1$ score
Dummy classifier	0.92	0.46	0.50	0.48
Logistic regression	0.92	0.46	0.50	0.48
Decision tree	0.85	0.55	0.55	0.55
Random forest	0.92	0.62	0.52	0.51
Support vector machine	0.92	0.46	0.50	0.48

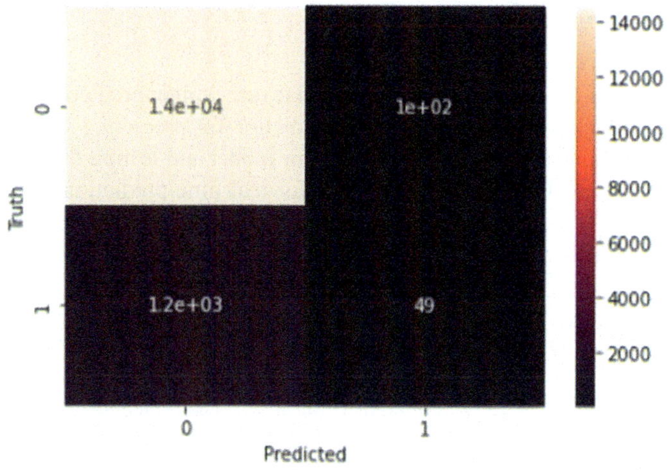

Fig. 4 Confusion matrix

problem, accuracy was not a suitable option. A basic model that always predicted "not win" was seen to have great accuracy, reducing the accuracy of predicting a win. Precision, Recall, and $F1$ score could have been used for comparison between models. For "Decision Tree," the accuracy was found to be 85% with a precision of 55% as compared to a precision of 46% for the Dummy Classifier and Logistic Regression. For "Random Forest," accuracy was 92% along with a high precision of 62% and for "Support Vector Machine" (SVM), the accuracy and precision were found to be the same as that of Logistic Regression.

After comparing all classification models, "Random Forest" was found to be the best algorithm for this classification problem because of its high precision and high accuracy. Figure 4 shows the confusion matrix.

The results from this particular study can be used effectively in various other important fields such as education, finance, industry, medicine, management, games, and businesses.

5 Future Scope

Hosting the model on the web along with creating a safe betting platform for the users. More accuracy on the prediction. Implementation of varied sports with a larger data set of different places. The way ML has been implemented for predicting the winner of a horse race, it can be used similarly for various sports.

6 Conclusion

In this research, a ML model is created for horse race winner prediction using Python and Sklearn. We developed a horse race prediction for predicting which horse will win. Training the model is the first half of our model and testing the trained model is as follow. Thus, the system can be used for real-time prediction of the winning horse.

References

1. Chung WC, Chang CY, Ko CC (2017) An SVM based committee for prediction of Hong-Kong horse racing. In: 2017 10th international conference on Ubi-media computing and workshops (Ubi-Media)
2. Nemdharry N, Ramma TRH, Mungroo R (2017) Impact of the variation of horse racing odds on the outcome of horse races at the champ de Mars. University Of Mauritius
3. Srinivas N, Atwal H, Rose DC, Mahalingam G, Ricanek K, Bolme DS (2017) Age, gender, and fine-grained ethnicity prediction using convolutional neural networks for the East Asian face dataset. In: 2017 IEEE 12th international conference on automatic face & gesture recognition
4. Chen H, Rindc PB, She L, Sutjahjo S, Sommer C, Neely D University of Arizona, Expert prediction symbolic learning and neural working an experiment on greyhound racing
5. Jain AK, Lad BK (2015) Speed milling cutters based on artificial neural network. In: International conference on robotics, automation, control and embedded systems–RACE 2015 18–20 February 2015. Hindustan University, Chennai, India
6. Verma A, Sharma A, Doshi S, Nair R (2018) House price prediction using machine learning and neural networks. In: International conference on inventive communication and computational technologies
7. Meena G, Sharma D, Mahrishi M (2020) Traffic prediction for intelligent transportation system using machine learning. In: 3rd international conference on emerging technologies in computer engineering: machine learning and internet of things (ICETCE-2020). 07–08 February 2020. IEEE Conference Record 48199
8. Chavan PA, Desai S (2023) Effective epileptic seizure detection by classifying focal and non-focal EEG signals using human learning optimization-based hidden markov model. Biomed Signal Process Control 83:104682. ISSN 1746-8094. https://doi.org/10.1016/j.bspc.2023.104682
9. Chavan P, Desai S (2021) A review on BCI emotions classification for EEG signals using deep learning. Rec Trends Intensive Comput, Adv Parallel Comput 544–551. 10.3233/ APC210241
10. Ganai MK Efficient data race prediction with incremental reasoning on time-stamped lock history

An Approach to Blockchain-Based E-Voting System Using Peer-To-Peer Protocol

Divya Boreda, Ashok Kumar Nanda, Chaitra Sai Jalda, V. Pradeep Kumar, and Kumar Gautam

Abstract It has been difficult for a long time to create a safe electronic voting system that delivers the openness and adaptability that come with electronic systems while maintaining the fairness and privacy of present voting procedures. Several forms of electronic voting have been tried in place of current procedures to increase effectiveness and decrease errors. Achieving such systems, however, presents numerous difficulties. In order to increase the viability of electronic voting systems, blockchain is a new disruptive technology. This study lays forth a strategy for using blockchain to build voting systems with greater effectiveness. As part of our effort to create a voting system that supports peer-to-peer protocols and other elements, we also make use of some of the fundamental characteristics that blockchain technology has to offer.

Keywords Blockchain · E-voting mechanism · Security

D. Boreda · A. K. Nanda (✉) · C. S. Jalda · V. P. Kumar
B V Raju Institute of Technology Vishnupur, Narsapur, Telangana 502313, India
e-mail: ashokkumarnanda@yahoo.com

D. Boreda
e-mail: boredadivya511@gmail.com

C. S. Jalda
e-mail: chaitrajalda278@gmail.com

V. P. Kumar
e-mail: pradeepkumar.v@bvrit.ac.in

K. Gautam
Gwangju Institute of Science and Technology (GIST), Gwangju, South Korea
e-mail: edumonk@ieee.org

B. R. Devi et al. (eds.), *Proceedings of Fifth International Conference on Computer and Communication Technologies*, Lecture Notes in Networks and Systems 897,
https://doi.org/10.1007/978-981-99-9704-6_31

1 Introduction

In a democratic country like India, voting plays a major role for forming a governing body by people. The traditional or paper-based polling method served to increase people's confidence in the selection by majority voting. It has helped to make the democratic process and the electoral system worthwhile for electing constituencies and governments more democratized. It is essential to ensure that assurance in voting does not diminish. A recent study revealed that the traditional voting process was not wholly hygienic, posing several questions, including fairness, equality, and people's will, and was not adequately quantified and understood in the form of government.

Technology introduced the new electronic voting techniques and methods, which are essential and have posed significant challenges to the democratic system. Electronic voting increases election reliability when compared to manual polling. In contrast to the conventional voting method, it has enhanced both the efficiency and the integrity of the process. Because of its flexibility, simplicity of use, and cheap cost compared to general elections, electronic voting is widely utilized in various decisions. The centralized voting system has many challenges where still many researchers are focusing on tackling the difficulties to build a safe, secure, reliable, and anonymous voting system using decentralized blockchain technology.

A decentralized node for online voting or electronic voting is provided by blockchain technology. Voting systems have therefore increasingly shifted away from paper ballots and polling booths in recent years. The advantages of end-to-end verification offered by distributed ledger technologies like blockchain have recently led to the development of electronic voting systems. Because of its decentralization, non-repudiation, and security protection features, blockchain offers a compelling alternative to traditional electronic voting systems. Using a decentralized system that allows voters to cast ballots whenever they want from anywhere, it is utilized to conduct boardroom and public voting, ensuring security. We must deal with well-known issues like maintenance and fraud notwithstanding the use of electronic voting. The blockchain is a relatively new technology, and academics are working to make use of features like openness, confidentiality, and non-repudiation that are crucial for voting applications. Recent initiatives to use blockchain technology to safeguard and correct elections have gotten a lot of attention due to the use of blockchain for electronic voting applications. There are numerous studies being done right now to develop a secure and trustworthy voting system while addressing anonymity and security concerns. With the help of the decentralized system, attention is shifting to putting the public's hands on an easy, safe, and anonymous voting process.

2 Related Works

The conventional voting procedures required a lot of time and effort and lacked credibility. Inefficiency is frequently caused by the inaccuracy of these systems. Blockchain was therefore employed to deliver reliable results for an effective and efficient e-voting system. Vote counting and end-to-end verification were its principal uses. Data tampering is prevented as well as the participation of third parties, which frequently leads to data breaches. By enabling voting from any location with an internet connection, it lowers organizational costs and makes voting more accessible. These attributes collectively aid in increasing user confidence in electronic voting systems. The contemporary techniques for the electronic polling that contributed to the creation of a reliable and secure system are relatively compared in Table 1 to one another.

The paper's importance in establishing trust in an online polling mechanism based on blockchain is highlighted in the summary that follows. The main elements that go into developing a reliable and secure online polling mechanism are offering privacy, voting anonymity, preventing stuffing of ballots, full audit trails, and the ability to vote from anywhere. The many online platforms, frameworks, consensus protocols, cryptographic algorithms, and system development technology are displayed in Table 1 along with their security features such as Auditability, Scalability, Accuracy, Affordability, and many others. These systems are not effective for large-scale or national election systems and have scaling concerns. However, they can be employed for small-scale systems.

The voting system that combined online electronic voting with blockchain technology was proposed by authors [6], to provide voters security. Everyone had online access to all additional voter data, including the voters' city, name, and number of votes. This approach allowed secure polling from any location, preventing the possibility of double counting.

Direct-Recording Electronic with Integrity and Enhanced Privacy (DRE-ip) system modification was proposed by authors [7] as a safe mechanism for a secret,

Table 1 Present blockchain-based online polling mechanisms differentiation

Consensus protocol	Cryptographic algorithms	Language	Framework	Online polling platform
POW	ECC	C++/Python	Bitcoin	Follow My Vote [1]
BFT-r	ElGamal	Python	Bitcoin	Agora [2]
PBFT	AES/GCM	Go/JavaScript	Hyperledger fabric	Voatz [3]
PET	ECC	NP	Private/local blockchain	Polyas [4]
PBFT	ECC/ElGamal	Go/JavaScript	Hyperledger Fabric	Luxoft [5]

authenticated, and end-to-end verifiable ballot election. This method prevents anyone from creating and secretly posting a valid ballot on a public bulletin board. To stop ballot stuffing attempts, the system offers a verifiable and secure framework for registering voters and verification. Blockchain-based voting system (BBVS) was introduced by authors [8] for the legislative elections in Jordan. The BBVS is a centralized, private blockchain that is deployed hierarchically, allowing voters to cast their ballots at two levels. There are two levels: one for the group and one for individual group members. The authors presented an innovative yet safe blockchain-based online polling mechanism that evaluates the efficacy, certainty, and integrity of the electoral process.

On the online platform where elections are conducted via a decentralized application, an electronic voting system (dApp) was presented by authors [9]. This system's simplicity, transparency, security, portability, reliability, and trustability were all guaranteed by the creators' use of blockchain technology. An analysis of blockchain-based voting systems was conducted by authors [10] and categorized them according to the number of participants, the types of blockchains used, and the consensus methods employed. The authors analyzed several voting systems in a methodical way and noted significant gaps in the literature as well as potential directions for future research.

The authors [11] looked at blockchain-based online polling and outlined the circumstances that lead to an attack on the system in order to discover transaction flexibility attacks within the system and develop appropriate defense measures. On a blockchain testbed hosting an online polling application, a transaction flexibility attack was successfully demonstrated. The studies determined the importance of block creation rate and network lag indicated the paths for further analysis in order to execute transaction malleability attacks successfully. Authors [12] presented a voting method for significant elections based on the blockchain that is safe, scalable, and useful. By designing the ballot using a randomizer token that serves as a user-facing black box, refusal of receipts and resistance to coercion are guaranteed. It was possible to verify everything thanks to the append-only structure.

A private blockchain network was created by authors [13] for polling software that made use of data analysis and storage technology and smart contracts to carry out transactions, and exchange data among peer blocks inside the network. Moreover, consistency, duration, and scalability characteristics were taken into consideration when analyzing the performance of the Ethereum clients Parity and Geth. Election fraud has long been a significant problem for the voting process. Authors [14] emphasized the key roles played by blockchain in the evolution of a reliable and secure polling system. Yet, because of how complicated the technology was, there were disputes in the voting process. To ascertain the technology's advantages and disadvantages, the authors conducted a qualitative analysis by speaking with blockchain specialists and election watchers.

The outcomes showed the value of human elements and the need for confidence in the electoral process. Proof of Credibility (PoC) and Proof of Stake are combined in the suggested hybrid consensus model (PSC-B chain) (PoS) by authors [15]. The PSC-B chain and the sharding method were integrated by the authors to ensure the scalable performance of the blockchain-based electronic polling system. The suggested model addressed the issues with an electronic voting system's scalability, effectiveness, and latency. A reliable and secure computer environment with smart contracts was implemented to guarantee the validity and safety of the voting procedures.

The proposed hybrid blockchain and traditional blockchain are then used to process it further in order to analyze the security aspect. Multi-agent and intelligent agent systems for ABVS were proposed by authors [16] as a way to improve security and safety at polling places and to create a sense of independence from client voting applications. To make it end-to-end verifiable, it was built using blockchain technology and used with an electronic voting procedure that was integrated into a controlled internet non-remote polling system. By combining fingerprint authentication, ring signature, and blockchain technology, a secure online polling system has been created by authors [17]. Because this method does not require many people, it is more verifiable and secure. Authors [18] examined and considered various uses for blockchain that were not already being discussed and assessed the suitability of blockchain from a quantitative standpoint. To reach clients for their benefits and market production, this was done to identify the applications' potential. The blockchain's advancement has sparked a quick growth in digital services and the creation of specialized knowledge in their legal and technological facets. Authors [19] provided examples of how blockchain might be applied to address safety and security-related issues. Voting on national television for singing competitions was made easier using blockchain, which also gives openness and a fair voting mechanism. Authors [20] provided an illustration of how blockchain can be used to solve problems with voting system transparency and confidence. Additionally, they strengthened democracy by promoting election openness and voter confidence and by looking at how technology affected issues relating to the e-voting system.

To increase voter confidence in the electoral process, a safe and online polling system with auditable cryptography has been proposed by authors [21] to replace the conventional voting techniques. The suggested system is based on the Preta Voter method of electronic voting using paper ballots. To create anonymous channels, the scheme used a combination of servers. Complex protocols were required by Mixnets to retain shared mix keys. Mixnets were also difficult to implement on a big scale and were susceptible to corruption. To guarantee verifiability, security, ballot secrecy, and trustworthiness, the suggested approach used paper ballots with robust cryptographic and tested security algorithmic characteristics.

3 Design Infrastructures

In this section, we have discussed about Java socket programming used in the implementation. Java socket programming is commonly used in various networked applications, including client–server systems, distributed computing, real-time messaging, and network protocols. It provides a flexible and powerful way to establish network communication between applications running on different machines.

4 Design Solutions

The proposed solution design is based on the peer-to-peer network Architecture using TCP/IP protocols. This system would allow for faster and more secure voting than traditional methods while providing greater transparency. This design utilizes the Java socket programming language to ensure the system is lightweight, quick, and reliable. In the experiment setup, as mentioned in Fig. 1, we initiate a connection to the devices through the web server and public network via a peer-to-peer network. The web server depicted in Fig. 1 uses Java socket programming language as the application logic to communicate between the devices and the blockchain.

To facilitate voting, our platform/setup offers users a Graphical User Interface (GUI) through the System Terminal or command prompt. The votes are then cast onto the machines via the nodes, and all the communication is tracked under the event listener, which listens to each node, captures the details like the transaction and voting details, and is securely recorded on the blockchain. The information or the details gathered by the user are encrypted using "AES" encryption standards, which help in protecting the data from tampering and privacy loss. The security of user data is paramount in any application or website. To ensure the safety of sensitive information, AES encryption standards encrypt the data collected from the user. This

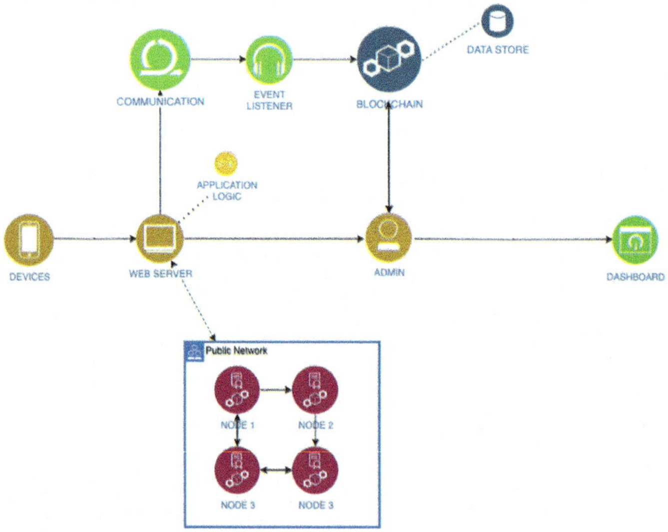

Fig. 1 Design solution for blockchain E-voting system

encryption standard helps to protect the data from tampering and privacy loss, thus providing a secure environment for users to store their data.

Blockchain details can be easily accessible for administrators to generate lists of transactions and votes given by users through multiple systems. This helps to ensure the accuracy and consistency of data. We guarantee that the log file remains tamper-proof. The data inside cannot be modified or altered, ensuring your user details are secure and safe. This also ensures readiness and prevents any potential data leakage. Also, it stores all records and transactions securely in a log file, making it easily accessible for administrators. Hence, after obtaining all the details of the vote pitched by the users, the admin can also produce the voting results over the dashboard.

5 Results

Blockchain-based polling mechanisms have the potential to revolutionize the way we conduct elections. Here are some key findings and insights on the topic. Examples: Several blockchain-based voting systems have been developed and tested, including Voatz, Horizon State, and Follow My Vote. However, these systems have faced challenges related to security, accessibility, and accuracy, which have raised concerns about their suitability for large-scale elections.

We have studied various examples of blockchain-based voting system. Based on the study, we have designed a prototype, which helps the voters cast their vote efficiently, without hesitation of his/her vote being tampered with. We also discuss the salient features in this paper, which help to decide the availability and the integrity of the solution design. The main objective of the solution design is to provide a lightweight, robust, and secure voting mechanisms via blockchain to the users, and also develop trust to use the decentralized system of voting. We also observe that the authenticity of the vote of the users remains intact and is immutable. We used the solution design on various platforms and it worked as expected. We also observed that the data stored after the voting were able to easily be accessed by the administrators, for declaring the results.

Overall, while blockchain-based voting systems have the feasible to address some of the challenges allied with traditional polling mechanisms, further research, development, and testing are needed to ensure their suitability for widespread adoption. The ability to produce cryptographic parameters with sufficient performance is crucial for an election which is depicted in Fig. 2 [22] with voting options, voter participation, memory usage, and runtime. The runtime should be linear in the number of voters, as shown in Fig. 2a. The complexity of the runtime should also be linear in the number of votes. With respect to the amount of voting alternatives, the time needed to generate the information behaves linearly. When the information is generated for various numbers of voters, Fig. 2b depicts the runtime's linear pattern. The necessary storage for storing vote transactions is shown in Fig. 2c. In relation to the quantity of transactions, storage is similarly linear. Figure 2d illustrates how long it

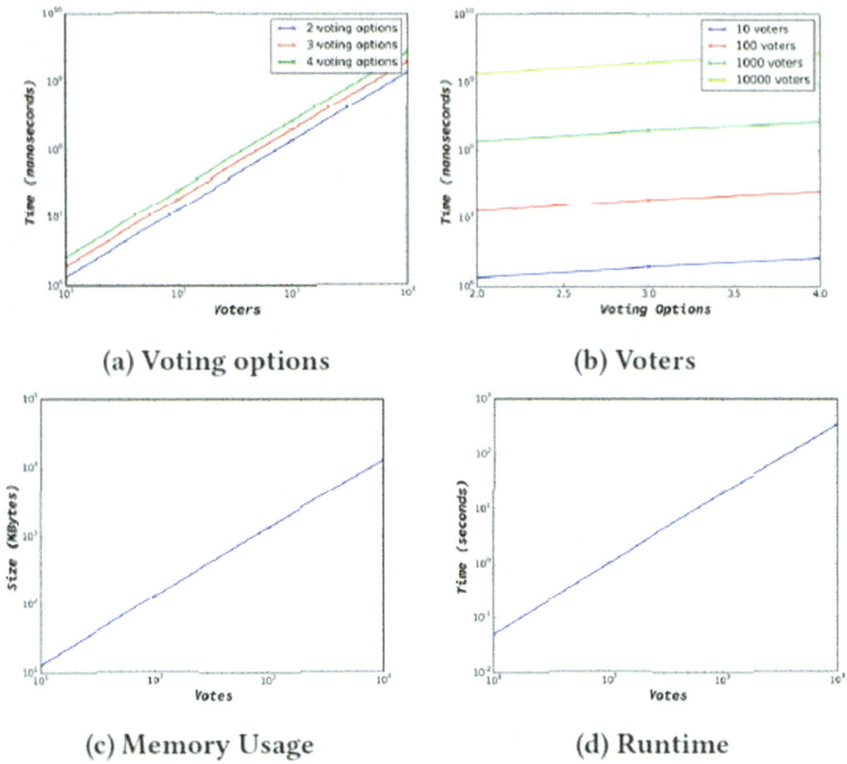

Fig. 2 Result analysis of blockchain E-voting system [22]

takes to process these votes, demonstrating that performance is also proportional to the quantity of operations.

Results of the code: First the vote officer would start the voting process and wait for clients (voting devices) to get connected. After the connectivity, the voting phase would start and the voting devices would collect the basic voter details like voter id, voter name, and party to which you want to vote and then the vote is registered. Please remember if the same person comes again to vote, the person who already voted and provided the same details then the vote will not be considered. After voting, the admin/server can be used to see the number of votes and the list of the members who voted for which party. All the data is encrypted and stored in the genesis block and is encrypted by AES. Here the blockchain data from the voting process is stored in a file in our machine in the form of log file. Here data is encrypted, which stores encrypted data in a hexadecimal format in a log file.

Thus, while exploring, we found that there are various methodologies that spoke about e-voting through blockchain technology, but this system lacked a comprehensive study; the implementation of other sources proved to be a very costly and time-taking affair. Hence, considering all the scenarios, we acknowledge that the proposed design solution would be beneficial for the users to cast their votes in a decentralized manner.

6 Conclusions and Future Work

This proof-of-concept study significantly contributes because it illustrates the problems with electronic voting systems and blockchain-based security requirements. It also assisted individuals in comprehending the issues with electronic voting and the opportunities and challenges presented by blockchain technology in this context. This proof-of-concept model for an electronic voting system based on blockchain technology would enable us to examine the concept of decentralized voting to implement Proof-of-work (POW) in the future to add a block to the blockchain, which would allow the blockchain to be used in the development of voting systems.

References

1. Vote FM (2021) The secure mobile voting platform of the future—follow my vote. Available online: https://followmyvote.com/. Accessed 26 July 2021
2. Agora (2020). Available online: https://www.agora.vote. Accessed 28 July 2020
3. Voatz—Voting Redefined ®® (2020). Available online: https://voatz.com. Accessed 28 July 2020
4. Polyas (2015). Available online: https://www.polyas.com. Accessed 28 July 2020
5. Luxoft. Available online: https://www.luxoft.com/. Accessed 28 July 2020
6. Divya K, Usha K (2022) Blockvoting: an online voting system using block chain. In: 2022 international conference on innovative trends in information technology (ICITIIT) 1–7. https://doi.org/10.1109/ICITIIT54346.2022.9744132
7. Panja S, Roy B (2021) A secure end-to-end verifiable e-voting system using blockchain and cloud server. J Inf Secur Appl 59(2021):102815. https://doi.org/10.1016/j.jisa.2021.102815
8. Malkawi M, Yassein MB, Bataineh A (2021) Blockchain based voting system for Jordan parliament elections. Int J Electr Comput Eng 11(10):4325–4335. https://doi.org/10.11591/ijece.v11i5.pp4325-4335
9. Kumar A, Goyal M (2021) Sustainable e-infrastructure for blockchain based voting system 221252. https://doi.org/10.1002/9781119792079.ch7
10. Huang J, He D, Obaidat M, Vijayakumar P, Luo M, Choo KKR (2021) The application of the blockchain technology in voting systems: a review. Comput Surv 54(04):1–28. https://doi.org/10.1145/3439725
11. Khan KM, Arshad J, Khan M (2021) Empirical analysis of transaction malleability within blockchain-based e-Voting. Comput Secur 100(01):102081. https://doi.org/10.1016/j.cose.2020.102081
12. Dimitriou T (2020) Efficient, coercion-free and universally verifiable blockchain-based voting. Comput Netw 174(04):107234. https: //doi.org/https://doi.org/10.1016/j.comnet.2020.107234
13. Dhulavvagol P, Bhajantri V, Totad S (2020) Blockchain ethereum clients performance analysis considering e-voting application. Proc Comput Sci 167(01):2506–2515. https://doi.org/10.1016/j.procs.2020.03.303
14. Baudier P, Kondrateva G, Ammi C, Seulliet E (2021) Peace engineering: the contribution of blockchain systems to the e-voting process. Technol Forecast Soc Change 162:120397. https://doi.org/10.1016/j.techfore.2020.120397

15. Abuidris Y, Kumar R, Yang T, Onginjo JO (2020) Secure large-scale E-voting system based on blockchain contract using a hybrid consensus model combined with sharding. ETRI J 43
16. Pawlak M, Poniszewska-Maranda A (2019) Blockchain e-voting system with the use of intelligent agent approach 145–154. https://doi.org/10.1145/3365921.3365927
17. Suralkar S, Udasi S, Gagnani S, Tekwani M, Bhatia M (2019) E-voting using blockchain with biometric authentication. Int J Res Anal Rev 6(1):72–81
18. Zeadally S, Abdo JB (2019) Blockchain: trends and future opportunities. Internet Technol Lett 2(09 2019). https://doi.org/10.1002/itl2.130
19. Desai S, Han M, Li L, Li Z, He J, Xu X (2019) Untampered electronic voting in entertainment industry: a blockchain based implementation. In: Proceedings of the 20th annual SIG conference on information technology education (SIGITE '19). Association for Computing Machinery, New York, NY, USA, pp 166. https://doi.org/10.1145/3349266.3351386
20. Moura T, Gomes A (2017) Blockchain voting and its effects on election transparency and voter confidence. In: Proceedings of the 18th annual international conference on digital government research (dg.o '17). ACM
21. AboSamra KM, AbdelHafez AA, Assassa GMR, Mursi MFM (2017) A practical, secure, and auditable e-voting system. J Inf Secur Appl 36(2017):69–89. https://doi.org/10.1016/j.jisa.2017.08.002
22. Killer C, Rodrigues B, Matile R, Scheid E, Stiller B (2020) Design and implementation of cast-as-intended verifiability for a blockchain-based voting system. In: SAC '20: proceedings of the 35th annual ACM symposium on applied computing. March 2020, pp 286–293. https://doi.org/https://doi.org/10.1145/3341105.3373884

Artificial Cognitive Intelligence and Information Technology in Cybersecurity

Tamanna Jena Singhdeo, Adyesha Singhdeo, J. R. Mohanty, and Suresh Satapathy

Abstract Cognitive intelligence is intelligence exhibited by the process of acquiring knowledge and understanding through sense and thought processes. Cognitive intelligence in cybersecurity includes understanding the prospect of protection of digital devices, computer networks, websites, and cloud computing from both external and internal threats. Cyber threats like phishing, smishing, and data theft cause threats to business processes. In today's day and age mobile, social networking and IoT are generating huge amounts of data. Analysis of the large volume of data helps in identifying the different attack vectors and their behavior pattern. It further can exceed the cognitive skills of security systems, rules, and protocols in organizations to keep the system more secure. Security analysts are now gathering huge amounts of data on users, networks, and transaction details. The captured data may reveal weak spots, hidden patterns, or disguised malicious activities. Evaluating and contemplating the underlying meaning of data is time-consuming and can be missed easily. In this paper, we are suggesting a few machine learning-based approaches to find the correlation between the data. This paper presents a cognitive classifier model that integrates machine learning with the data obtained from networks, web servers, logs, and cognitive processes to generate insight and knowledge which can be used to make the response actions better and more secure.

Keywords Cognitive intelligence · Machine learning · Weka · Cybersecurity

T. J. Singhdeo (✉)
Fairleigh Dickinson University, Vancouver, Canada
e-mail: tamannasinghdeo@gmail.com

A. Singhdeo
University of British Columbia, Vancouver, Canada

J. R. Mohanty · S. Satapathy
KIIT University, Bhubaneswar, India

B. R. Devi et al. (eds.), *Proceedings of Fifth International Conference on Computer and Communication Technologies*, Lecture Notes in Networks and Systems 897,
https://doi.org/10.1007/978-981-99-9704-6_32

1 Introduction

According to [1, 2] in the year 2017, there were almost 8.3 billion digital devices that were connected which increased to 11.19 billion in 2018, and it is expected to increase to 20.4 billion devices in 2020. However, Statista suggested that there were 30.73 billion connected devices by the year 2020 which will increase to 75.44 billion by the year 2025 [3]. The interconnection of digital devices generates a huge volume of data. According to Gartner 2023, in strategic technology trends, the observable data is the most precious monetizable asset for any organization [4, 5]. If the data and metadata are used using proper AI-based models to extract business capabilities, then companies can use the obtained knowledge to gain competitive advantage and security. Additionally, by 2026, the organizations that operationalize AI for transparency, trust, and security were likely to see a 50% improvement in terms of business goals and user acceptance [4].

2 Cognitive Information and Cognitive Computing

The brain and natural intelligence extract information are categorized at the level of data, information, knowledge, and intelligence [6, 7]. In this paper, we are trying to follow the same path. The data level represents the raw data. At the information level, we clean the data and define a scope. At the knowledge level, we try to get insight. Finally, at the intelligence level, we try to apply the gained knowledge in the system to make it more robust and efficient.

Data and information processing have been studied for a long time. However, with the emergence of Big data and powerful data processing tools, researchers and academicians are non-stop working to application-oriented data processing. The research progress in theories, mathematical approaches, systematic studies in cognitive informatics, and cognitive computing is yet to be mainstream. The basic approach is to invent cognitive computers, cognitive robots, and cognitive systems that extend human learning ability, wisdom, and creativity [8–12]. In cybersecurity, machine learning can be beneficial to understand the weak spot, identify the attackers, and categorize the actors of intrusion detection systems [13–16]. Applications like cognitive robots, cognitive learning engines, cognitive Internets, cognitive translators, cognitive control systems, and cognitive automobiles are a few functional aspects of cognitive computing [17, 18]. Implementation of cognitive computing will enhance the security of IT in businesses of all sizes [19].

3 Cognitive Classifier Model

We proposed a cognitive classifier model, where we collected datasets from various authentic websites shared for educational and research purposes. The flowchart of the model is represented in Fig. 1. It is pivotal in research to use datasets of high quality and authenticity. It enhances the research findings. The obtained dataset was cleaned and decluttered. We used Jupyter Notebook, where both R studio and Python kernels are used for normalizing and cleaning data. Additionally, we identified the important features which were relevant to the scope. We used Weka, the web-based graphical user interface (GUI) for machine learning. The classifiers offered by Weka are intuitive and effective. The algorithms we implemented are Na¨ıve Bayes, treeJ48, and Decision Stump on the dataset. The dataset on which we worked is a large dataset, having more than 1.5 M instances. Each instance of the dataset has more than 130 features and functionalities.

Businesses are now using cyber specialists or data analysts to evaluate every action of the users. It is found that actions are basically categorized into malicious or non-malicious. If the motivation of any action or pattern is found to be malicious for others or inclined to extract information to gain competitive advantage, then the user is flagged as the attacker. On the basis of the outcome of the machine learning on the data and metadata of the attacker, user profiles are created. Table 1 outlines the categories of the attackers along with the impact of the attack. The important decision factors in identifying the category of the attacker are the type of attack, the extent of the impact, behavior pattern, and cultural characteristics.

The cognitive classifier model includes the following processes:

1. Collection of data
2. Cleaning of data
3. Generation of hypothesis
4. Research of Hypothesis
    - Apply correlation methods to investigate the intelligence in the incident/ transaction

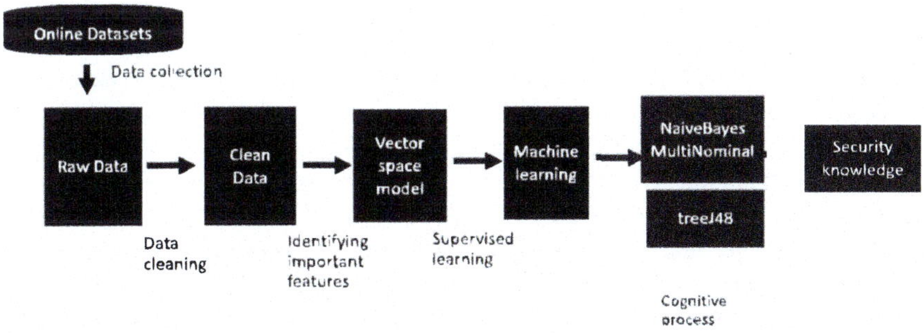

Fig. 1 Design flow of the cognitive classifier model

Table 1 Categories of cyber actors

Category	Motivations	Adverse Impact
Script Kiddies	Curiosity	Low
Hactivist	Defacement of individual or group for political reasons	High
Cyber punks	Exploring and engaging malicious attacks	High
Coders	Write/use automated tools	Low
Insiders	Displeased employer	High
Cyber terrorists	Spreading fear and instability	High
Hackers	Leaking data	High
Penstesters	Finding/identifying vulnerability	Low

- Identify the infected IPs
- Quantify the motive behind the incident
- State the profile of the attacker
- Analyzed other incidents and the pattern of the attack

4 Methodology and Experimental Setup

The dataset used is exclusively curated. We experimented using a dataset by the Canadian Institute for Cybersecurity. The datasets are of diverse types, including DNS datasets [4], IDS datasets, malware datasets, and Darknet datasets. The dataset which we used most for our experiment is collected from CICDarknet2020 [20], CIC-Bell-DNS2021 [20–22], and CIC-Bell-DNS-EXP-2021 [23]. The former dataset consists of audio-stream, browsing, chat, email, video-stream, VOIP [24]. The later datasets consist of benign, malware, phishing, and spam datasets. These datasets are collected during 2015–2021 in collaboration with CIC and Bell [22, 23]. After collecting datasets, we cleaned the data using Jupyter Notebook. Some data were balanced (60/40%, benign/malicious) and some were unbalanced (90/10%, benign/malicious) and used Python. After cleaning the data, we used the functions to date time in pandas to make the data into a more readable form. After cleaning data, we used Weka, the GUI to classify using different types of machine learning algorithms. Table 2 gives the cleaned data which is used as an input for the cognitive classifier model.

Table 3 outlines the Darknet network traffic details. It was having more than 120 features. We decided to make the data specific to understand. So, we identified 27 features we would like to consider. It further helped us to decide vector space model [25]. Figure 1 outlines the design flow of the cognitive classifier model outlines the design outlines the vector space model, where each row corresponds to each row of the dataset. However, as the data obtained is in raw form, we identified 27 important features as the column of matrix X. Y is another matrix that consists of a type of traffic category explained in Table 3. We used multiple types of classifiers to categorize the dataset into classes based on the types mentioned in the above Table 3.

Table 2 Relations of data and knowledge in cognitive classifier model

	Flow ID	Src IP	Src port	Dst IP	Dst port	Protocol	Timestamp	Flow duration	Total fwd pkt	Total bwd pkt
0	1,057,216	10.152.152.11	57,158	216.58.220.99	443	6	24/7/15,14:09:40	229	1	1
1	105,774	10.152.152.11	57,159	74.125.136.120	443	6	24/7/15,14:10:48	407	1	1
2	1,057,173	10.152.152.11	57,160	173.194.65.127	19,305	6	24/7/15,14:09:45	431	596	400

Table 3 Details of darknet network traffic

Traffic category	Applications	Shortlisted features
Audio-stream	YouTube, Netflix	Flow ID, Source IP,
Browsing	Firefox and Chrome	Source port, destination port
Chat	CQ, AIM, Skype, Facebook, and Hangouts	Protocol, timestamps
Email	SMTPS, POP3S, and IMAPS	Flow duration, total length
P2P	uTorrent and Transmission (BitTorrent)	Forward urgent flag
Transfer	Skype, FTP over SSH (SFTP) and FTP over SSL (FTPS) Vimeo and YouTube	SYN flag down/up
Video-stream	Facebook, Skype, and Hangouts voice calls	Ratio average packet
VOIP	YouTube, Netflix	Label

The outcome of the cognitive classifier model allows organizations to get insight into the existing risks, vulnerabilities, possible attacks, hidden attacks, patterns of attackers, and gaps. Based on the knowledge obtained, the organizations can project policies, conditions, and strategies in the near future. The outcomes are categorized primarily into two levels of security insights. The first/lower level is to find the technical solution, which can be further used in automation or responsive action. The second level/higher level of awareness helps to establish modified strategic decisions. The latter mainly involves human intervention. It is risky and needs a visionary approach. However, the scope of this paper is to compare classifiers on the relevant dataset and obtain security knowledge. We used Na¨ıve Bayes classifier [26–28] J48 algorithms [28–30], and Decision Stump [30] as classifiers. The former one is mainly based on probability, the middle one is based on multiple-level decision trees and the latter one consists of a one-level decision tree. Table 4 outlines the result and compares the output of the algorithms.

It is found that tree J48 algorithm is able to classify the dataset to 95% accuracy. Output of decision stump is found to be 51% and Naive Bayes performance is 34%. The confusion matrix obtained using treeJ48 is a diagonal matrix, which shows that the classifier is clearly able to categorize the classes. However, the confusion matrix

Table 4 Comparison of classifiers

	Total instances	Precision (%)	Relative absolute error (%)	Correctly classified	RMS error	Cross validation	Time taken to build model (s)
Naive Bayes	141,530	34.8397	89	48,520	0.2689	10	0.08
TreeJ48	141,530	95.085	16.59	135,818	0.0741	10	29.97
Decision stump	141,530	51.81	80.63	73,333	0.2414	10	09.62

obtained is not diagonal but vertical in the case of decision stump and Naive Bayes, which means it is poorly classifying the data.

5 Conclusion

The cognitive classifier model proposed used three classifiers on the datasets. The performance of the Treej48 algorithm was better than the decision stump and Naive¨ Bayes algorithm. The obtained confusion matrix during the experiment validates the result. Hence, the cognitive model discussed in this paper is able to gain knowledge from datasets. The web-based application Weka was found to be intuitive to implement multiple classifiers. This paper additionally outlines the datasets, and platforms which are beneficial for machine learning research. Future work will include cognitive computing on various datasets to extract knowledge and obtain knowledge on security concerns.

References

1. Carretero J, Garcia JD (2014) The internet of things: connecting the world. Int J Pers Ubiq Comput 18:445–447
2. Renu N (2021) Technological advancement in the era of COVID-19, SAGE Open Medicine, vol 9. SAGE Publications Sage UK, London, England
3. Andriole S (2015) What C-suite executives need to know about digital strategy and emerging technologies. https://www.forbes.com/sites/steveandriole/2015/10/07/analytics-iot-social-location-security-how-to-all-get-along-a-note-to-c-suiters/?sh=22cf5cc1ab27accessed on 4th Aug 2023
4. Gartner 2023 Recent. https://www.gartner.com/en/articles/top-strategic-cybersecurity-trends-for-2023. Last accessed 18 Aug 2023
5. Top Strategic Technology Trends 2023 https://emtemp.gcom.cloud/ngw/globalassets/intlca/information-technology/documents/2023-gartner-top-strategic-technology-trends-ebook-ca.pdf accessed on 5th July 2023
6. Xin Y, Kong L, Liu Z, Chen Y, Li Y, Zhu H, Gao M, Hou H, Wang C (2018) Machine learning and deep learning methods for cybersecurity. IEEE Access 6:35365–35381
7. Russell I, Markov Z (2006) An introduction to the Weka data mining system. ACM SIGCSE Bull 38(3):367–368
8. Andrade RO, Yoo SG (2003) Cognitive security: a comprehensive study of cognitive science in cybersecurity. J Inf Secur Appl 48:102352 2019 Oct 1
9. Buczak AL, Guven E (2015) A survey of data mining and ma- chine learning methods for cyber security intrusion detection. IEEE Commun Surv Tutorials 18:1153–1176
10. Kantarcioglu M, Xi B (2016) Adversarial data mining: Big data meets cyber security. In: Proceedings of the 2016 ACM SIGSAC conference on computer and communications security, pp 1866–1867. 24 Oct 2016
11. Dua S, Du X (2016) Data mining and machine learning in cybersecurity 2016. CRC press
12. Sarker IH, Kayes ASM, Badsha S, Alqahtani H, Watters P, Ng A (2015) Cybersecurity data science: an overview from machine learning perspective. J Big Data 7:1–29
13. Thomas T, Vijayaraghavan AP, Emmanuel S (2020) Machine learning approaches in cyber security analytics 2020. Springer

14. Shaukat K, Luo S, Varadharajan V, Hameed IA, Xu M (2015) A survey on machine learning techniques for cyber security in the last decade. IEEE Access 8:222310–222354
15. Alqahtani H, Sarker IH, Kalim A, Minhaz Hossain SM, Ikhlaq S, Hossain S (2020) Cyber intrusion detection using machine learning classification techniques. In: Computing science, communication and security: first international conference, COMS2 2020, Gujarat, India. Springer Singapore, pp 121–131. March 26–27, 2020
16. Lakshminarasimman S, Ruswin S, Sundarakantham K (2017) Detecting DDoS at- tacks using decision tree algorithm. In: 2017 fourth international conference on signal processing, communication and networking (ICSCN), pp 1–6
17. Makhdoom I, Abolhasan M, Lipman J, Liu RP, Ni W (2018) Anatomy of threats to the internet of things. IEEE J Commun Surv Tutorials 21(2):1636–1675 IEEE
18. Reddy YH, Ali A, Kumar PV, Srinivas MH, Netra K, Achari VJ, Varaprasad RA (2022) Comprehensive survey of internet of things applications, threats, and security issues. South Asian Res J Eng Tech 4(4):63–77
19. Rizvi S, Scanlon M, Mcgibney J, Sheppard J, Yoo SG (2003) Application of artificial intelligence to network forensics: survey, challenges and future directions. IEEE Access 2022 10:110362–110384
20. CIC dataset. https://www.unb.ca/cic/datasets/dns-2021.html
21. Samaneh M, Nasim M, Arash HL, Matt B, Amir HR (2021) Classifying malicious domains using DNS traffic analysis. In: The 19th IEEE international conference on dependable, autonomic, and secure computing (DASC), Oct. 25–28, 2021. Calgary, Canada
22. Samaneh M, Amgad HS, Princy V, Miguel G, Amir HR, Natasha H, Arash HL (2021) Lightweight hybrid detection of data exfiltration using DNS based on machine learning. In: The 19th IEEE international conference on dependable, the 11th IEEE international conference on communication and network security (ICCNS), Dec 3–5, 2021. Beijing Jiaotong University, Weihai, China
23. Yavanoglu O, Aydos M (2017) A review on cyber security datasets for machine learning algorithms In 2017 IEEE international conference on big data (big data) 2017 Dec 11. IEEE, pp 2186–2193
24. Arash HL, Gurdip K, Abir R () DIDarknet: a contemporary approach to detect and characterize the darknet traffic using deep image learning. In: 10th international conference on communication and network security. Tokyo, Japan, November 2020.
25. Shahmirzadi O, Lugowski A, Younge K (2019) Text similarity in vector space models: a comparative study. In: 2019 18th IEEE international conference on machine learning and applications (ICMLA). IEEE, pp 659–666
26. Webb GI, Keogh E, Miikkulainen R (2010) Na¨ıve bayes, encyclopedia of machine learning 15(1):713–714
27. Taheri S, Mammadov M (2013) Learning the naive bayes classifier with optimization models. Int J Appl Math Comput Sci 23(4):787–795
28. Razdan S, Gupta H, Seth A (2021) Performance analysis of network intrusion detection systems using j48 and naive bayes algorithms. In: 6th International conference for convergence in technology (I2CT). IEEE, pp 1–7, 2021 Apr 2
29. Bhargava N, Sharma G, Bhargava R, Mathuria M (2013) Decision tree analysis on j48 algorithm for data mining. Proc Int J Adv Res Comput Sci Softw Eng 3(6):713–714
30. Ahishakiye E, Taremwa D, Omulo EO, Niyonzima I (2017) Crime prediction using decision tree (J48) classification algorithm. Int J Comput Inf Technol 6(3):188–195

Analyzing the Counterfeit Medicines Based on Classification Using Machine Learning Techniques

Binitha S. Thomson and W. Rose Varuna

Abstract Medicines play a vital role in the pharmaceutical industry, dealing in the hospital and treating patients to cure their diseases. Counterfeit medicines are produced and distributed worldwide, increasing the risk to public health. To find counterfeit medicines in drug discovery by using ML algorithms. By analyzing the images in the Machine Learning (ML) algorithms identify the pattern of medicine for the safe supply of drugs that are Random Forest (RF), Naive Bayesian (NB), and Support Vector Machine (SVM). A better result will be obtained by finding the dataset samples applied in the SVM ML algorithms, which gives the accuracy of around 94.5% of the counterfeit medicine.

Keywords Artificial intelligence · Counterfeit medicines · Machine learning · Pharmaceutical industry · Supply chain management

1 Introduction

Artificial Intelligence (AI) and Machine Learning (ML)

Artificial Intelligence (AI) in the field of medicine for discovering data and understanding health to improve health experiences. AI makes clinical research reach the outcome to support medical professionals to give the best solution for healthcare problems. Machine Learning (ML) is the subgroup of AI and DL is the subgroup of ML [1]. ML which works on analyzing counterfeit medicines in supply chain management (SCM). The medicines are produced in the pharmaceutical industry with unique technologies or types of equipment which is collaborated in the field of chemical and mechanical engineering [2]. Producing and distributing the complex

B. S. Thomson (✉) · W. R. Varuna
Department of Information Technology, Bharathiar University, Coimbatore 641046, India
e-mail: bini2796tha@gmail.com

W. R. Varuna
e-mail: rosevaruna@buc.edu.in

B. R. Devi et al. (eds.), *Proceedings of Fifth International Conference on Computer and Communication Technologies*, Lecture Notes in Networks and Systems 897,
https://doi.org/10.1007/978-981-99-9704-6_33

medicines that are identified and characterized in the pharmaceutical industry by dealing with SCM. Counterfeit medicines are recognized by their set of patterns, determined by the method of AI by collecting the data of the medicines [1].

A report from World Health Organization (WHO) in the year 2017 stated that 10.5% of counterfeit medicines are sold in low and middle countries which have less income, which leads to serious issues for patient safety. Counterfeit medicines may cause severe damage to health conditions like the spread of diseases and cause side effects [2]. Numerous amount of counterfeit medicines is sold all over the world, which can determine the medicine whether is counterfeit or not. This method helps to find counterfeit medicines with their uniqueness of the medicinal formula or the molecules of the drug which is pre-defined in the system using ML [4].

AI technology has increased and data training is carried out. AI is used in four main areas to function in the pharmaceutical industry: (1) By assessing and predicting the severe diseases and the treatment for the particular patient will be successful by all possible means. (2) Preventing the complications of diseases while treating the patients. (3) The important function of dependable technology during the treatment or operation procedures on the patients. (4) The last function is used for chemicals and drugs to deal with the instruments or treat the chemicals for patients which can improve safety and efficiency [5].

ML is used for various benefits in healthcare and producing better results in classifying medicines in the pharmaceutical industry. ML algorithms have significant importance in medical findings. To analyze the counterfeit medicines, the ML algorithms are used such as, Random Forest (RF), Naive Bayesian (NB), k-Nearest Neighbor (KNN), Support Vector Machine (SVM) in the medicines findings. The ML algorithms can be functioned for developing various models for identifying chemicals and features of physical compounds of the medicines [6].

2 Literature Survey

Ham [7], Rahman et al. [8] stated that counterfeit medicines are produced and distributed by changing their origin, uniqueness, and benefits of the packages falsely the information represented. Consumption of these harmful or replaced formulae of counterfeit medicines leads to health issues and sudden changes, side effects in the body, or damage in the internal organs even causing death [7, 8].

Uppa et al. [9] state that the significant problem is determined in the pharmaceutical industry for counterfeit medicine in developing and developed countries. It affects people's lives and causes severe damage to their health. This leads to treatment failure, cannot find accurate therapy for treatment, and even cause death. Counterfeit medicines are manufactured which brings a negative impact on society. Serious damage and risk in the lifestyle are caused by the outcome of the pharmaceutical industry. The steps are determined to reduce the counterfeit medicine production and distribution in the pharmaceutical industry [9].

Table 1 Sample datasets of medicines

Sample medicines	Strength (mg)
Tylenol	325
Doxycycline hyclate	100
Aspirin	325
Bystolic	2.5
Gilenya	0.5

Amusa et al. [10] discuss the development of preventing counterfeit medicine for structured and useful information of the system in the world. This author depicts the Information Communication Technology-based telecommunication tools for assisting the pharmaceutical industry's response to control counterfeit medicine and health information. The approach states that the manufacturing and distribution of counterfeit medicines play an important role in rules and regulations for the healthcare system [10].

Pathak et al. [11] state that counterfeit medicines are manufactured and sold in the pharmaceutical industry by their unique, efficient, and easy way to transfer them. Preventing counterfeit medicines will make the public reduce the causes of the effects. Detecting counterfeit medicines must be taken action according to Government law and Act so that counterfeit medicines will not be supplied in the pharmaceutical industry [11].

3 Dataset

The counterfeit medicines testing samples are manufactured mostly in China and India. The datasets are collected from the University of California Irvine (UCI) Machine Learning datasets repository [12] (Table 1).

4 Methodology

4.1 Random Forest (RF)

The ML algorithm that is widely used for large data collection is RF which can function with multiple features.

This works by removing the exceptions of the datasets based on the specific feature by applying a particular algorithm. The input data and variables are trained for a number of datasets that are collected from multiple databases. The exception data works for accessing the attribute missing value by classifying its characteristics. By analyzing the data, false positive occurs in RF along with NB and SVM, which is compared with other algorithms that have a small amount of errors are identified. From analyzing the feature selection and classifier, the training process of the datasets will predict the molecule descriptors based on the training sets [13].

In the training process of the RF function, a few missing parameters are identified and it collaborates with the nonparametric data. The multifunctional approach in RF includes sustaining the genome of the drug sequencing by achieving its drug responses in the feature processes. The input of the data is combined with the drug classifier and framework to identify and predict the data from the medicine's response time [14]. The main functionality of the medicines in RF is to improve the prediction based on the response time for the input data framework is given in (1).

$$\begin{aligned}\text{Gini Index} &= 1 - \sum_{i=1}^{n} (B_i)^2 \\ &= 1 - \left[(B_+)^2 + \left(B_-\right)^2\right]\end{aligned} \tag{1}$$

where

B_+ represents the probability of a positive class
B_- Represents the probability of a negative class.

4.2 *Naive Bayesian (NB)*

NB algorithm in ML function is a supervised learning method which is one of the important tools for predicting the classification of the datasets. The standard NB algorithm works for features like classifying the datasets from the characteristics of input fields, factor correlation, and the data size, which is one easiest way to perform the task. This technique helps to enhance the accuracy of the original data and retrieved data which is large in number [6]. In counterfeit medicine discovery, the main aim of this process is to function for a large set of data that has been collected [13].

This technique will classify the several important functions of NB which is incorporated with the medicine discovery process. The selection of the medicine compounds might be active or inactive, which have possible functions of classifying the data, often incorporated with the random noise of the datasets. Mainly, it processes the datasets for classifier from the original data and can find the accuracy from the level of the collected dataset with excellent output. This NB technique is very useful in finding the medical process [5]. To determine the classification of medicine and to find the accuracy is given in (2).

$$P(B|I) = \frac{P(I|B)P(B)}{P(I)} \tag{2}$$

$P(B|I)$ represents the posterior probability of class B—denotes target given predictor I—denotes attributes
$P(B)$ represents the prior probability of class
$P(I|B)$ represents the probability of the predictor given class
$P(I)$ represents the prior probability of the predictor.

4.3 Support Vector Machine (SVM)

SVM is the supervised ML algorithm that is used for the classification of high structural data. This is a linear algorithm with a feature selector of the hyperplane. SVM is a difficult task for medicine discovery which has the capability of active and inactive compounds to function in the process with a regression training set [15]. The regression method predicts the relationship between the medicine and complex data to predict the dataset that is determined in the training dataset [5]. The functions of SVM predict the medicine with multiple data of bioactivities. SVM can process several numbers of root functions such as radial basis, function, linear, and polynomial. The main aim of the SVM function is the nonlinear space with high structured data of medicines [14]. By implementing the sequential minimal function in the training process, it can determine the support vector classifier in SVM. To determine the linear algorithm in SVM in (3):

$$h(p) = q + \sum_{n=1}^{N} y_t b_t K\left(p_, p_t\right) \tag{3}$$

Here,

$h(p)$ denotes the value of the distance from the decision edge
q denotes bias weight
b denotes the coefficient maximum of correct classification from the training dataset elements
N denotes the features
K denotes the kernel
p denotes the features vector.

5 Result and Discussions

The data of 200 samples used for the set to find the accuracy of counterfeit medicines are Tylenol 325 mg, Doxycycline Hyclate 100 mg, aspirin 325 mg, Bystolic 2.5 mg, and Gilenya 0.5 mg. The sample dataset has been tested from the ML algorithms by training the dataset to the machine to give an accurate result. The medicines are validated by using ML algorithms. RF algorithm and NB algorithm give lower accuracy prediction results for classifying counterfeit medicine when compared with

Table 2 Accuracy of the counterfeit medicines

Algorithms	Number of samples tested	Accuracy of the counterfeit medicines (%)
RF	200	84.2
NB	200	87.01
KNN	200	88.05
SVM	200	94.5

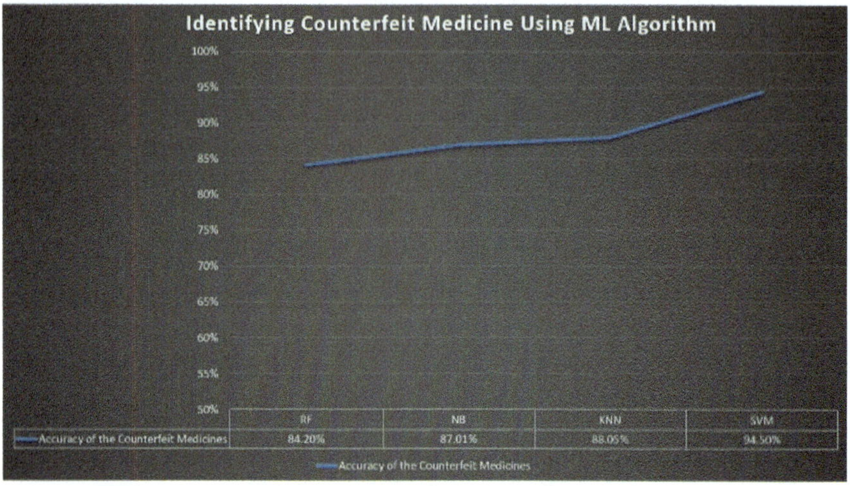

Graph 1 Graphical representation of identifying counterfeit medicine using ML algorithms

SVM. RF algorithm which performs the slow functioning of the large training sets when predicting the data, and NB algorithm gives the probability of zero frequency that cannot be predicted and gives the wrong estimation when compared with SVM.

The dataset samples are active elements with the genuine sample which is classified by comparing with four different ML algorithms RF, NB, and SVM and can be classified the images by using these ML techniques. By obtaining this, the classifier of the medicine with the below-mentioned accuracy of the counterfeit medicine gives the better accuracy level by using SVM algorithm mentioned in Table 2 and Graph 1 shown the counterfeit medicine accuracy level.

6 Conclusion

The main aim of ML has widely benefited health care. Through ML techniques the treatment can be given in a proper way, identification of treatment, suggesting a personal review to the patient, prediction of disease, etc. From the analyzed dataset, the supervised ML algorithm SVM gives better accuracy when it is compared with other ML algorithms such as RF and NB. The SVM is able to detect counterfeit medicines based on analyzing and predicting the data values which is classified with the accuracy of 94.5%. ML has the outcome to predict and analyze different fields like temperature, weather, and other details of major issues happening around the world.

References

1. Patel V, Shah M (2022) Artificial intelligence and machine learning in drug discovery and development. Intell Med 2(134–140)
2. "ETHealthWorld (2021) Counterfeit drugs: threat to India's Reputation of Indian Self-reliant Model. https://health.economictimes.indiatimes.com/news/pharma/counterfeit-drugs-threat-to-Indiasreputation-of-Indian-self-reliant.model/81342614. May 2021
3. Gupta P, Singhal K, Pandey A (2012) Counterfeit (fake) drugs & new technologies to identify it in India. Int J pharm Sci Res 3(11):4057–4064
4. Paulz D,Sanapz G, Shenoyz S, Kalyane D, Kalia K, Tekade RK (2021) Artificial intelligence in drug discovery and development. Drug Discov Today 26
5. Adsul KB, Kosbatwar SP (2020) A novel approach for traceability & detection of counterfeit medicınes through blockchain. EasyChair Preprint:№ 2539, Feb 2020
6. Patel L, Shukla T, Huang X, Ussery DW, Wang S (2020) Review machine learning methods in drug discovery. Molecules 25:5277
7. Ham MT (2003) Health risks of counterfeit pharmaceuticals. Drug Saf 26(14):991–997
8. Rahman MS, Yoshida N, Tsuboi H, Tomizu N, Endo J, Miyu O, Akimoto Y, Kimura K (2018) The health consequences of falsified medicines—a study of the published literature. Tropical Med Int Health 23(12):1294–1303
9. Uppal P, Bansal T, Tyagi HV (2023) Fighting counterfeit medicines: a review. Int J Curr Sci (IJCSPUB) 13(2). ISSN: 2250-1770
10. Amusa M, Oluwade B (2020) A historical background of some basic ICT tools used in counterfeit drug control. Afr J Comput 13(1):52–61
11. Pathak R, Gaur V, Sankrityayan H, Gogtay J (2023) Tackling counterfeit drugs: the challenges and possibilities, pp 281–290
12. UCI. Drug Review Dataset. Available online: https://archive.ics.uci.edu/ml/datasets/Drug+Review+Dataset+%28Drugs.com%29/
13. Lo YC, Rensi SE, Torng W, Altman RB (2018) Machine learning in chemoinformatics and drug discovery. Drug Discov 23:1538–1546
14. Talevi A, Morales JF, Hather G, Podichetty JT, Kim S, Bloomingdale PC, Kim S, Burton J, Brown JD, Winterstein AG et al (2020) Machine Learning in drug discovery and development part 1: a primer. CPT Pharmacomet Syst Pharmacol 9:129–142
15. Alsallal M, Sharif MS (2018) A machine learning technique to detect counterfeit medicine based on X-Ray fluorescence analyser. IEEE,ISBN: 978-1-5386-4904-6

Water Quality Classification Using Multi-cell RNN in Aquaculture Ponds for Catla Fish

Peda Gopi Arepalli and Jairam Naik Khetavath

Abstract In aquaculture, water quality has a crucial role in fish development and survival. The link between different water quality sections is not well captured by the water quality prediction systems now in use, and their capacity to take advantage of the temporal relationships and interactions between various physical and chemical properties of water parameters is constrained. In this study, we present a multi-cell RNN for catla fish that combines the benefits of LSTM and GRU to capture these temporal linkages and interactions in a deep learning model for water quality prediction in aquaculture catla fishponds. Compared to other models, our suggested approach performs better at categorizing data on water quality for catla fish. The experimental findings show that the suggested multi-cell RNN achieves a high accuracy of 99.88% and a low loss of 0.0192. This study offers a viable method for forecasting aquaculture water quality, which may help fish develop and survive and support sustainable aquaculture methods.

Keywords Water quality · Aquaculture · Deep learning · Multi-cell RNN · LSTM · GRU · Catla fish · Classification

1 Introduction

Aquaculture is a rapidly growing industry that is crucial in meeting the demand for seafood as the world population continues to increase. To ensure the success and sustainability of this industry, water quality [1] monitoring is key. Water quality parameters, such as temperature, dissolved oxygen, pH, turbidity, and salinity, are

P. G. Arepalli (✉) · J. N. Khetavath
Department of Computer Science and Engineering, National Institute of Technology Raipur, Raipur, India
e-mail: apgopi.phd2020.cse@nitrr.ac.in

J. N. Khetavath
e-mail: Jnaik.cse@nitrr.ac.in

B. R. Devi et al. (eds.), *Proceedings of Fifth International Conference on Computer and Communication Technologies*, Lecture Notes in Networks and Systems 897,
https://doi.org/10.1007/978-981-99-9704-6_34

crucial in determining the health and growth of aquatic organisms. However, traditional manual testing can be time-consuming and inaccurate, as water quality parameters can fluctuate with time. To address this issue, modern technologies like the IoT and ML [2] can be used in aquaculture to constantly monitor water quality parameters. Existing water quality prediction methods often fail to capture the correlations between water quality sections, limiting their ability to exploit the temporal relations and interaction among different physical and chemical characteristics of water parameters.

In this research, a DL model for water quality prediction in aquaculture using a multi-cell RNN for catla fish is proposed. The model combines the advantages of LSTM and GRU [3] to capture temporal relations and interactions among different water parameters. The proposed model outperforms existing models in classifying water quality data for catla fish. Multi-cell RNN is a variant of recurrent neural network that has multiple cells instead of a single cell. Each cell in a multi-cell RNN can have a different architecture, allowing it to capture different aspects of the temporal dynamics [4] of the input sequence. In the proposed model, the LSTM cell is used to capture long-term dependencies, while the GRU cell is used to capture short-term dependencies. By combining these two cells, the model can capture both long-term and short-term temporal dependencies in the input sequence.

The suggested model predicts catla fish water quality [5] given a series of water quality time-series samples. The model learns the correlation among the quality of the water and the development of fish by being exposed to a huge dataset of water quality factors and their associated fish growth rates. The experimental findings show that the suggested model provides more accurate predictions of water quality for catla fish than the current models. The model's high level of precision means it may be used to boost aquaculture's efficiency and longevity. Here is how the remainder of the article is structured: The study is divided into five sections: Sect. 2 a literature review, Sect. 3 a presentation of the suggested model, Sect. 4 a presentation of the findings and discussions, and Sect. 5 a summary and conclusion.

2 Literature Survey

Aqua animals in ponds are subject to the impact of water, and therefore Rana [5] suggested a machine learning technique to study this phenomenon. The leading objective of the study is to determine the factors prompting water quality and its consequences on aquatic stock and food production. In order to calculate an approximation of the Water Quality Index, Venkataramana [6] conducted a study using several distinct Deep Learning models. These included an ANN, RNN, and LSTM. Parameters affecting water quality are thoroughly analyzed. When compared to other approaches, the execution time of the models utilized here is quite quick while still producing accurate and exact results. However, additional variables that might alter water quality are ignored in this analysis. Kalaivanan [7] conducted a survival analysis of several machine learning approaches to predicting water quality. Methods of reducing dimensionality, such as the ANN linked discriminate analytic approach and the SVM model. However, not all water resources can benefit from machine learning methods owing to variations in environmental conditions or other reasons.

Khatri et al. [8] ANN-FFBP, DFFBP, and DCFBP are effluent quality models that use ANNs. At 26 kg d1, UASB-FP was able to remove > 84% of organics. The COD, AN, TSS, BOD, TKN, and TP values were used by each model to make predictions about the water quality. The models were taught, tested, and validated using 180 samples spanning one year.

3 Proposed Methodology

A multi-cell RNN is used to predict fish growth and mortality based on water quality data. The model is trained on historical data, learning patterns between water quality parameters and fish outcomes. During inference, it processes a sequence of water quality readings, capturing temporal dependencies. The RNN's output is used for classification, but it is crucial to validate the model on new data before real-world use.

Figure 1 illustrates a multi-cell RNN used for classifying water quality data. Initially, water quality parameters are inputted and preprocessed with the drop method. The data is then split into a 70:30 training and validation ratio. Multi-cell RNNs are employed for classifying data, especially for assessing the impact of water quality on fish mortality. These RNNs enhance sequential data analysis by incorporating multiple memory cells in the hidden layer to capture long-term dependencies. Each cell manages information flow using input, output, and forget gates. The mathematical model is represented in the content.

$f_t, i_t, O_{gt}, z_t, r_{gt}$ represent the forget, input, output, update, reset gates.

$wf_x, wg_x, wi_x, w_{ox}, w_{ux}, w_{rx}$ represent the weight matrices of all gates, b_f, b_i, b_g, b_o represent the biases, σ—represents the sigmoid function, x_t—represents the input, h_t—represents the hidden state.

The input x_t is fed into the first memory cell, where it is processed using the Eqs. (1) to (6) to obtain the output (O_t).

$$f_t = \sigma\left[(wf_x * x_t) + (wf_h * h(t-1)) + b_f\right] \tag{1}$$

$$O_{tf} = O(t-1) * f_t \tag{2}$$

The information that must be remembered and the information that must be lost is determined by forget gate (f_t), and forget gate is calculated as (1). And the output of forget gate is calculated as (2).

$$i_t = \sigma[(wi_x * x_t) + (wi_h * h(t-1)) + b_i] \tag{3}$$

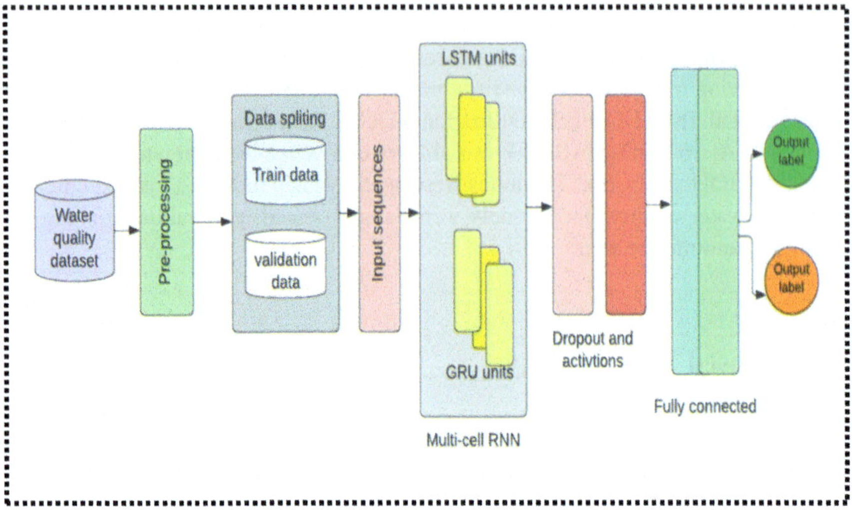

Fig. 1 Multi-cell RNN for water quality data classification

$$i_d = \tanh\left[(wg_x * x_t) + (wg_h * h(t-1)) + b_g\right] \quad (4)$$

$$O_{ti} = (i_t * i_d) \quad (5)$$

$$O_t = \left(O_{ti} + O_{tf}\right) \quad (6)$$

The succeeding operations to update the cell status are performed by the input gate (i_t) and calculated as Eqs. (3), (4). And the output (O_t) of the inputgate is calculated as Eqs. (5), (6). In a multi-cell RNN, information flows between memory cells and hidden states, enabling the capture of long-term dependencies. This sequential process involves updating cell states, calculating hidden states using Eqs. (8)–(10), and transmitting output from one cell to the next. Backpropagation through time (BPTT) optimizes gate and memory cell weights. Equation (11) updates the hidden state using reset and Eq. (12) combines it with the previous state via the update gate. These mechanisms enable the model to process and retain relevant information effectively.

$$O_{gt} = \sigma[(w_{ox} * x_t) + (w_{oh} * h(t-1)) + b_o] \quad (7)$$

The value of the following hidden state is determined by output gate (O_{gt}) and is calculated as Eq. (7)

$$h_t = \tanh(O_t) + O_{gt} \quad (8)$$

The tanh function is applied to the transformed cell state that resulted from the original cell state, and the result is found using the formula Eq. (8)

$$z_t = \sigma[w_{ux} * x_t + w_{uh} * h(t-1)] \tag{9}$$

The update gate is denoted by z_t It is computed as Eq. (9) and indicates the amount of past's information must be handed on to the future.

$$r_{gt} = \sigma[w_{rx} * x_t + w_{rh} * h(t-1)] \tag{10}$$

The reset gate is denoted by r_{gt}, it reveals how much of the past intelligence to forget and is calculated as Eq. (10).

$$h_t = \tanh\left[w_{xt} + r_{gt} \oplus wh(t-1)\right] \tag{11}$$

$$h_t = \left[z_t \oplus h(t-1) + (1-z_t) \oplus h't\right] \tag{12}$$

The final output is determined from Eqs. (11) and (12).

Algorithm: Water Quality Classification Using Multi-Cell RNN

Input: Water quality features [B.O. D(mg/l), Temp, D. O (mg/l), PH, Nitrate.

Output: predict the rate of dissolved oxygen.

1. Take the input values and declare (h_{t-1}) and current input x_t.
2. Declare σ () =sigmoid & tanh () =hyperbolic tangent are Activation functions.
   // LSTM Cell structure.
3. Initialize the network parameters, calculate the f_t from Eq. (1)
4. Compute the O_{tf} from Eq. 2).
5. Determine the i_t, i_d from the Eqs. 3 and 4.
6. Calculate the output O_{ti}, O_t from the Eqs. 5 and 6
7. Compute the O_{gt}, from the Eq. (8)
8. Calculate output H_t from Eq. (8)
   // GRU Cell structure.
9. Take the past information as input, Calculate the z_t from the Eq. (9)
10. Calculate the r_{gt} from the Eq. (10)
11. Compute the memory content using the Eq. (11)
12. Determine H_t—Vector which holds the Information, z_t computed using Eq. (12).
13. End

4 Experimental Analysis

4.1 Dataset

The dataset was gathered from three fishponds located in the Indian state of Andhra Pradesh. To collect information about the quality of the water in these ponds, seven different types of water quality sensors were employed [9, 10]. These sensors measured various parameters including pH, dissolved oxygen (DO), ammonia, temperature, nitrate, turbidity, and manganese. The data was collected over a period of 12 months, spanning from February 2022 to January 2023. In total, 74,759 data records were obtained and used for analysis. This dataset was then used to train and test various machine learning models, including the proposed multi-cell RNN model, in order to classify the water quality based on its parameters for fish.

4.2 Results and Discussion

This section discusses about the investigational results of proposed multi-cell RNN model and state-of-art models. The results show that the performance of proposed model with accuracy and loss. And next made comparison with classification metrics as follows:

Figure 2 depicts a comparison between the accuracy and validation accuracy of a multi-cell RNN model used to classify water quality data for fish. The training accuracy and validation accuracy both achieved a score of approximately 99.5%, indicating that the model performed well in accurately classifying the data. This suggests that the model is well-suited to fit the given dataset.

In Fig. 3, a comparison is made between the loss and validation loss of a multi-cell RNN model that was used for classifying water quality data related to fish.

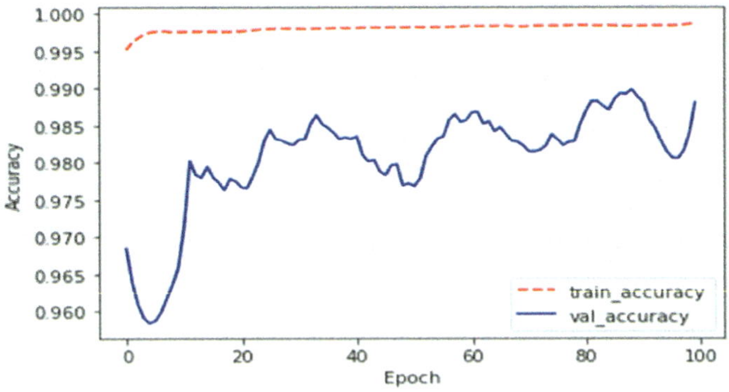

Fig. 2 Accuracy

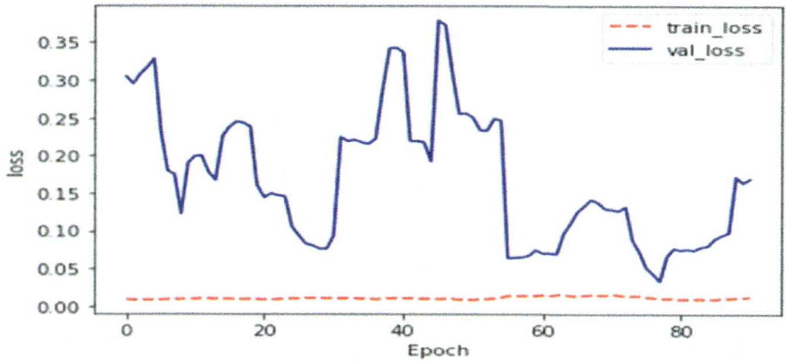

Fig. 3 Loss

The training loss and validation loss attained scores of around 0.0195 and 0.1982, respectively, which indicates that the model was able to classify the data with a high level of accuracy. This performance suggests that the model is well-suited to the dataset at hand. It is important to note that lower loss values in this context are generally considered better, as they suggest that the model is making fewer errors in its predictions. Thus, achieving a low loss value is a significant performance metric for the model.

Table 1 compares the proposed multi-cell RNN with LSTM and ANN models in terms of accuracy and loss. The multi-cell RNN achieved superior accuracy at 99.25%, thanks to its increased depth and ability to handle data complexity and time-series sequences effectively. In contrast, LSTM's accuracy was lower at 59%, likely due to its limitations with complex data. ANN performed better with 90.1% accuracy but couldn't match the proposed multi-cell RNN's ability to capture intricate data features.

Table 2 compares the proposed multi-cell RNN model with LSTM and ANN models based on precision, recall, and F1-score. The multi-cell RNN outperformed LSTM and ANN with scores of 0.982, 0.974, and 0.961, respectively. This improvement is due to the multi-cell RNN's greater depth, enhancing its ability to handle complex data and effectively capture time-series sequences. In contrast, LSTM scored lower (0.761, 0.896, 0.793) due to limitations in handling data complexity. ANN performed better (0.901, 0.890, 0.912) but still lagged behind the multi-cell RNN in capturing intricate data features.

Table 1 Comparison of proposed multi-cell RNN with stat-of-art models with accuracy and loss

Model	Loss	Val loss	Accuracy	Val accuracy
ANN	0.1601	0.1878	0.9133	0.8991
LSTM	0.0486	0.8832	0.9766	0.5457
Proposed multi-cell RNN	0.0192	0.0712	0.9988	0.9907

Table 2 Comparison of proposed multi-cell RNN with stat-of-art models with classification metrics

Model	Precision	Recall	*F*-score
ANN	0.901	0.890	0.912
LSTM	0.761	0.896	0.793
Proposed multi-cell RNN	0.982	0.974	0.961

5 Conclusion

Maintaining optimal water quality is vital for fish health and survival. Traditional machine learning struggles with complex, dynamic water quality data. To tackle this, we introduced a multi-cell RNN model for fish water quality classification, comparing it to LSTM and ANN models. Our proposed RNN effectively handles data complexity and time-series sequences, achieving impressive metrics: 0.985 accuracy, 0.0195 loss, 0.982 precision, 0.974 recall, and 0.961 F1-score. Significantly, it outperforms LSTM and ANN models due to its depth and sequence memory capabilities. This study underscores the potential of multi-cell RNNs in handling complex temporal data, surpassing conventional machine learning methods.

References

1. Pehme K-M, Burlakovs J, Kriipsalu M, Pilecka J, Grinfelde I, Tamm T, Jani Y, Hogland W (2019) Urban hydrology research fundamentals for waste management practices. Res Rural Dev 1:160–167
2. Hameed M, Sharqi SS, Yaseen ZM, Afan HA, Hussain A, Elshafie A (2017) Application of artificial intelligence (AI) techniques in water quality index prediction: a case study in tropical region, Malaysia. Neural Comput Appl 28:893–905
3. Khan UT, Valeo C (2015) A new fuzzy linear regression approach for dissolved oxygen prediction. Hydrol Sci J 60"1096–1119
4. Kumara A, Sharmab M, Raic SP (2017) A novel approach for river health assessment of Chambal using fuzzy modeling, India. Desalination Water Treat 58:72–79
5. Rana M, Rahman A, Dabrowski J, Arnold S, McCulloch J, Pais B (2021) Machine learning approach to investigate the influence of water quality on aquatic livestock in freshwater ponds. Biosys Eng 208:164–175
6. Venkataramana Y (2022) Water quality analysis in a lake using deep learning methodology: prediction and validation. Int J Environ Anal Chem 102(17):5641–5656
7. Kalaivanan K, Vellingiri J (2022) Survival study on different water quality prediction methods using machine learning. Nat Environ Pollut Technol 21(3)
8. Khatri N, Vyas AK, Abdul-Qawy ASH, Rene ER (2023) Artificial neural network based models for predicting the effluent quality of a combined upflow anaerobic sludge blanket and facultative pond: Performance evaluation and comparison of different algorithms. Environ Res 217:114843
9. Arepalli PG, Khetavath JN (2023) An IoT framework for quality analysis of aquatic water data using time-series convolutional neural network. Environ Sci Pollut Res Int. https://doi.org/10.1007/s11356-023-27922-1. Epub ahead of print. PMID: 37284950
10. Arepalli PG, Naik KJ (2023) A deep learning-enabled IoT framework for early hypoxia detection in aqua water using light weight spatially shared attention-LSTM network. J Supercomput. https://doi.org/10.1007/s11227-023-05580-x

Advanced Techniques for Color Image Blind Deconvolution to Restore Blurred Images

Jonnadula Narasimharao, P. Deepthi, Bolla Aditya, C. R. Shruthi Reddy, Avala Raji Reddy, and Ganpat Joshi

Abstract Due to improper camera settings and atmospheric turbulence, captured scene images are usually degraded by blurring. Image restoration is the process of extracting a blurred image's original image. Image restoration is an articulate technique that facilitates the retrieval of original images in a one-step manner. First blind deconvolution is among the most significant restoration methods in which the blurring operator is not identified. Blind system deconvolution is a very complicated process that restores objects without knowing what they were like before they started to deteriorate PSF. The PSF is an impulse response of a point source. This paper introduces a novel approach of blind deconvolution for blurred color images. Here, first perform the degradation process, then use the Wiener filter to reduce the noise from the input blurred image using the degradation function. Then the restoration process is performed using blind deconvolution, involving the estimation of PSF kernels through a thresholding mechanism. Following this, deblurring can be carried out for different sizes of PSF images. The deblurred images are restored with some ringing effects. In conclusion, the Canny edge detector is utilized to reduce noise and achieve the optimal restored image. The cost of MSE and PSNR was calculated for each method and compared in the results.

Keywords Blind deconvolution · Blurred color image · PSF (point spread function) kernels · Wiener filter · Restored image

---

J. Narasimharao (✉) · C. R. S. Reddy
Department of CSE, CMR Technical Campus, Hyderabad, Telangana, India
e-mail: jonnadula.narasimharao@gmail.com

P. Deepthi
Department of CSE (AI&ML), CMR Institute of Technology, Hyderabad, Telangana, India

B. Aditya
Department of IT, CMR Technical Campus, Hyderabad, Telangana, India

A. R. Reddy
Department of Mechanical Engineering, CMR Technical Campus, Hyderabad, Telangana, India

G. Joshi
Department of Computer Science and Engineering, Madhav University, Rajasthan, India

B. R. Devi et al. (eds.), *Proceedings of Fifth International Conference on Computer and Communication Technologies*, Lecture Notes in Networks and Systems 897,
https://doi.org/10.1007/978-981-99-9704-6_35

1 Introduction

In comparison to other human perceptions, vision is the most trusted source of information. An image is a fundamental container of pictorial data. The primary areas of image processing involve enhancing this data for human understanding and application along with computer-based image processing for extracting and analyzing information. One of the key areas of image processing is enhancing the visual quality of images and optimizing the extraction of information present in an image. However, in the domain of image processing and vision-based systems, blurring has been a critical issue [1]. Evaluating image quality is an important aspect that needs to be considered to assess the usability of images in various vision-based applications. In various domains such as product photography, robot vision, medical imaging, astronomical imagery, satellite photography, forensic applications, and military applications, the need for image restoration has been evident [2]. Blurry images contain less information and create difficulties in scene interpretation as well as image analysis.

Image blurring results in information loss and degradation of the image. The extraction of information from a blurred image and restoring it is a crucial task when there is no forehand knowledge about the scene being imaged or the limitations in the image formation and sensing mechanism [3]. From a technical perspective, image degradation becomes possible as the original image is combined with a blurring function, and this information becomes necessary to restore the original image from the degraded version. When there is no knowledge about the blur or the original image, restoring the image becomes a challenging task, and the image is to be restored using blind deconvolution techniques [4]. Blind deconvolution is used in various image processing applications due to its many motivations. Collecting prior information about the scene being imaged often comes with an associated cost. To provide a few examples, in applications such as remote sensing, medicine, and astronomy, specific information or the original image itself is challenging to model statistically because the scenes have never been imaged before. Additionally, it is not possible to accurately determine the degradation caused by blurring. In medical video-conferencing, a real-time imaging activity, the PSF parameters cannot be pre-determined. Classical image restoration algorithms are not well-suited for tackling practical blurred images because they depend on knowledge of the imaging system's PSF. If information about the PSF parameters can be estimated in advance, the image can be restored using these methods. By identifying the type of blurring and subsequently applying parameter estimation techniques, the necessity for prior information about the blur is reduced, and the necessity of simultaneous estimation of both image and PSF in the image restoration process is eliminated. The classification of blur types and the determination of PSF for restoration remain challenging tasks for researchers. This challenge has inspired further research in this field [5–7].

2 Literature Survey

Blind deconvolution seeks both the connected blur kernel and the latent picture. Due to the occurrence of several pairs of solutions, blind deconvolution is regarded as an ill-posed issue. To address the well-posedness of the blind deblurring problem, the majority of blind picture deblurring solutions commonly define the challenge as a minimization problem. An alternate technique is commonly used to solve the latent image and the blur kernel. When a proper and effective initialization is carefully chosen, the method swiftly converges. In their work within the subject of Applied Computational Intelligence at Hindawi, Liu et al. [7] first-order and second-order image gradients' L0-regularized norms were combined to standardize the estimation results. Various academics have presented various picture priors as well as optimization strategies to solve the ill-posed problem in generic photos.

Pan et al. [8] expanded their research by working with generic images, incorporating modifications to a dark channel concept initially proposed by He et al. [12]. The majority of natural scene patches include at least one color channel with pixels with intensities near to zero, which is the foundation for our method. Our strategy beats existing methods by increasing sparsity in the dark channel of latent images during kernel estimation. However, if the images are without dark pixels and instead prominently feature bright pixels, the dark channel may not contribute to the estimation of intermediate latent images. Extreme Channels Prior (ECP), a brand-new natural image prior that Yan et al. [9] presented, combines the advantages of Bright Channel Prior (BCP) and Dark Channel Prior (DCP) to enhance natural images. The robustness and favorability of this method surpass state-of-the-art picture deblurring techniques for both real-world and artificial photos. However, this method has its limitations and may not be well-suited for kernel estimation. Kernel estimation is a complex process, and its performance is primarily dependent on the intensity attributes present in images. To enhance the precision of kernel estimation, a number of researchers have chosen to utilize a gradient prior.

Different regularizations or assumptions, however, are necessary. Pan et al. [10] proposed a gradient and L0-regularized intensity prior for text picture deblurring. While this method is less successful when dealing with a complex backdrop, it is still appropriate for text graphics with a clear background. Cho et al. [11] proposed a method for including text-specific features in the deblurring process. Although this method worked better in two-tone shots, it performed less well in generic images. The authors of [13] employed a hyper-Laplacian prior to mimic the gradient distribution in real-world images, with the purpose of avoiding local solutions. In order to address kernel estimation difficulties, Shan et al. [14] used a sparse picture prior and produced a unified probabilistic model that matched the gradient distribution of natural images. Fergus et al. [15] suggested a method for obtaining the prior of picture gradients by utilizing a zero-mean mixture of Gaussian distributions, with the goal of imitating the gradient distribution in genuine photographs.

3 Blind Deconvolution of Blurred Color Images

The main steps of blind deconvolution and deblurring of blurred color images can be explained as (shown in Fig. 1).

Step 1: Initially, the blurred color image is taken as an input.
Step 2: Apply the degradation process to obtain the degradation function.
Step 3: The ideal image is initially processed using a Wiener filter to eliminate noise.
Step 4: Then restoration is performed with blind deconvolution to reduce or eliminate the degradation that occurred in the image.
Step 5: Next, kernel estimation is conducted to obtain accurate PSF parameters.
Step 6: Next, the restored blurred image PSFs of various sizes are performed. This performs three restorations such as Deblurring with undersized PSF, Deblurring with initial PSF, and Deblurring with oversized PSF.
Step 7: Analyze these restored PSFs and generate a restored image with ringing effects at its edges.

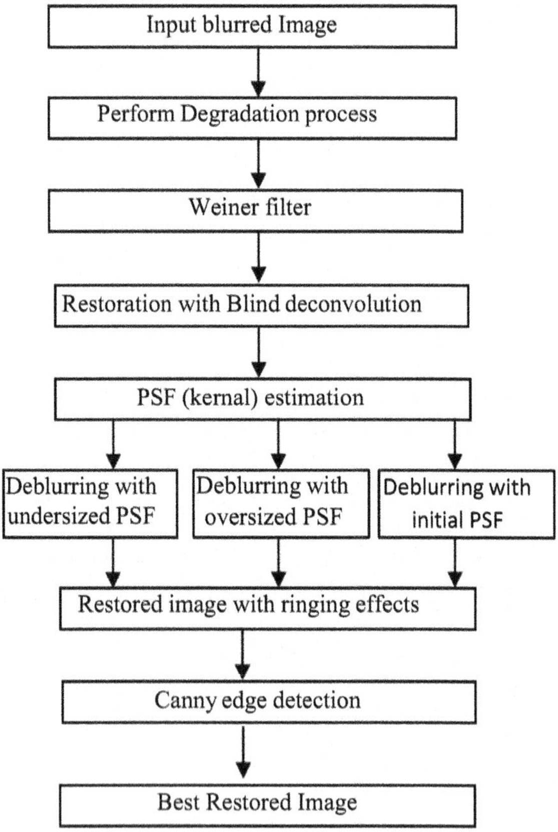

Fig. 1 Blind deconvolution process of blurred color images

Step 8: The Canny edge detection technique is utilized to eliminate the ringing effect at the edges.
Step 9: Finally, the highest-quality restored image is obtained from the degraded image.

3.1 Degradation Process

Image degradation can occur because of the presence of noise and external disturbances in an image during its transmission from one location to another through cable, wireless, or satellite channels. Images taken by surveillance cameras, consumer cameras, satellite cameras, and MRI machines may undergo degradation due to sensor limitations, as well as turbulence caused by camera defocusing, noise, and motion. Restoration of such degraded images is necessary to enhance analysis and vision. The more accurate the degradation model, the better the restored image will be when applying the inverse imaging process. In this model, the image is blurred by H degraded function and noise is added. The process of image degradation is represented as

$$g(x, y) = H(x, y).(x, y) + (x, y) \tag{1}$$

The original or input image is represented by $f(x, y)$, the degraded or consequent picture is represented by $g(x, y)$, the degradation function is represented by $H(x, y)$, and the additive noise is represented by $n(x, y)$. The H defines a convolution matrix which models blurring that introduced by multiple imaging systems. For instance, the H function will model the defocus of the camera, imperfections in lenses, and motion blur.

3.2 Weiner Filter

The Wiener filter is one of the linear filters used to filter degraded images affected by blurring and additive noise. For this type of filter calculation, it is necessary to assume that both the noise and signal processes are second-order stationary. This is commonly used in the frequency domain. During the data collection or transmission, noise can corrupt the image. Image denoising is used to eliminate additive noise while preserving the important signal features as much as possible. This filter handles both the noise and degradation functions. The main objective of the Wiener filter is to reduce the mean square error between the unaltered and altered images. The Wiener filter determines an estimate $\hat{f}$ that minimizes the statistical error function as

$$e^2 = E(f - \hat{f})^2 \tag{2}$$

where f represents the under graded image and E stands for the expected value operator. The minimum of the error function provided in the above equation is achieved in the frequency domain.

3.3 Restoration with Blind Deconvolution

The technique of reconstructing the original image from a degraded reproduction is known as image deconvolution. Blind image deconvolution refers to the combination of blur detection and image restoration. The original image and the degradation function are unknown in blind image deconvolution, leaving only the degraded image to be known. Non-blind picture deconvolution is well recognized for using degradation functions, such as the blur function, blur kernel, or point spread function. It is a very effective approach for picture restoration that requires little or no prior information of a point spread function (PSF) that produced image blurring. Blind deconvolution of images primarily involves the approach of identifying the PSFs from the true image and storing them for future utilization with one of the well-known techniques to restore the original image. Determining the PSF and identifying the true image are two separate processes.

3.4 PSF (Kernal) Estimation

A known PSF is used in both regular linear and non-linear deconvolution procedures. The PSF is calculated from the picture or a group of images in blind deconvolution, which speeds up the deconvolution process. Blind deconvolution can be performed iteratively, with each step improving PSF estimate. By using kernel estimation, the blur kernel k is retrieved from the source image. With $M_i = 1$ if $\underline{k} >$, where t is a threshold and $k_{\max}$ is the greatest kernel value, a kernel mask M is created at each level by thresholding the kernel values. We determine two masks, M_{low} and M_{high}, using two thresholds, t_{low} and t_{high}. M_{low} contains M and is bigger.

To reduce noise at level l after the kernel estimate is finished, we give zero values to all Kl elements outside of the mask M_{high}. Then, to further reduce noise, at the finer level $1 + 1$, those components of $Kl + 1$ that are not part of $M_{\text{low's}}$ upsampled mask are set to zero.

3.5 Deblurring with Various Sizes of PSFs

The blurred picture is first read, and the procedure then employs the point spread function for various sizes to reconstruct the blurred image. After that, the recovered PSF is analyzed, and the restoration method is improved. The initial stage is to extract references from the colored picture for detection. The image's point spread function (PSF) is then determined in the next phase. Following that, the colored picture is converted into an undersized PSF that is four times shorter in terms of pixels. Following that, the colored image is turned into an enlarged PSF, which is four times larger in terms of pixels than the original one. The colored picture is then transformed into the original PSF while retaining the original pixel size. The PSF pictures that have been saved are then analyzed. The Sobel filter is used to detect edges in the colored image's Red, Green, and Blue channels. In the last stage, Blind deconvolution is applied to the colored picture for each Red, Green, and Blue. These are the stages involved in deblurring a picture using the blind deconvolution technique.

3.6 Canny Edge Detection

The Canny edge detection technique involves calculating the vertical gradient G_y and horizontal gradient G_x by convolving the image pixel with a provided mask, calculating the gradient magnitude G and angle Θ for the given pixel, implementing non-maximal suppression, and facing challenges in real-time implementation because of the complexity of the arc tangent function and its dependency on floating-point calculations. A comparison is performed between the actual pixel and its neighbor along the gradient direction. If a pixel does not classify as a local maximum, the pixel is removed, and interpolation is used to determine the values of neighboring pixels.

4 Results

During the result analysis, this type of color image to color image reconstruction method utilizes a total of 5 datasets. Various methods are utilized to restore the image based on different MSE and PSNR values.

MSE: The mean square error, abbreviated as MSE, is an estimator that calculates the "errors" average squares, representing the variation between the estimator and the predicted values. Mean Square Error may be measured in a variety of methods to assess the gap between estimated and verified real dependability.

PSNR: Signal-to-noise ratio (PSNR) represents a scientific concept indicating the ratio of a signal's maximum power to the corrupting noise power, and this ratio has an impact on the accuracy of its representation. PSNR is usually represented as a decibel value. The PSNR is a widely used metric for evaluating image reconstruction performance. A high PSNR value is indicative of high image quality (Figs. 2, 3, 4, 5, 6, 7, 8, 9, 10, 11, 12, 13, 14, 15, 16, 17, 18, 19, 20 and 21).

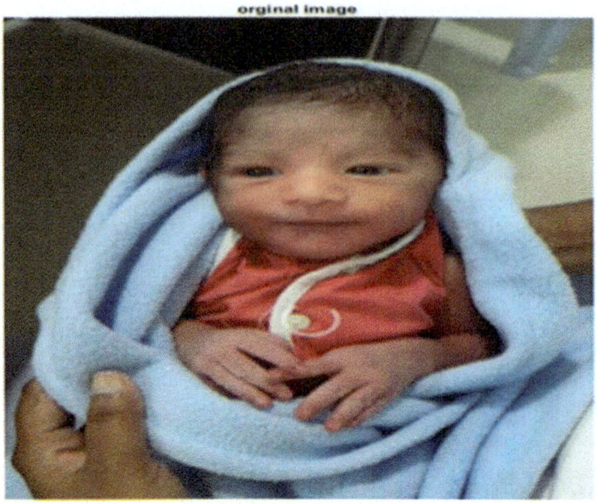

Fig. 2 Original input image

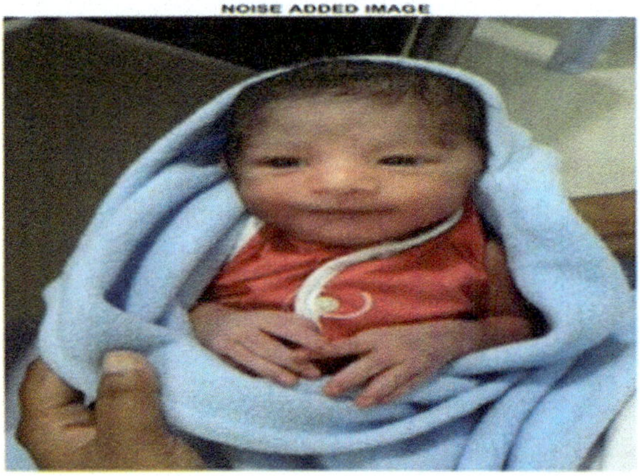

Fig. 3 *N* added image

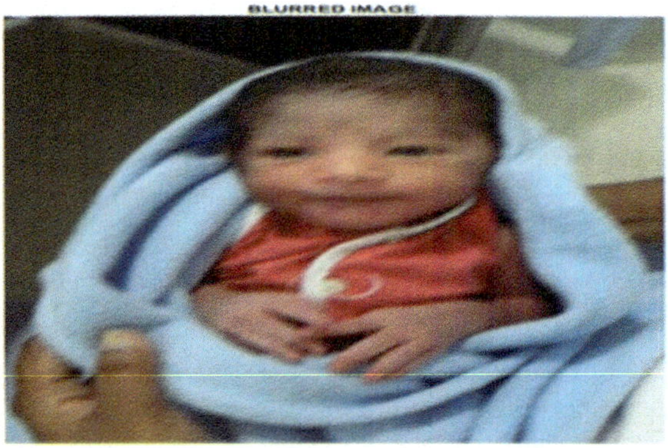

Fig. 4 Blurred image

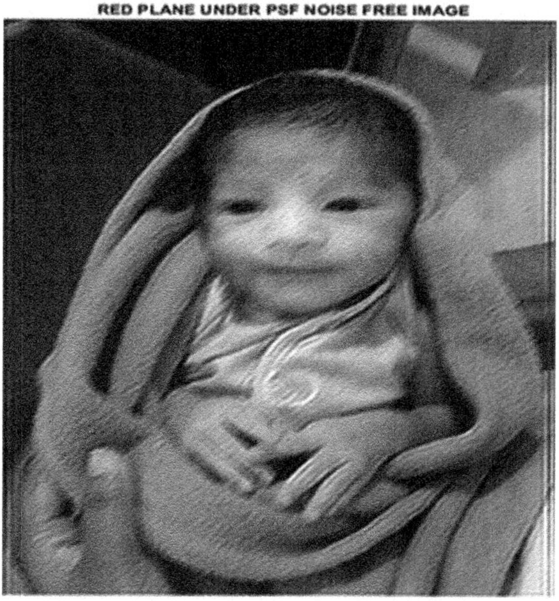

Fig. 5 Red plane under PSF noise free image

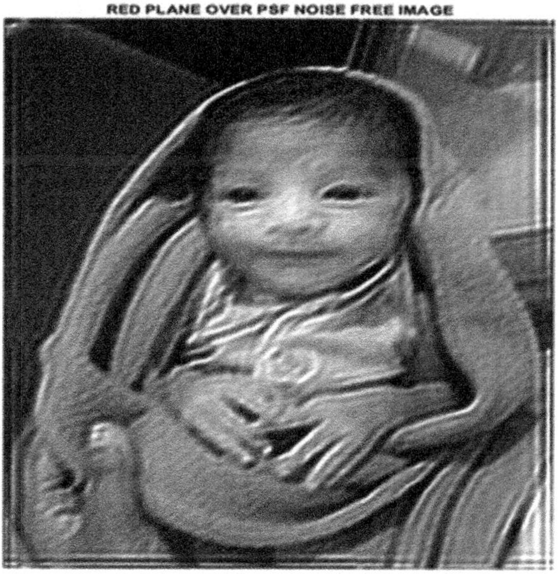

Fig. 6 Red plane over PSF noise free image

(continued)

(continued)

Images/ data sets	Methodology (Color Image to Color Image)					
	PSNR	MSE	Reconstructed using under PSF	Reconstructed using over PSF	Reconstructed using init PSF	Best restored image
Images/ data sets	Methodology (Color Image to Color Image)					
	PSNR	MSE	Reconstructed using under PSF	Reconstructed using over PSF	Reconstructed using init PSF	Best restored image
1	62.333	0.150	✓			✓
	48.542	1.181		✓		
	61.449	3.165			✓	
2	58.552	6.215	✓			
	39.902	8.235		✓		
	74.645	2.213			✓	✓
3	62.346	1.149	✓			✓
	18.519	7.182		✓		
	61.462	15.164			✓	
4	25.390	0.192	✓			✓
	19.501	6.215		✓		
	17.307	4.194			✓	
5	59.036	0.216	✓			✓
	38.545	2.228		✓		
	19.036	4.215			✓	

5 Conclusion

This paper presents advanced techniques for color Image Bling Deconvolution to Restore Blurred Image. In this, a proposed technique is introduced for effectively restoring blurred images using the filter and PSF estimation. The Wiener Filter is implemented efficiently, facilitating the creation of a highly blurred image in a relatively short amount of time. Our blind deconvolution provides the advantage of deblurring the degraded image without the need for prior knowledge of additive noise and PSF. The filter is applied to remove the texture components, while the Canny edge detector is utilized to detect edges in the blurred image. Clear images can be reconstructed with significantly less ringing and less noise, thereby achieving a more effective deconvolution. The above experimental results that are MSE, PSNR values indicate that this method is highly efficient for deblurring motion blur.

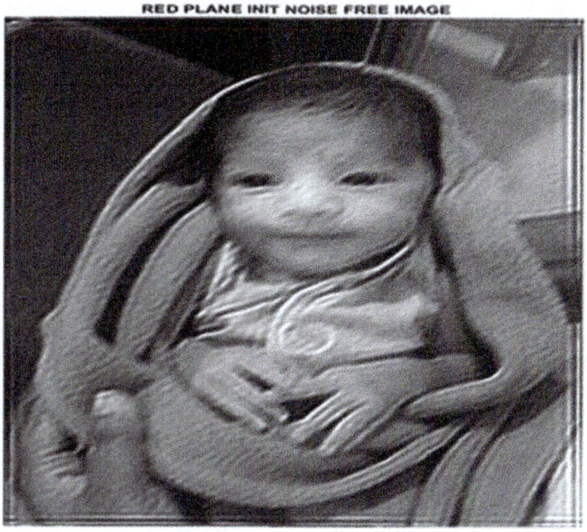

Fig. 7 Red planet init noise free image

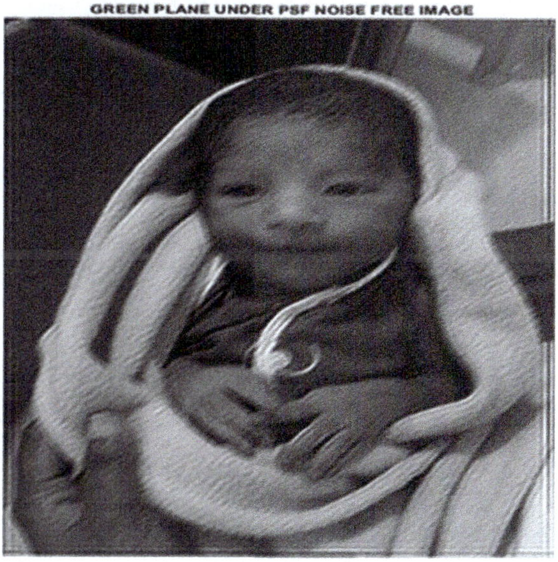

Fig. 8 Green plane under PSF noise free image

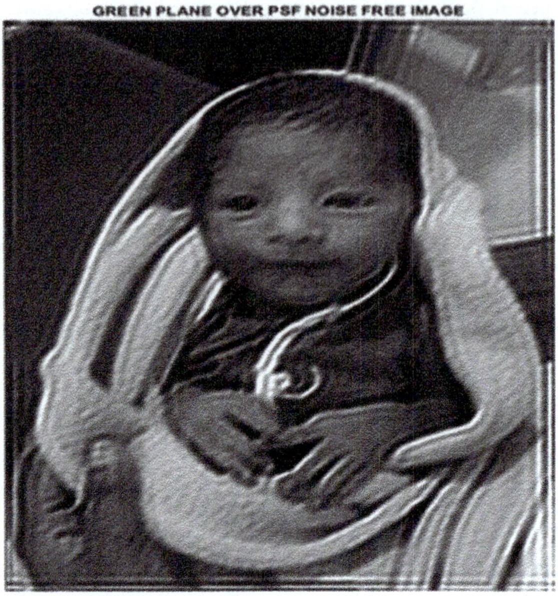

Fig. 9 Green plane over noise free image

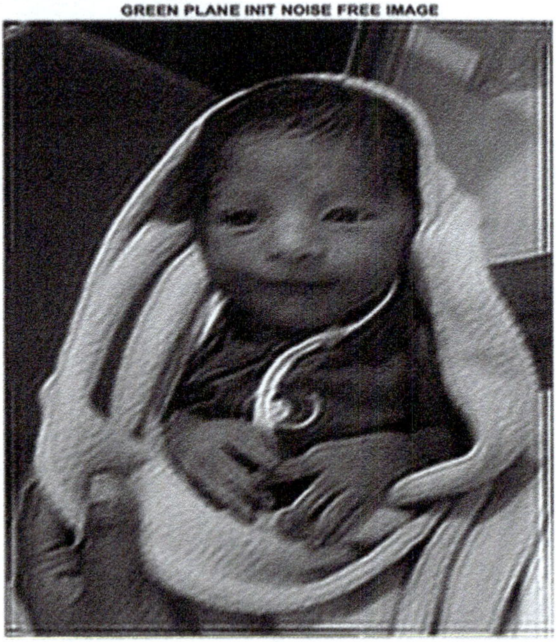

Fig. 10 Green plane init noise free image

Fig. 11 Blue plane under PSF noise free image

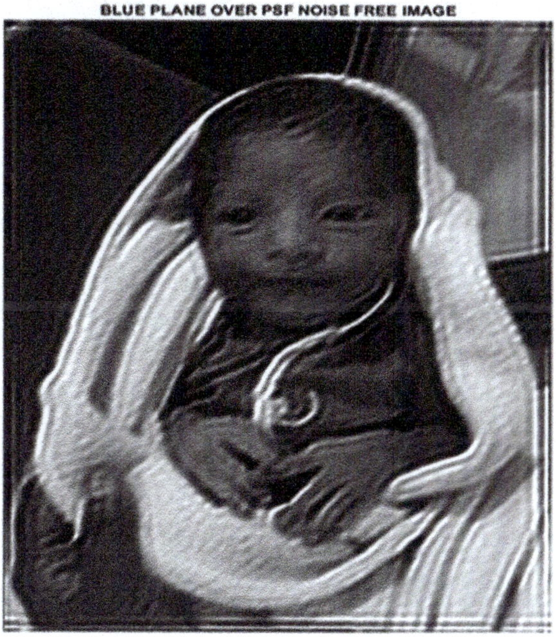

Fig. 12 Blue plane over PSF noise free image

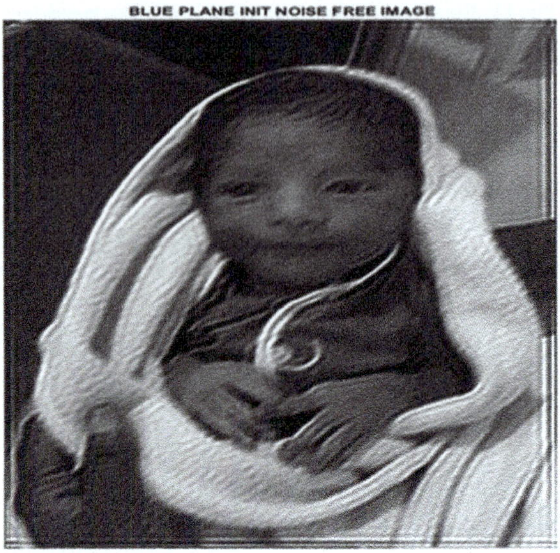

Fig. 13 Blue plane init noise free image

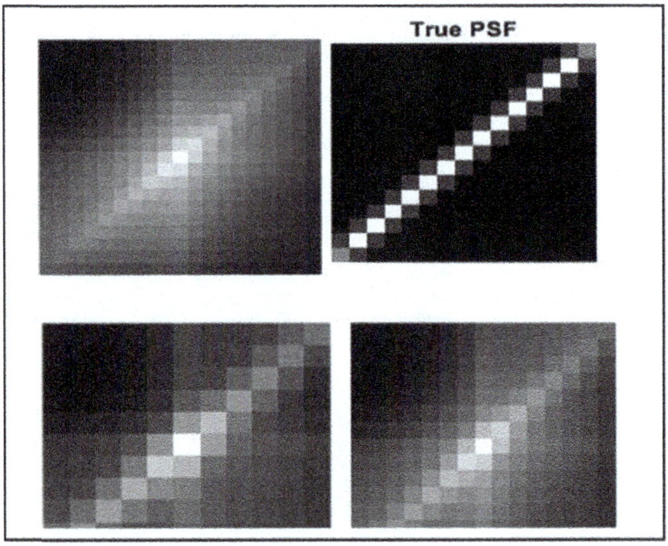

Fig. 14 All PSF 1st plane valued images

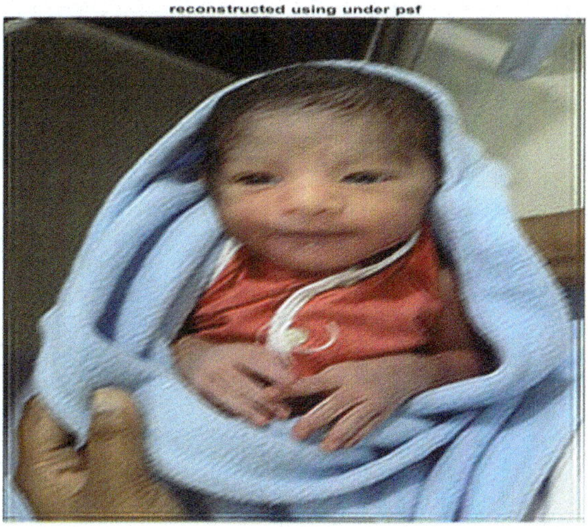

Fig. 15 All PSF 1st plane valued images

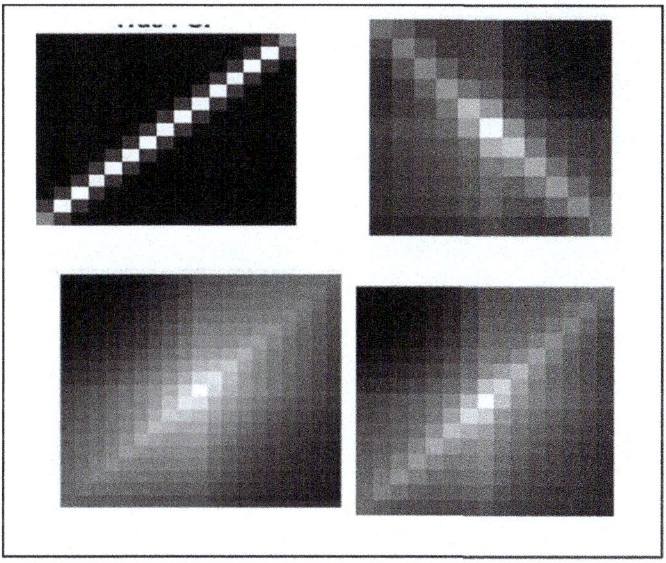

Fig. 16 All PSF 2nd plane valued images

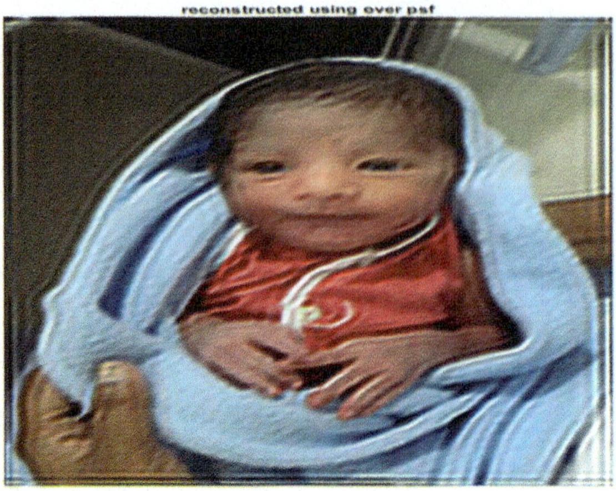

Fig. 17 Reconstructed image using over PSF

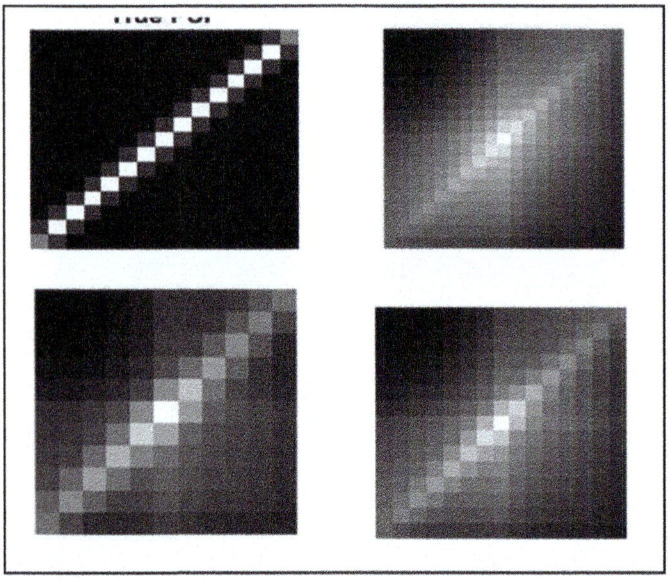

Fig. 18 All PSF 3rd plane valued images

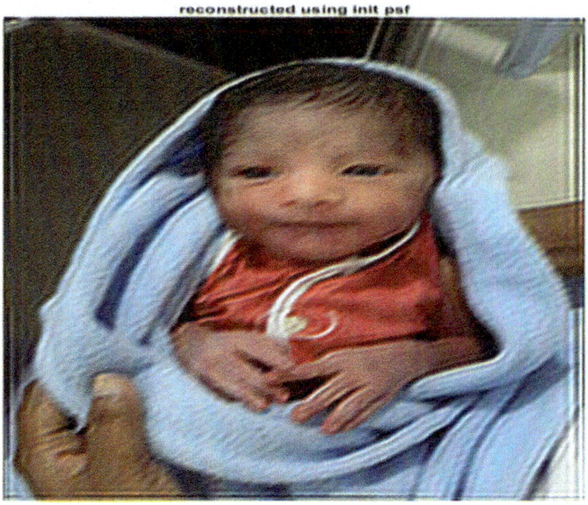

Fig. 19 Reconstructed image using init PSF

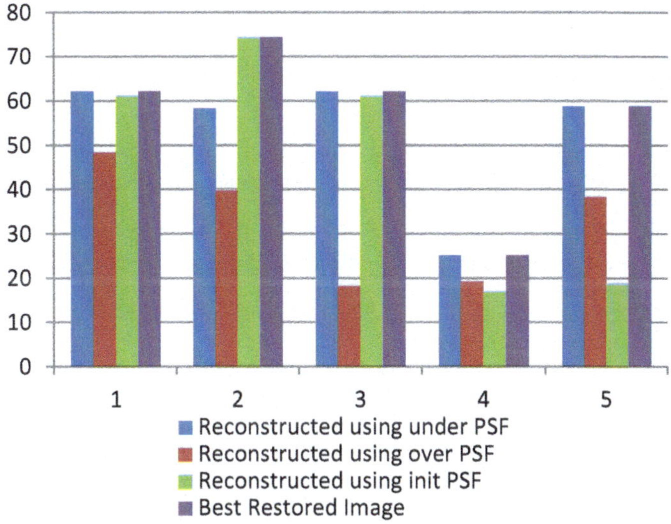

Fig. 20 Comparison plot of PSNR values of different datasets for different methods

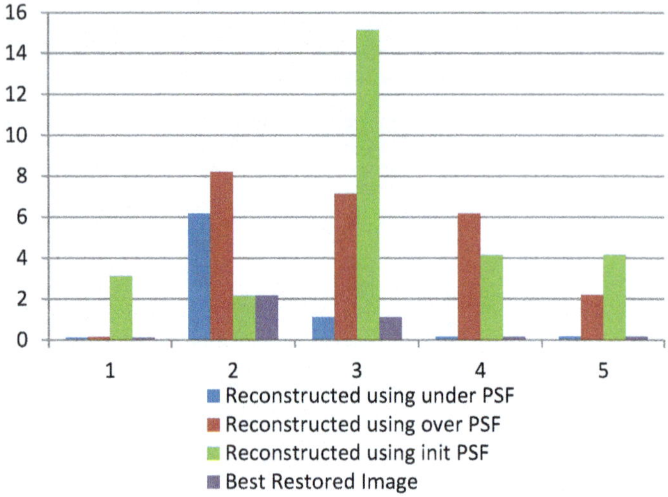

Fig. 21 Comparison plot of MSE values of different datasets for different methods

References

1. Khan A, Ali I, Waleed M (2020) An online platform based on google collaboration for quick estimation of real blur in blind deblurring of a single image. In: Electronics, computers, and AI 2020: 12th international conference
2. Shajkofci A, Liebling L (2020) Blind deconvolution and depth estimation using spatially variable CNN-based point spread function estimation in optical microscopy. IEEE Trans Image Process 29:5849–5862
3. Ning M, Zhang S. Wang S (2018) Image restoration of motion blurred stars using mems gyroscope aid and blur kernel correction. 18(8):2663
4. Tabib RA, Mudenagudi U, Patil U, Dhanakshirur RR (2019) Learning-based PSF estimation for motion deblurring: evidence-based feature selection and collaborative representation. In: International conference on computer vision by the IEEE
5. Khongkraphan K, Phonon A, Yusoh M (2018) Using the gradient descent method to estimate motion blur parameters. In: 22nd International conference on computer science and engineering. Tao X et al (2018) Deep Image deblurring using a scalable recurrent network. In: 2018 IEEE Conference on pattern recognition and computer vision
6. Xiang S, Peng S, Yin W, Liu RW (2018) L0 regularized priors based on hybrid gradient sparsity for robust single image blind deblurring. In: IEEE conference proceedings on acoustics, speech signal processing, Seoul, Korea, April 2018, pp 1349–1353
7. Pfister H, Yang MH, Sun D, Pan J (2018) Image blurring using a dark channel before. IEEE Trans Pattern Recogn Mach Learn 40:2316–2329
8. Ren W, Cao X, Guo Y, Yan Y, Wang R (2017) Prior to deblurring images using extreme channels. In: The IEEE conference on computer vision and pattern recognition, Honolulu, July 2017, pp 6976–6984
9. Yang M, Su Z, Pan J, Hu Z (2014) Text image deblurring using L0 regularised gradients and intensities. In: The IEEE Conference on pattern recognition and computer vision proceedings, Columbus, Sept 2014, pp 2902–2909
10. Lee S, Wang J, Cho H (2012) Deblurring of text images utilizing text-specific characteristics. J Eur Comput Vis Conf 525–538

11. Tang X, Sun J, He K (2011) Using the dark channel to remove haze from a single image. J Pattern Recogn Mach Learn 33:2342–2354
12. Durand F, Levin A, Freeman W, Weiss Y (2009) Understanding and evaluating blind deconvolution techniques. In: IEEE Conference on computer vision and pattern recognition, Miami, Florida, pp 1965–1972, Mar 2009
13. Jia J, Shan Q, Agarwala A (2008) High-quality motion blurs removal with only one image. ACM Trans Graph 27:1–11
14. Hertzmann A, Fergus R, Freeman WT, Roweis ST, Singh B (2006) Removing camera shake from a single shot. ACM's Trans Graph 25:788–795

CRNN-Based Eye Behavior Analysis for Drowsiness Detection

J. Sasi Kiran, Gurram Sunitha, Marri Sireesha, U. Mahender, K. Reddy Madhavi, Swathi Rudra, and Voruganti Naresh Kumar

Abstract Deep learning plays a critical role in enhancing road safety through various applications and technologies. Deep learning is at the core of Advanced Driver Assistance Systems, which provide real-time feedback and warnings to drivers. Drowsiness detection is a critical aspect of ensuring road safety and preventing accidents, especially in long-haul driving and transportation sectors. Deep learning models can analyze driver behavior, including eye movement and facial expressions, to detect signs of drowsiness. Drowsiness detection systems can then issue warnings suggesting rest breaks to prevent accidents caused by drowsy driving. This research explores the development and evaluation of a Convolutional Recurrent Neural Network architecture for drowsiness detection from eye images. The CRNN-based design utilizes a synergistic combination of CNN and LSTM layers, offering a comprehensive solution for drowsiness detection that considers both spatial and temporal aspects of eye behavior. The proposed model leverages the Eye Aspect Ratio and Eye Closure Ratio as key features for accurate detection. The proposed CRNN-based model is evaluated against traditional machine learning and deep learning models. Experimental results reveal that the CRNN-based model achieved an impressive accuracy of 93.2%, outperforming compared models and showcasing its potential for real-world applications.

J. S. Kiran
Department of CSE, Lords Institute of Engineering and Technology (Autonomous), Himayathsagar, Hyderabad, Telangana, India

G. Sunitha · K. R. Madhavi
Department of AI & ML, School of Computing, Mohan Babu University, Tirupati, Andhra Pradesh, India

M. Sireesha · V. N. Kumar (✉)
Department of CSE, CMR Technical Campus, Hyderabad, Telangana, India
e-mail: nareshkumar99890@gmail.com

U. Mahender
Department of CSE, CMR ENGINEERING COLLEGE, Hyderabad, Telangana, India

S. Rudra
Department of CSE(AI & ML), CMR Technical Campus, Hyderabad, Telangana, India

B. R. Devi et al. (eds.), *Proceedings of Fifth International Conference on Computer and Communication Technologies*, Lecture Notes in Networks and Systems 897,
https://doi.org/10.1007/978-981-99-9704-6_36

Keywords Road safety · Drowsiness · Deep learning · Digital imaging · CRNN

1 Introduction

Digital imaging classification has undergone a profound transformation in recent years. This transformation is led by rapid advancement of deep learning techniques. Deep learning has revolutionized the way complex problems are interpreted and solved in various domains, including computer vision and image classification [1–3].

Deep learning is at the core of Advanced Driver Assistance Systems, which provide real-time feedback and warnings to drivers. Drowsiness detection is a critical aspect of ensuring road safety and preventing accidents, especially in long-haul driving and transportation sectors [4]. Deep learning models can analyze driver behavior, including eye movement and facial expressions, to detect signs of drowsiness. Drowsiness detection systems can then issue warnings suggesting rest breaks to prevent accidents caused by drowsy driving. Drowsiness-related accidents on the road have become a significant concern, leading to substantial loss of lives and property. Detecting drowsiness in drivers is of paramount importance for road safety [5]. One of the early indicators of drowsiness is the closure of the eyes. In recent years, advancements in deep learning and computer vision have opened up new avenues for drowsiness detection, with a focus on eye image analysis [6].

This research introduces a comprehensive approach to drowsiness detection from eye images. It investigates the development and evaluation of a Convolutional Recurrent Neural Network (CRNN)-based model for drowsiness detection. This research uniquely emphasizes on integrating two critical features—Eye Aspect Ratio (EAR) and Eye Closure Ratio (ECR). EAR captures geometric aspects of the eye, while ECR accounts for temporal changes in eye closure. The synergy between CNN-based image analysis and RNN-based temporal modeling is harnessed to effectively distinguish between drowsy and alert states.

2 Literature Review

The importance of deep learning in digital imaging classification lies in its ability to automatically learn and represent intricate features from images, its scalability, robustness, and adaptability [7]. As digital imaging continues to play a vital role in various domains, including healthcare, entertainment, and scientific research, deep learning remains at the forefront of innovation, enabling accurate and efficient classification of digital imagery. Its contributions to computer vision are transformative and hold immense promise for the future.

Deep learning's impact extends beyond computer science. It has found applications in fields such as healthcare (medical image diagnosis), automotive (autonomous driving), agriculture (crop monitoring), and security (video surveillance) [8–10]. Its versatility makes it an invaluable tool for solving cross-disciplinary challenges.

The field of drowsiness detection from eye images has witnessed significant advancements, with a multitude of techniques and models proposed in recent years [11]. Early approaches predominantly relied on handcrafted features and traditional machine learning algorithms. However, these approaches often struggled with complex feature extraction and modeling temporal dependencies.

The emergence of deep learning transformed the landscape of drowsiness detection. CNNs exhibited remarkable capabilities in image feature extraction, allowing for improved representation of eye images [12]. Additionally, RNNs proved invaluable in capturing temporal patterns [13]. The proposed CRNN-based model amalgamates these strengths, offering a holistic approach.

Various datasets have provided researchers with labeled eye image sequences for training and evaluation [14, 15]. EAR and ECR metrics have gained popularity due to their effectiveness in quantifying eye openness and closure [16].

However, despite these advancements, the challenges of real-world deployment persist. Variability in lighting conditions, head poses, and the presence of eyeglasses are among the complexities that necessitate robust models.

3 CRNN-Based Deep Model for Drowsiness Detection

A brief description of the CRNN-based design created for drowsiness detection from eye images using EAR and ECR is presented in this section.

3.1 Architecture

The proposed CRNN-based architecture seamlessly combines the strengths of CNN and RNN to tackle drowsiness detection with precision. Table 1 presents configuration of CRNN-based model. It starts by extracting spatial features from input eye image sequences using convolutional layers, capturing essential details and patterns. The convolutional layers employ 3×3 kernels with ReLU activation functions to extract relevant image features efficiently. Subsequent max-pooling layers reduce spatial dimensions, preserving essential information while reducing computational complexity.

Following the CNN component, the model incorporates recurrent layers. These layers capture temporal dependencies across a sequence of images. Long Short-Term Memory (LSTM) units are used in the proposed model for temporal modeling. This enables the network to capture sequential dependencies and patterns in the eye image sequences over time, a critical aspect of drowsiness detection. The fully connected layers process the extracted features and make classifications. Classification is based on EAR and ECR values. They quantify eye openness and closure.

Table 1 Proposed CRNN-based model configuration

Layer	Hyperparameter	Configuration
Input layer	Sequence length	5
	Height and width	224 × 224
	Channels	3
CNN-convolutional layers 1 and 2	Type	2D Convolutional layer
	No. of filters	Conv 1—32, Conv2—64
	Kernel size	(3, 3)
	Activation function	ReLU
	Padding	Same
CNN—pooling layers 1 and 2	Pool type	Max
	Pool size	(2, 2)
	Normalization	Batch
RNN—LSTM layers 1 and 2	No. of units	64
	Activation function	tanh
	Dropout	0.3
	Return sequences	False
Dense layers 1 and 2	No. of units	DL 1—128, DL2—3
	Activation function	DL1—ReLU, DL2—Sigmoid
	Dropout	DL1—0.5

3.2 Custom Components for Eye Features

EAR and ECR are common metrics used in facial landmark analysis and drowsiness detection. They are calculated using the coordinates of specific facial landmarks around the eyes. Two features EAR and ECR are calculated separately for each image to detect whether the eye is open or closed.

EAR is a metric used to analyze the geometry of an individual eye image, typically used for detecting blinks or assessing eye openness. In the context of static eye images, EAR is calculated using the coordinates of specific landmarks on the eye usually the inner and outer corners of the eye.

$$\text{EAR} = \frac{2 \times |P_2 - P_6|}{|P_3 - P_5| + |P_4 - P_1|} \tag{1}$$

where P_1 to P_6 are coordinates of specific eye landmarks in image; P_1 and P_6 represent leftmost and rightmost points on the horizontal axis of eye; P_2 and P_5 represent topmost and bottommost points on the vertical axis of eye; P_3 and P_4 represent inner corner and outer corner of eye.

In this context, EAR is used to quantify the degree of openness of the eye depicted in a single eye image. A lower EAR value suggests a more closed eye (e.g., during a blink), while a higher EAR value indicates a more open eye.

The vertical distance $|P_2 - P_6|$ represents the height of the eye, and the horizontal distance $|P_3 - P_5| + |P_4 - P_1|$ represents the width of the eye. EAR measures how open or closed the eye is, with a lower value indicating more closed eyes (e.g., during a blink) and a higher value indicating more open eyes.

ECR is a metric used for drowsiness detection by analyzing the closure of the eyes over a sequence of frames. It assesses the degree of eye closure based on EAR calculations across a sequence of eye images. In this context, ECR is calculated as follows.

$$\text{ECR} = \frac{C}{TE} \tag{2}$$

where C is count of images in sequence where the EAR falls below a specified threshold, indicating closed eyes; E is total number of eye images in analyzed sequence.

To use ECR effectively for drowsiness detection, the system would monitor the ECR value over time, and if it exceeds a certain threshold for a specified duration, it may trigger a drowsiness alert.

4 Experimental Results and Evaluation

In this section, a comprehensive overview of the experimental setup, the performance evaluation metrics used, a comparison with baseline methods, and the presentation of results are presented. The experiments were conducted to assess the effectiveness of the proposed CRNN-based architecture for drowsiness detection from eye images.

Three models were implemented and evaluated—A CNN model designed for image feature extraction, RNN model used for temporal feature capture, proposed CRNN-based architecture that combines both CNN and RNN elements, with EAR and ECR features incorporated. All models were trained using standard techniques, such as mini-batch stochastic gradient descent with Adam optimization.

4.1 RLDD Dataset Description

The RLDD dataset was created for multistage drowsiness detection, focusing on both evident and subtle cases of drowsiness [17]. The dataset comprises of videos approximately from 60 healthy participants. The participants were all adults, with diverse demographics and ages. Some participants wore glasses, and others had facial hair in videos. All videos are self-recorded by participants from different angles and environments. Each video represents driver in three conditions: alertness, low vigilance, and drowsiness, totaling 180 videos. The frame rate in the videos is consistent with typical consumer cameras, always less than 30 fps. The three classes in the dataset were defined as follows:

Alert: Participants were considered alert when they exhibited no signs of sleepiness.
Low Vigilant: This category represented subtle instances where some signs of sleepiness were present, but individuals did not exert effort to stay awake.
Drowsy: In this state, participants actively had to make an effort to avoid falling asleep, indicating a significant level of sleepiness or drowsiness.

A sequence of five images from each video were extracted for the purpose of experimentation. Human faces are detected in each image and are automatically cropped from the images. OpenCV is used for the process. A total of 4000 images were created (1371 drowsy images, 1436 non-drowsy images, 1193 low vigilant images).

4.2 Experimental Results

Figure 1 presents the performance comparison chart of the experimentation on various deep models. The proposed CRNN-based model achieved the highest accuracy of 93.2%, demonstrating its superior ability to classify drowsy, low vigilant, non-drowsy states accurately. Both CNN and RNN models also performed well, achieving accuracies of 87.5% and 88.9%, respectively. CRNN-based model achieved a balanced precision and recall, ensuring both false positives and false negatives are minimized. CNN and RNN models also exhibited commendable precision and recall values. F1-Score of CRNN-based model is 0.931, indicating a robust balance between precision and recall. CRNN-based model demonstrating the highest AUC, emphasizing its superior discriminatory power. Confusion matrix (Fig. 2) visually illustrate the performance. CRNN-based model has fewer misclassifications compared to SVM and Random Forest. CRNN-based model outperformed traditional baseline models (SVM and Random Forest) across all performance metrics. This superior performance highlights the effectiveness of the proposed architecture in utilizing EAR and ECR features for drowsiness detection (Fig. 2).

Proposed CRNN-based architecture, incorporating EAR and ECR features, exhibited superior performance in drowsiness detection from eye images. While both CNN and RNN models showed competitive performance, CRNN-based model outperformed them and traditional baseline models. This suggests that the combination of convolutional and recurrent layers in the CRNN architecture, along with the integration of EAR and ECR features, offers an effective solution for accurate drowsiness detection.

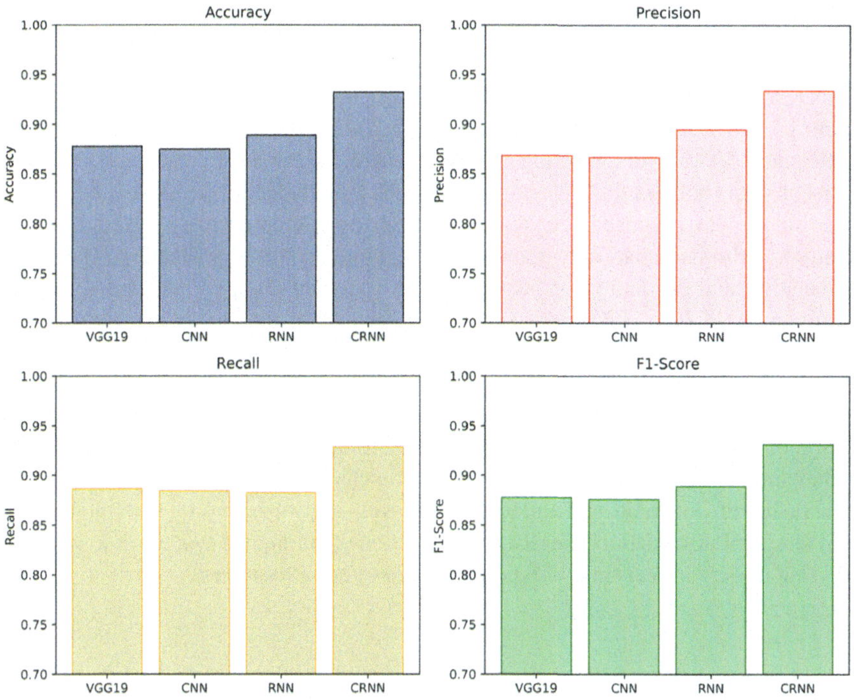

Fig. 1 Experimental results of CRNN-based model for drowsiness detection

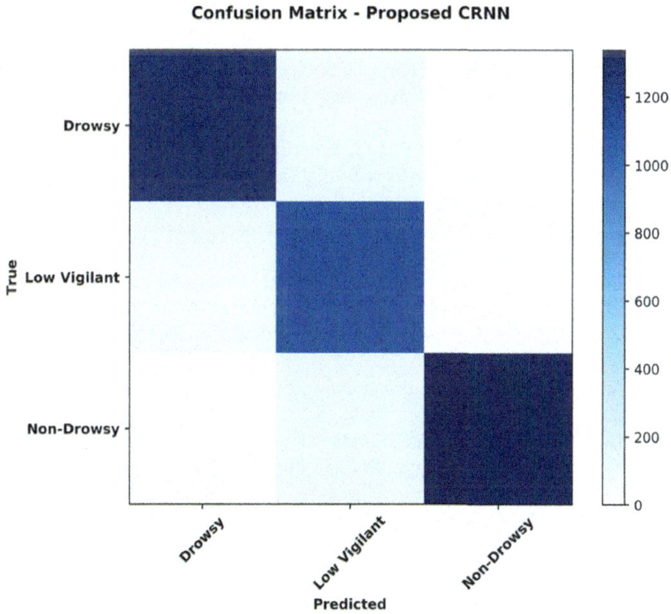

Fig. 2 Performance of CRNN-based Model for drowsiness detection

5 Conclusion

This research endeavors to address the pressing issue of drowsiness detection in the context of road safety. The CRNN-based design utilizes a synergistic combination of CNN and LSTM layers, offering a comprehensive solution for drowsiness detection that considers both spatial and temporal aspects of eye behavior, ultimately enhancing road safety. The proposed CRNN-based architecture, empowered by EAR and ECR features, exhibits remarkable potential for accurate and timely detection. The experimentation and evaluation of our model have yielded compelling results, with an accuracy of 93.2%. CRNN-based models superior performance, when compared to traditional models SVM and Random Forest, underscores the significance of deep learning in this domain. The achieved precision, recall, and F1-score emphasize the model's ability to minimize both false alarms and missed detections. The incorporation of EAR and ECR as crucial features facilitates a deeper understanding of eye behavior, enabling the model to discern subtle changes associated with drowsiness. The fusion of Convolutional and Recurrent layers allows the model to exploit spatial and temporal information effectively. Integration of automated systems into vehicles, alerting drivers in real-time will potentially save countless lives.

References

1. Arunachalam R, Sunitha G, Shukla SK, Urooj S, Rawat S (2023) A smart Alzheimer's patient monitoring system with IoT-assisted technology through enhanced deep learning approach. Knowl Inform Syst 1–39
2. Charan NS, Narasimhulu T, Bhanu Kiran G, Sudharshan Reddy T, Shivangini Singh T, Sunitha G (2023) Solid waste management using deep learning. In: 14th International conference on soft computing and pattern, vol 648, pp 44–51. Springer, Berlin
3. Reddy Madhavi K, Jarugula J, Karuna G, Kaleru S, Srujan Raju K, Sunitha G (2023) Bird species recognition using deep transfer learning. In: 4th International conference on computer and communication technologies, vol 606, Springer, Berlin
4. Albadawi Y, Takruri M, Awad M (2022) A review of recent developments in driver drowsiness detection systems. Sensors 22(5):2069
5. Ngxande M, Tapamo JR, Burke M (2017) Driver drowsiness detection using behavioral measures and machine learning techniques: a review of state-of-art techniques. In: 2017 pattern recognition Association of South Africa and Robotics and mechatronics (PRASA-RobMech), pp 156–161
6. Deng W, Wu R (2019) Real-time driver-drowsiness detection system using facial features. IEEE Access 7:118727–118738
7. Sunitha G, Geetha K, Neelakandan S, Pundir AKS, Hemalatha S, Kumar V (2022) Intelligent deep learning based ethnicity recognition and classification using facial images. Image Vis Comput 121:104404
8. Karthikeyan C, Sunitha G, Avanija J, Reddy Madhavi K, Madhan ES (2021) Prediction of climate change using SVM and Naïve Bayes machine learning algorithms. Turk J Comput Math Educ 12(2):2134–2139
9. Suma KG, Sunitha G, Avanija J (2019) Accident prevention and traffic control by Otsu Method and Haar-cascade hand detector. Int J Eng and Ad Technol 9(2):5328–5333

10. Reddy Madhavi K, Sunitha G, Avanija J, Viswanadha Raju S, Sindhooja A (2021) Impact analysis of hydration and sleep intervention using regression techniques. Turk J Comput Math Educ 12(2):2129–2133
11. Siddiqui HUR, Saleem AA, Brown R, Bademci B, Lee E, Rustam F, Dudley S (2021) Non-invasive driver drowsiness detection system. Sensors 21(14):4833
12. McDonald AD, Lee JD, Schwarz C, Brown TL (2018) A contextual and temporal algorithm for driver drowsiness detection. Accid Anal Prev 113:25–37
13. Pandey NN, Muppalaneni NB (2021) Temporal and spatial feature based approaches in drowsiness detection using deep learning technique. J Real-Time Image Proc 18(6):2287–2299
14. Drowsiness Detection Computer Vision Project (2022) RoboFlow universe. https://universe.roboflow.com/augmented-startups/drowsiness-detection-cntmz
15. Weng CH, Lai YH, Lai SH (2017) Driver drowsiness detection via a hierarchical temporal deep belief network. In: Computer Vision—ACCV 2016 workshops: ACCV 2016 international workshops, Taipei, Taiwan, 20–24 Nov 2016. Revised Selected Papers, Part III 13. Springer International Publishing, Berlin, pp 117–133
16. Maior CBS, das Chagas Moura MJ, Santana JMM, Lins ID (2020) Real-time classification for autonomous drowsiness detection using eye aspect ratio. Expert Syst Appl 158:113505
17. Ghoddoosian R, Galib M, Athitsos V (2019) A realistic dataset and baseline temporal model for early drowsiness detection. In: Proceedings of the IEEE/CVF conference on computer vision and pattern recognition workshops

Sentiment Analysis Principle Technical Approach on Online Social Network Data Using CNN for Detection of Stress

K. Srujan Raju, Nagu Chandra Sekhar Reddy, Matoori Praveen, V. S. Manoj Kumar Chenna, Sandhyarani, and Bollampelly Chandana

Abstract According to this research, stress is typically a human response to various dangers or goals. When it works well, this reaction can help us stay focused, motivated, and cognitively engaged. However, when it goes out of control, it can be hazardous and result in depression, anxiety, hypertension, and a variety of other serious problems. Cyberspace serves as a vast soap box for people to communicate everything they come across in their daily lives. Based on the posts and status updates the person shares, it may then be utilized as a very effective way to determine how stressed out they are. This is a proposal for a website that receives the user's Twitter username, uses sentiment analysis to scan and analyze the topic's profile, and then provides findings. These results give an overview of the subject's mental and emotional health as well as its overall stress levels. The CNN deep learning architecture is the focus of this study since it aims to address the stress of evolution in online social networks like Twitter.

Keywords Social media · Factor graph model · Social interaction · Stress detection · Healthcare

1 Introduction

Data mining, also known as information exploration, is the computer-assisted process of sifting through and analyzing large amounts of information and extracting the meaning from the data. Tools for forecasting preferences and future trends in data

K. S. Raju (✉) · V. S. M. K. Chenna
Department of CSE, CMR Technical Campus, Hyderabad, Telangana, India
e-mail: ksrujanraju@gmail.com

N. C. S. Reddy
Department of CSE, MLR Institution of Technology, Hyderabad, Telangana, India

M. Praveen · Sandhyarani · B. Chandana
Department of CSE (Data Science), CMR Technical Campus, Hyderabad, Telangana, India

B. R. Devi et al. (eds.), *Proceedings of Fifth International Conference on Computer and Communication Technologies*, Lecture Notes in Networks and Systems 897,
https://doi.org/10.1007/978-981-99-9704-6_37

mining, enabling experienced-led businesses to make well-informed strategic decisions. Data mining software provides answers to market problems that have traditionally taken a long time to resolve. They explore hidden trend databases, uncovering predicting data that experts might overlook because it falls outside of their assumptions. Data mining is typically defined as the process of evaluating and summarizing data from many perspectives to produce meaningful information [1, 2] that may be used to generate income, reduce costs, or do both. Data mining software is one of several analytical approaches to data analysis. It allows users to examine, categorize, and summarize data-related relationships from a variety of perspectives [3, 4].

Data mining is the process of identifying commonalities [5] or patterns between thousands of fields in large databases. Despite the fact that data mining may yield similar results in the long run, the innovation organization is not. For a long time, organizations have used powerful machines to sift through large amounts of general store scanner data and deconstruct statistical surveying reports. However, nonstop advancements to workstation preparing power, disc storage, and statistical methods [6], while driving down the cost, are dramatically increasing research accuracy. Even though data mining is still in its infancy, data mining tools and techniques are already being used by companies in a wide range of industries to take advantage of historical data, including retail, finance, healthcare, manufacturing, transport, and aerospace. Data mining serves examiners with comprehend vital information, relationships, trends, patterns [7, 8], and exceptions. Furthermore, anomalies that might overall try unnoticed with filter through warehoused learning. Eventually, Tom's perusing utilizing example distinguishment advances Furthermore measurable Furthermore scientific systems [9]. Information mining will be used to recognize examples. Furthermore, correlations in the information for organizations in place should help in making educated business choices. Information mining will encourage with spot bargains trends, create smarter offering campaigns, what is more faultlessly foresee customer devotion the natural provisions for learning mining.

2 Literature Review

The detection of psychological is associated with the topic of sentiment analysis and emotion detection [10].

In recent years, computer-aided identification, analysis, and application of emotions, especially on social networks, have attracted a great deal of attention [11, 12]. Relationships between stress and psychology [13, 14]. Personality characteristics could be an important topic to consider. For instance, Refs. [15, 16] provides proof that Daily 44 based on behavior metrics, stress may be accurately identified. Service by users on mobile phones. Huge numbers of social media-based feeling examination investigations would be on the tweet level, utilizing text-based semantics. Characteristics have also established arrangement methodologies and [17] recommended an arrangement called temperament lens to perform a feeling investigation on the

Chinese micro-blog stage Weibo, classifying the feeling classifications under four types, i.e., angry, gross, happy, and sad.

References [12, 18] thought on the problem of passionate transmission for social media and additionally discovered that resentment needs a more robust bond between disparate clients than happiness, indicating that unfavorable emotions may spread more widely and efficiently among the group. This inference can help us aggregate the social impact of people who identify stress because stress is typically perceived as a negative emotion. Nevertheless, the majority of these jobs make use of social network textual material. Social network data is fundamentally cross-media data since it typically comprises sequential and related items from several sources and modalities.

Although the identification of tweet-level emotions reflects the intermediate emotion exhibited in a single tweet, people's psychological stress states or emotions are usually more persistent and change over time. In recent years, a lot of research has begun to focus on user-level identification of emotions on social networks [19, 20]. Our latest job [20] proposed using a deep convolution network to learn user-level presentations on successive tweet series over a predetermined amount of time in order to detect psychological stress situations on social media. Inspired by the hemophilia theory, Ref. [7] strengthened Twitter's user-level sentiment analysis by integrating social interactions [21]. Although certain user-level feeling discovery tasks have been completed, the role that social connections have in an individual's mental states of stretch and the ability to integrate such information into push discovery have not yet been investigated. Social networking is among the most important features of social media websites. These days, a lot of scholars are concentrating on leveraging social interaction information to improve the effectiveness of social media analysis. References [22, 23] investigated the connections between social interactions and users' ideas and activities and found that conversations on Twitter can lead to insightful conclusions. References [24, 25] used comments on Flickr to anticipate the emotions portrayed in images posted there. Printed remarks content, for example, were important aspects of social interactions conveyed by these works, but they ignored the systemic nuances that underlie, say, how customers are interfaced [26].

3 Problem Identification

Stress is mainly how people react to different challenges or goals. When this response is under check, it can help us stay alert, focused, and mentally engaged while working effectively. However, when it goes out of control, it can lead to dangerous conditions like depression, anxiety, hypertension, and a variety of other conditions. People can share everything and everything they go through in their daily lives on the vast expanse of the internet. Consequently, it may be used as a highly helpful way to ascertain an individual's stress levels based on the posts and status updates that they share. This is a proposal for a website that scans and analyzes the subject's profile using Sentiment Analysis, provides results, and accepts the Twitter username of the topic as input. These findings provide a description of the subject's mental and emotional state as well as the overall stress level of the individual [27].

4 Implementation

Difficulties a rise in psychological stress recognition.

(1) How to extract attributes for the user level from the tweeting sequence and user attribute tackling the question of lack of modality in tweets.
(2) We proposed a process in which we gathered messages from Twitter and assigned distinct moods to each detail. The structure and class of each tweet can be used to identify them. It was nostalgic after categorizing all of the tweets for each of the words. It is simple to edit each of the tweets with the help of emotion extraction so that each of the stress rate levels may be identified.

5 Proposed System

Machine learning algorithms can be used to understand the cardinal indication of psychological stress: excessive negative thoughts and a lack of good sentiments. Optical Character Recognition (OCR) for image processing, Natural Language Processing (NLP), and Convolutional Neural Network (CNN) for text content processing are all included in the proposed system. Image processing, which detects and extracts text tweets from photographs, preprocessing, attribute extraction, categorization, and linking to NGOs are all critical components of the proposed system, as shown in Fig. 1.

The data was acquired via social media platforms such as Twitter and Facebook [28]. The user's stress level was estimated using a data selection of graphics and text. The photo dataset is extracted and analyzed using OCR, which extracts the text. After that, the text tweet material dataset and image extracted text dataset were used as preprocessing input and feature extraction using NLP. CNN plays a crucial role in categorizing favorable and negative tweet content. Finally, users' disparaging tweets and information are compiled and given to the NGO for advice. The stressed user information (i.e., user ID) is extracted and will be given to the CLASSIFER. This is one of the ways in which the suicidal person is detected in the social network, and it is very likely to decrease the measure of depression and results indicates a non-stressed user and a negative (0) anxious user. When the CNN classification of the social media dropout layer is eventually applied to the network regularization and to avoid over-fitting concerns value, which gives the value of likelihood as an output, CLASSIFER that interacts with the projected outcome can be characterized as positive (1) in order to alert (Figs. 2 and 3).

The initial layer of CNN is made up of one contraction. The layer and the temporal convolution operation are executed using the layer's kernel size and zero padding.

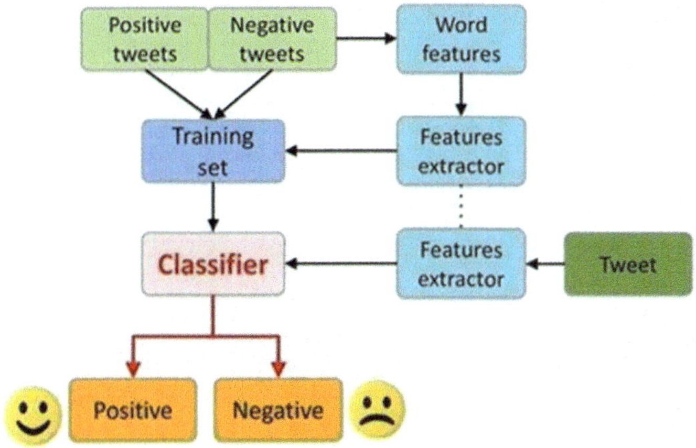

Fig. 1 How all the knowledge that is initially obtained can be analyzed and how all the sentence is extracted using emotions

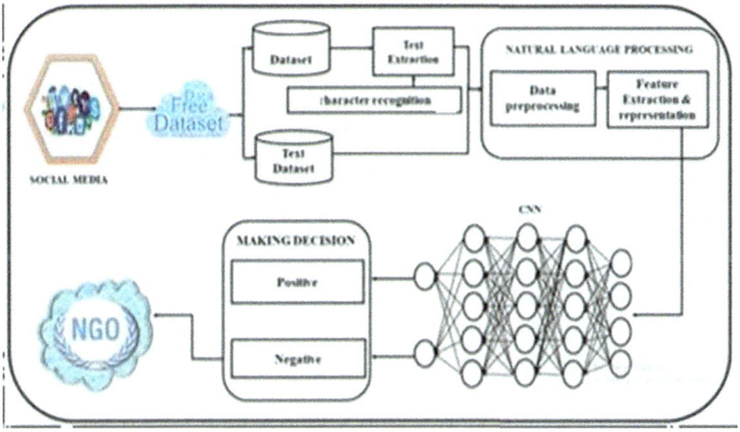

Fig. 2 Architecture of Psychological stress prediction

Next, the layer is subjected to the functionality of relu activation, and the pooling procedure is carried out. Here, global maximum pooling is used to reduce the dimensionality of the data. To generate a single value, the fully linked convolution layer receives the pooling output as an input. With this, the sigmoid activation function is carried out (Figs. 4, 5, 6, 7, 8 and 9; Table 1).

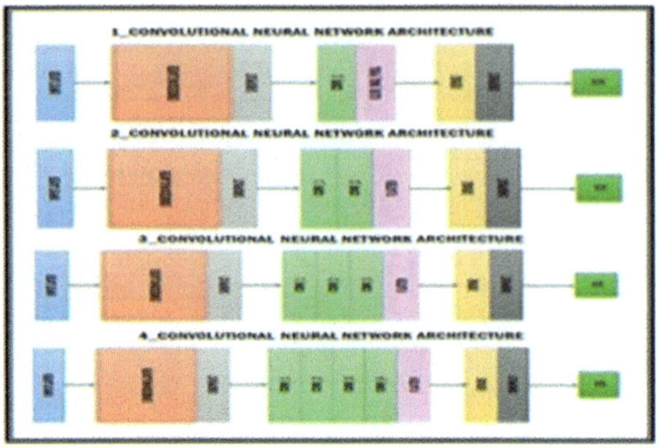

Fig. 3 Layered Convolution neural networks for stress classification

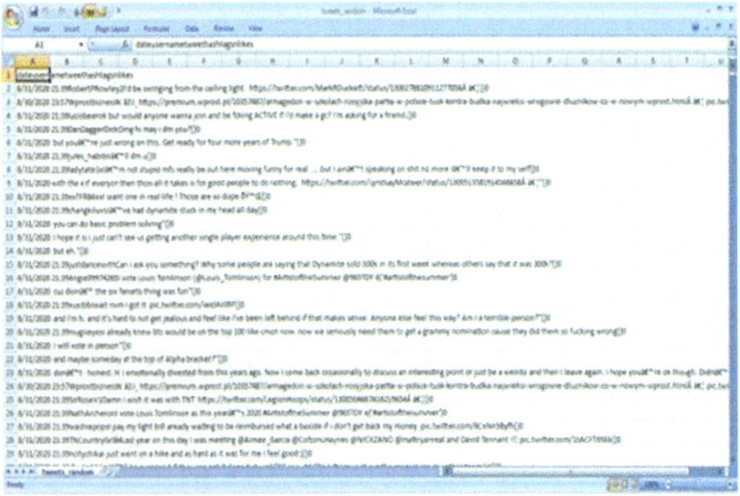

Fig. 4 Scrapping tweets from the online tweets

6 Results and Analysis

7 Conclusion

In today's world, where young people and almost all of the population mostly suffer from overcoming stress, whether due to peer pressure, workload, or other domestic tensions, it is very important to have a reality check on how stressed a person really is. It is for this reason that prompt stress diagnosis and prevention is a dire necessity. We developed this project to enable individuals to investigate the problem of stress. This project would be especially helpful for people who find it difficult to talk to others about their problems. It can provide these people with a reality check and

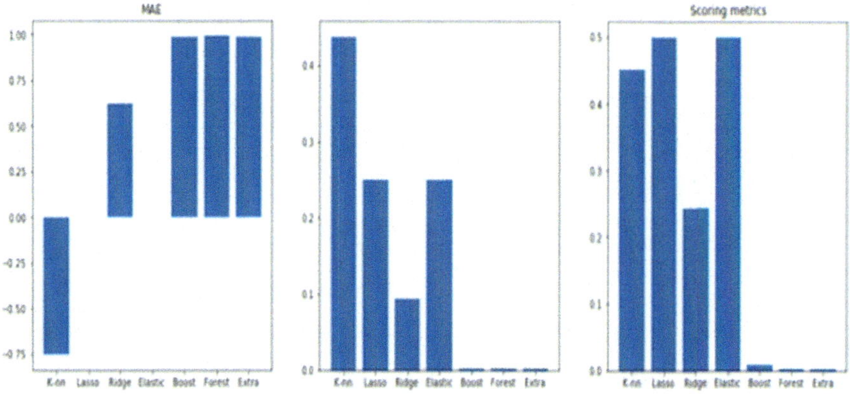

Fig. 5 Scoring metrics like mean absolute error and scoring metrics results graph

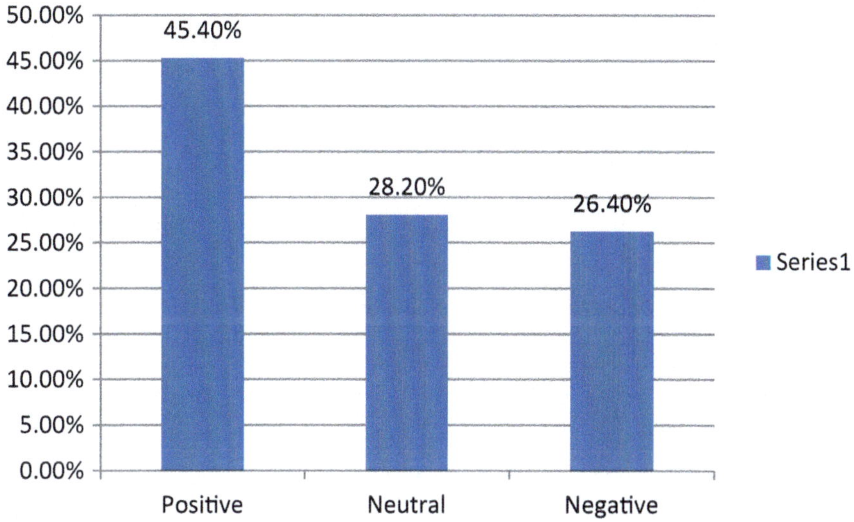

Fig. 6 Scoring metrics results using machine learning approach

empower them to seek medical attention based only on their social connections. We have utilized the concepts of sentiment analysis and used both human and machine learning. Compared to earlier methods, the technique's primary advantage is its non-invasiveness and fast-oriented execution in stress detection.

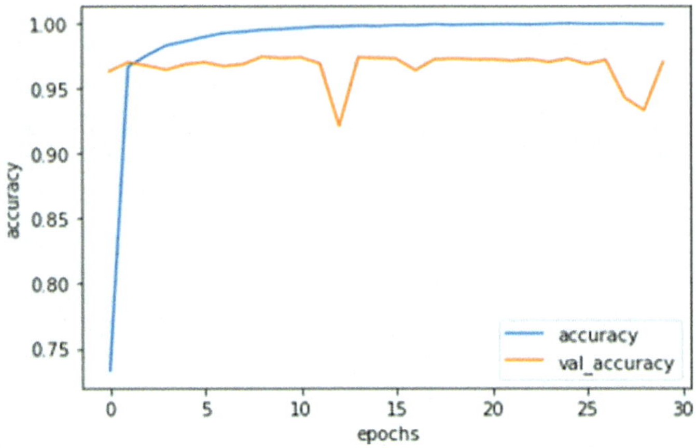

Fig. 7 Training and validation accuracy graph

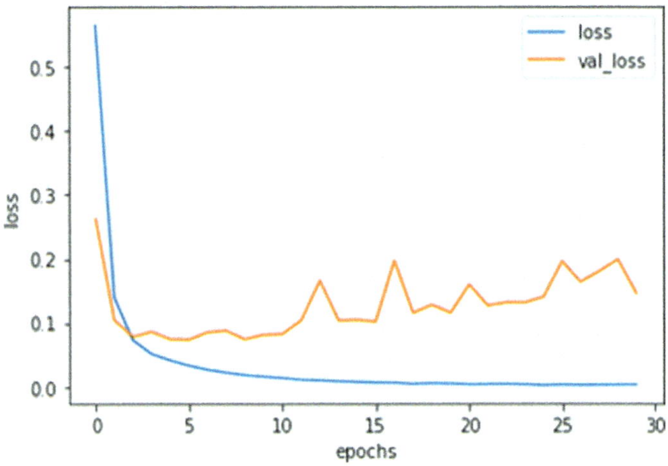

Fig. 8 Training and validation loss graph

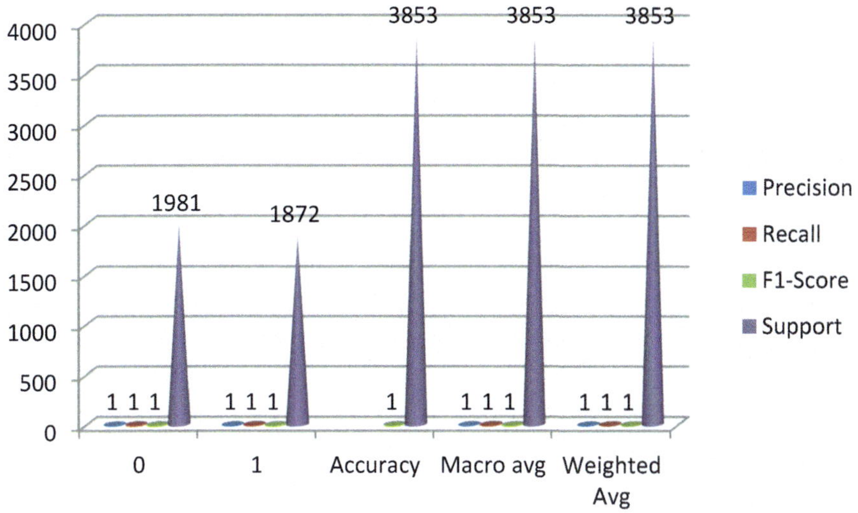

Fig. 9 Evolution metrics results

Table 1 Fraction of each tweet for stress classification

S. No.	Model	MSE	MAE	Score
0	K-nn	0.4	0.45	− 75.05
1	Lasso	0.2	0.50	− 0.02
2	Ridge	0.1	0.24	62.26
3	Elastic	0.2	0.50	− 0.02
4	Boost	0.0	0.01	99.17
5	Forest	0.0	0.00	99.19
6	Extra	0.0	0.00	99.15

References

1. Buckley C, Voorhees EM (2004) Retrieval evaluation within complete information. In: Proceedings of the 27th annual international ACM SIGIR conference on Research and development in information retrieval, pp 25–32
2. Ciresan DC, Meier U, Masci J, Gambardella LM, Schmidhuber J (2011) Flexible, high performance convolutional neural networks for image classification. In: Proceedings of International joint conference on artificial intelligence, pp 1237–1242
3. Lee K, Agrawal A, Choudhary A (2013) Real-time disease surveillance using twitter data: demonstration on flu and cancer. In: Proceedings of ACM SIGKDD international conference on Knowledge discovery and data mining, pp 1474–1477
4. Lin H, Jia J, Guo Q, Xue Y, Huang J, Cai L, Feng L (2014) Psychological stress detection from cross-media microblog data using deep sparse neural network. In: Proceedings of IEEE international conference on multimedia & expo
5. Golbeck J, Robles C, Edmondson M, Turner K (2011) Predicting personality from Twitter. In: Passat/social.com 2011, privacy, security, risk and trust, pp 149–156

6. Friedman JH (1999) Greedy function approximation: a gradient boosting machine. Ann Statist 29(5):1189–1232
7. Granovetter MS (1973) The strength of weak ties. Am J Sociol 1973
8. Kschischang FR, Frey BJ, Loeliger H-A (2001) Factor graphs and the sum-product algorithm. IEEE Trans Inform Theory 47(2):498–519
9. Chang C, Lin C-J (2001) Libsvm: a library for support vector machines. ACM Trans Intell Syst Technol 2(3):389–396
10. Fan E, Zhao J, Chen Y, Xu K (2014) Anger is more influential than joy: sentiment correlation in weibo. PLoSONE 9:e110184
11. Cohen S, Thomas AW (1985) Stress, social support, and the buffering hypothesis. Psychol Bull 98(2):310–357
12. Coppersmith G, Harman C, Dredze M (2014) Measuring post traumatic stress disorder in twitter. In: Proceedings of the International conference on weblogs and social media, pp 579–582
13. Fang Z, Zhou X, Tang J, Shao W, Fong ACM, Sun L, Ding Y, Zhou L, Luo J (2014) Modeling paying behavior in game social networks. In: Proceedings of the twenty-third conference on information and knowledge management (CIKM'14), pp 411–420
14. Gettinger J, Koeszegi ST (2015). More than words: the effect of emoticons in electronic negotiations.
15. Verhoeven B, Daelemans W, Plank B (2016) Twisty: a multi lingual twitter stylometry corpus for gender and personality profiling. In: Proceedings of the Tenth international conference on language resources and evaluation LREC, pp 1632–1637
16. Bogomolov A, Lepri B, Ferron M, Pianesi F, Pentland A (2014) Daily stress recognition from mobile phone data, weather conditions and individual traits. In: ACM International conference on multimedia, pp 477–486
17. Kelman HC (1958) Compliance, identification, and internalization: three processes of attitude change. General Inform 1(1):51–60
18. Gao R, Hao B, Li H, Gao Y, Zhu T (2013) Developing simplified Chinese psychological linguistic analysis dictionary for microblog, pp 359–368
19. Hosmer DW, Lemeshow S, Sturdivant RX (2013) Appliedlogistic regression. In: Wiley series in probability and mathematical statistics
20. Cun YL, Bengio Y (1995) Convolutional networks for images, speech, and time series. In: The handbook of brain theory and neural networks, p 3361
21. Chang X, Yang Y, Hauptmann AG, Xing EP, Yu Y-L (2015) Semantic concept discovery for large-scale zero-shot event detection. In: Proceedings of international joint conference on artificial intelligence, pp 2234–2240
22. Farnadi G, Sitaraman G, Sushmita S, Celli F, Kosinski M, Stillwell D, Davalos S, Moens MF, DeCock M (2016) Computational personality recognition in social media. In: User modeling and user adapted interaction, pp 1–34
23. Kamvar SD (2011) We feel fine and searching the emotional web. In: Proceedings of WSDM, pp 117–126
24. Fischer E, Rebecca Reuber A (2011) Social interaction via new social media: (how) can interactions on twitter affect effectual thinking and behavior? J Bus Venturing 26(1):1–18
25. Guo Q, Jia J, Shen G, Zhang L, Cai L, Yi Z (2016) Learning robust uniform features for cross-media social data by using crossauto encoders. Knowl Based Syst 102:64–75
26. Kobayashi S (1981) The aim and method of the color image scale. Color Res Appl 6(2):93–107
27. Kralj NP, Smailovi J, Sluban B, Mozeti I (2015) Sentiment of emojis. PlosOne 10(12)
28. Che W, Li Z, Liu T (2010) LTP: a Chinese language technology platform. In: Proceedings of international conference on computational linguistics, pp 13–16
29. Hwang SJ (2013) Discriminative object categorization with external semantic knowledge

Performance Study of Three Different Types of Data Center Network Architectures, NovaCube, BCube, and FatTree Using NS-3 Simulator

A. Ranjith, P. Subhash, G. Nagaraju, M. Srinivas, and A. Prashanth

Abstract In Data Centers, the Data Center Network (DCN) is an important component, which is an interconnection of servers and switches with high-speed communication links. As demands of cloud-based services are increasing, the scale of the DCN has been increasing and cost and power consumption have became one of the major concerns. Existing fundamental DCN architectures such as FatTree, BCube use external switches to interconnect servers, which are cost inefficient and more power-consuming. In order to deal with these problems, a novel torus based DCN architecture named NovaCube and Probabilistic Oblivious Routing Algorithm PORA for NovaCube was recently proposed. However, effectiveness of this proposed architecture of NovaCube and PORA algorithm had been given only theoretically. In order to take better business decisions, improve the architecture, or to introduce an entirely new architecture, in this paper, we presented performance studies of three different types of DCN Architectures named FatTree, BCube, and NovaCube. NS-3 platform had been used to implement these architectures. Performances of these DCN architectures are measured in terms of average throughput, average latency, cost, power, and scalability.

Keywords FatTree · BCube · NovaCube · Data center network · PORA

A. Ranjith
School of Computer Engineering, KIIT University, Bhubaneswar, India

P. Subhash (✉)
Department of CSE (CyS, DS) and AI&DS, VNRVJIET, Hyderabad, India
e-mail: subhash_p@vnrvjiet.in

G. Nagaraju
Department of CSE (AIML & IOT), VNRVJET, Hyderabad, India

M. Srinivas
Department of CSE, Vaageswari College of Engineering, Karimnagar, India

A. Prashanth
Bandenburg University of Technology (BTU), Cottbus, Senftenberg, Germany
e-mail: akkaldev@b-tu.de

B. R. Devi et al. (eds.), *Proceedings of Fifth International Conference on Computer and Communication Technologies*, Lecture Notes in Networks and Systems 897,
https://doi.org/10.1007/978-981-99-9704-6_38

1 Introduction

The Data Center Network (DCN) architecture is noticed as one of the most important factors of network performance in Data Centers and plays an important role in fulfilling the cloud computing services such as on-demand access of information, social network websites, hypertextual web search engines [1] and for large clusters, platforms of data processing. As demands of cloud-based services are rising, a scalable Data Center Network is required to support those services. In order to take better business, decisions cost becomes a major issue. To understand the completely functional DCN of 50,000 [2] (Fifty Thousand) servers, the approximate estimated cost incurred is around $125 million. To be specific, almost 45% of the costs were spent into computational servers and storage system, 25% of the costs went into power distribution and cooling, and the rest of the 30% into network equipment [3] and electrical utility. Data center plays an important role, in providing on-demand access of information and services in the cloud-based applications. In order to deliver peak performance to end users, DCN architecture must be effective in terms of throughput and latency as performance metrics. By keeping these parameters in mind, many DCN architectures, such as BCube [4], FatTree [5], DCell, through [6], Helios [7], FlatNet [8], Hypercube [9], CamCube [10], Sprintnet [11] have been proposed. DCN topologies are mainly classified into three types:

1. Tree-based DCN Topology (e.g., FatTree)
2. Recursive defined Topology (e.g., BCube), and
3. Direct network Topology (e.g., CamCube).

These three types of topologies have their own disadvantage and advantages. FatTree and Clos examples of Tree-based Topologies which can provide full bisection bandwidth, poor in providing scalability, and its building complexity and cost are comparatively high. Whereas, recursive defined topology provides scalability and cost-effective but full bisection bandwidth may not be provided. In the direct network, topology servers are directly connected to other servers without any routers and switches in between servers to connect. The direct network topology is usually designed based on torus topology. Torus is an attractive interconnect network architecture in high performance system. But existing torus based DCN topology such as CamCube design is suffering from long network diameter; network diameter is defined as the longest distance in topology to reach any node to any other node in topology. SmalWorld suffers from poor load balance and low worst case throughput. These basic torus based network topologies usually suffer from long network diameter when the scalability is high. Due to this long network diameter, latency is high and suffers from poor throughput. To overcome these drawbacks, the work introduced NovaCube [12] DCN architecture and Probabilistic Routing Algorithm called PORA [12]. But, in the work, they have presented theoretical explanation of NovaCube such as how it is better when compared with other torus based DCN architectures. In order to take better decisions, we came up with the simulated result of comparison between NovaCube with BCube and FatTree.

1.1 Motivation

Why compare with torus based DCN architecture? The torus based DCN architecture has unique advantages compared with other DCN architectures, some of those are presented below. Firstly, the infrastructure cost of the torus based design is very low since it does not need any external switches to interconnect servers. So it is cost-effective. Secondly, since it does not use external switches to interconnect the systems, the power utilized by the switches and cooling power associated with switches also can be reduced. Thirdly, fault tolerance. Existing popular DCN architectures are generally built with a large number of switches. If there is any failure in interconnected switch, then the performance of network and interconnected system reliability can be impacted. Fourthly, the scalability. Since torus based DCN architecture is symmetrical in nature, it allows optimal scalability as well, and adding and removing systems in the network does not impact of existing architecture and routing algorithm. So, it also allows flexibility.

2 Related Work

2.1 FatTree Based DCN Architecture

Generally, FatTree [5] data center network architecture interconnects the servers using external switches and high capacity data cable links. These switches and servers in Fat-Tee are arranged in hierarchical architecture, which is having 4 layers: end host layer, edge switch layer, aggregation layer, and core switch layer. In m-ary, FatTree core switch layer contains $(m/2)2m$-port switches. There are m-pods; these are arranged beneath the core switch layer; these m-pods are named pod 0 to pod $m - 1$. Each port of each switch of core switch layer is connected to one of the m-pods. Each pod, below the core switch layer, consists of two layers. One is an edge switch layer and the other one is an aggregation switch layer. Each of these two layers contains m/2 m-port switches. Each $m/2$ server of end host layer is connected to one of the $m/2$ switches of edge layer. Among the m-port of each switch of edge layer, $m/2$ ports are connected to $m/2$ servers of end host layer and $m/2$ ports of each and every switch that is further connected to $m/2$ switches of Aggregation layer. Among the m-port of each switch of Aggregation layer, $m/2$ ports are connected to $m/2$ servers of core switch layer and $m/2$ ports of each switch are connected to $m/2$ switches of edge layer. As m-ary FatTree has m-pods, each pod is having $m/2$ switches and each $m/2$ switch is connected to $m/2$ servers in end host layer, so the total number of servers in m-ary FatTree is $m3/4$ servers. Figure 1 shows the FatTree DCN architecture.

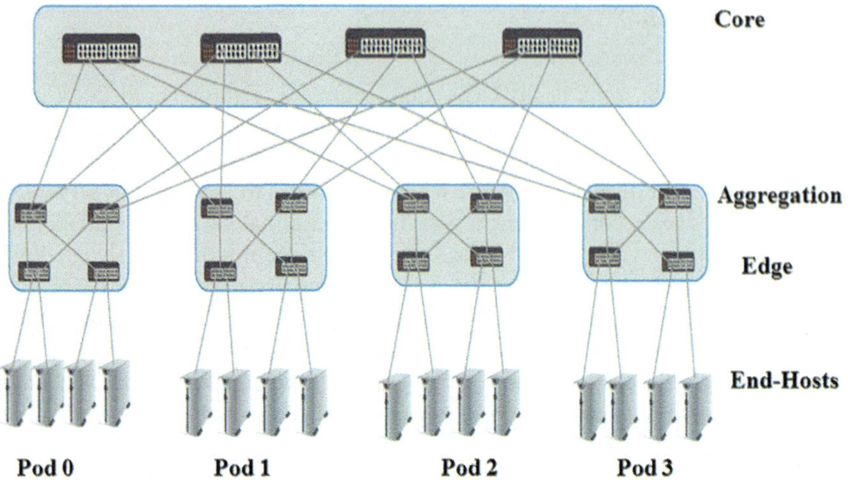

Fig. 1 A m-FatTree DCN architecture with $m = 4$

2.2 BCube

One of the fundamental recursive data center network architectures is BCube. In a more general form, a BCubem ($k \geq 1$) is built from n BCubem−1s and m-port switches. In the base, a BCube0 is formed by an n-port switch connected to n servers. A Becube1 is structured using n BCube0 n n-port switches. Every server of BCubem has $m + 1$ ports, labeled from level0 to level-m, each of these $m + 1$ ports is connected to exactly one switch of each level. In BCubem, there are $m + 1$ level of switches and each level has n-port switches, so each level can connect to servers. A BCubem exactly has n servers and each server has $m + 1$ ports. All servers have number of ports. As their $m + 1$ level of switches, each level has n-port switches so all switches together have the number of ports. So these server ports and switche ports are connected using the following addressing scheme. Notice that as shown in Fig. 2, no two switches are directly connected through link. The addressing for switches is enumerated as follows Lev, Sm − 1 Sm − 2. S0, where Lev($0 \leq$ Lev $\leq m$) is denoting the level of switch and next part of address Sp ∈ [0, $n - 1$], where $p \in [0, k - 1]$ denotes the unique identification for each switch. In BCubek to address each server uniquely, every server is enumerated as rm, rm − 1 … r1, r0 where ri ∈ [0, $n - 1$], i ∈ [0, k]. In this, addressing scheme of servers rm denotes unique enumeration for n BCubek − 1 from 0 to n1. In n BCubem − 1 s, each BCubem − 1 has n m servers rest of rm − 1, rm − 2 … r0 uniquely enumerates the server of each BCubem − 1 from 0 to n $m - 1$. To connect the switches to the servers, the level m-port of the ith server in the jth BCube $m - 1$($j \in [0, n - 1]$) is getting connected to jth port of ith level-m switch. A typical architecture illustration is shown in. Figure 2, with $m = 1$ and $n = 4$.

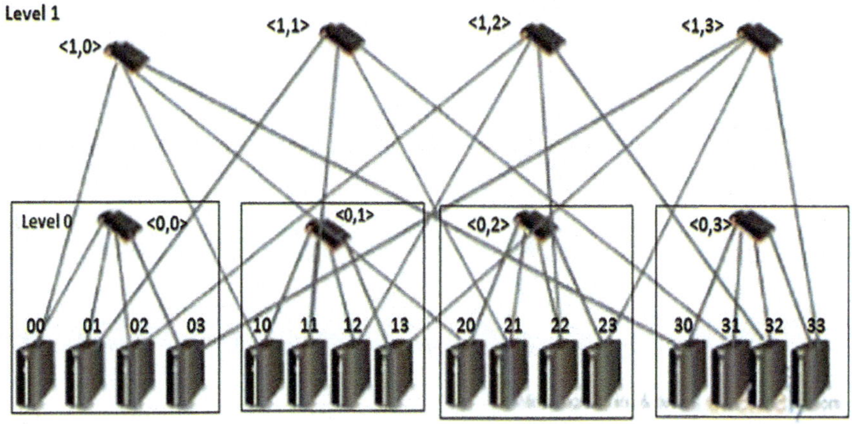

Fig. 2 A BCube_m architecture with $m = 1$ and $n = 4$

2.3 *NovaCube Torus Based DCN Architecture*

Torus based DCN architecture has many advantages such as it allows high path diversity, which offers many different paths from source node to destination node. Its symmetrical architecture allows the nodes to connect closely and it is cost and energy effective as it does not need external switches to interconnect servers in DCN architecture. Also, it is very flexible to increase scalability as it does not damage existing architecture and routing algorithm. But as we are increasing the scalability, network diameter increases. As network diameter increases latency and throughput increase, then the efficacy of network performance will be reduced. As shown in Fig. 3, a k-ary n-cube torus has bk/2c × n network diameter. So as the number of nodes increases, the network diameter increases, which big disadvantage of torus based DCN architecture. To overcome this problem, the work [12] proposed a new torus based DCN architecture named NovaCube. As shown in Fig. 4, NovaCube is constructed based on the regular torus topology by adding extra jump-over link between all longest distance node pairs throughout the network.

3 Experimental Setup

We have used NTU-DSI-DCN framework which builds on top of a common open-source simulator called ns-3. In this framework, to get fair comparison results, the network configuration devices, protocols, and traffic flow generation are made consistent. To run simulation, we have to change only the size of network topologies to be compared. Table 1 gives the simulation settings to compare the performances of the NovaCube with BCube and FatTree. The number of servers in all four architectures starts from 16 and ends with 4000 servers. Ethernet switches capable of

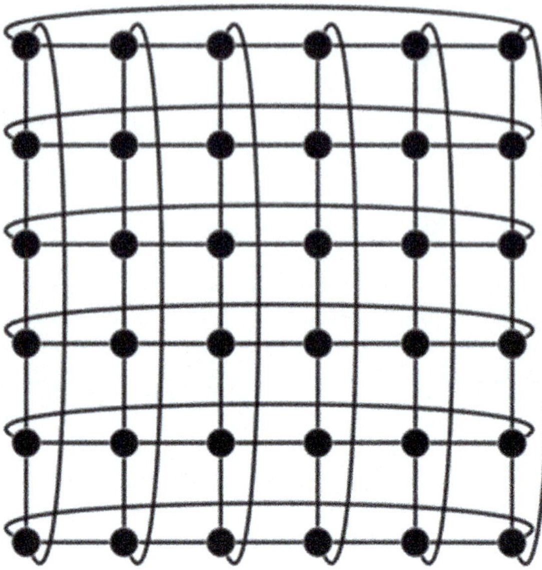

Fig. 3 A 6-ary 2 cube torus topology

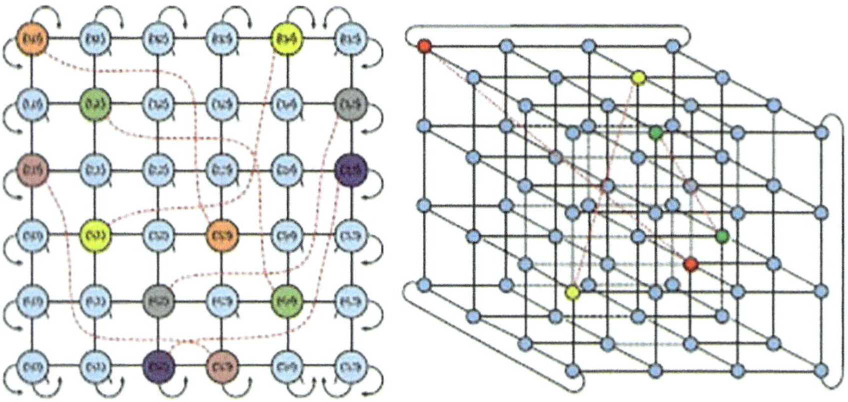

Fig. 4 A 2D 6 × 6-node and 3D 4 × 4 × 4-node NovaCube (for simplicity, not all jump-over links are shown) [12]

1000 Mbps with a real IP address and Nix-Vector Routing [13] technique were used. Sender and receiving communication pairs are selected randomly for generating traffic flow. Simulation running time is 100 s and each sender sends 1Mbps flow of data simultaneously.

In this framework, to get fair comparison results, the network configuration devices, protocols, and traffic flow generation are made consistent. To run simulation, we have to change only the size of network topologies to be compared. Table 1 gives the simulation settings to compare the performances of the NovaCube with BCube

Table 1 Simulation setting for the NovaCube, BCube, and FatTree architectures

	Nova Cube	BCube	FatTree
Number of pods (k)	–	3	4–24
No. of BCube levels	–	3 ($m = 2$)	–
No. of nodes	16–4000	64–3375	16–3456
Running time of simulation (s)	100	100	100
Size of packet (bytes)	1024	1024	1024
Device channel capacity (Mbps)	1000	1000	1000
Selection of communication pairs	Uniform random selection	Uniform random selection	Uniform random selection
Traffic pattern	Exponential random traffic	Exponential random traffic	Exponential random traffic
Routing protocol	Nix-vector	Nix-vector	Nix-vector

and FatTree. The number of servers in all four architectures starts from 16 and ends with 4000 servers. Sender and receiving communication pairs are selected randomly for generating traffic flow. Simulation running time is 100 s and each sender sends 1Mbps flow of data simultaneously.

The flow of traffic patterns for all architectures is an exponential random distribution with on–off behavior, which is a realistic traffic flow in data centers as proved in [14]. With these consistent simulation settings, we moved to construct 2D NovaCube, 3D NovaCube, BCube, and FatTree. We used Flow Monitor module [15] in ns-3 to fetch performance statistics and analyze them. To study the performances of architectures, we have taken two main performance metrics: Average throughput and the other one are average packet delay. The average packet delay is calculated as below:

$$\mathrm{DL}_{\mathrm{avg}} = 1/N \sum_{j=1}^{n} D_j \tag{1}$$

where $\mathrm{DL}_{\mathrm{avg}}$ is the average packet delay, N is the number of total packets successfully received in the network and D_j is the delay of the received packet.

The average throughput was calculated as below:

$$T_{\mathrm{avg}} = \frac{\sum_j^n (P \times \Theta i)}{\sum_j^n \mathrm{DL}_j} \tag{2}$$

where average throughput of the network is denoted by T_{avg}, $Pj \in [0,1]$ with $Pj = 0$ refers to the loss of packet j, $Pj = 1$ denotes the receiving packet j, θj denotes the received packet size in bits, Dj denotes a delay of a received packet j and the total number of received packets in an entire network is denoted by N.

4 Result and Analysis

By observing the result of average throughput vs number of nodes in Fig. 5., up to 350 nodes 3D NovaCube gives higher throughput than BCube and FatTree. While the number of nodes in 3D NovaCube is increasing from 350 to 1000, it is giving slightly better through than widely used DCN architecture called FatTree and slightly lower performance than BCube. When the number of nodes increases from 1000 to 3500, consistently small reduction in NovaCube average throughput, whereas FatTree architecture maintains steady throughput. When comparing BCube with FatTree where number of nodes ranges from 64 to 3500 nodes, BCube average throughput changes from 230 to 174Mbps, and FatTree Architecture average throughput ranges from 117 to 125Mbps. By the above observation, Bcube is performing slightly better than NovaCube and FatTree with respect to average throughput as a performance metric.

From Fig. 6., average latency versus number of nodes, we can observe that when the number of nodes increases, average latency of FatTree and BCube increases moderately, whereas NovaCube average latency increases a little quickly. When the number of nodes increases from 16 to 350 nodes, the average latency of NovaCube is lesser than Bcube. From 350 to 1000 nodes, average latency of NovaCube is slightly higher than BCube and a little lower than FatTree. From 16 to 350 nodes, average latency of NovaCube is in between 0.02 and 0.10 ms, average latency of FatTree is 0.065–0.073 ms, and average latency of BCube is 0.035–0.047 ms.

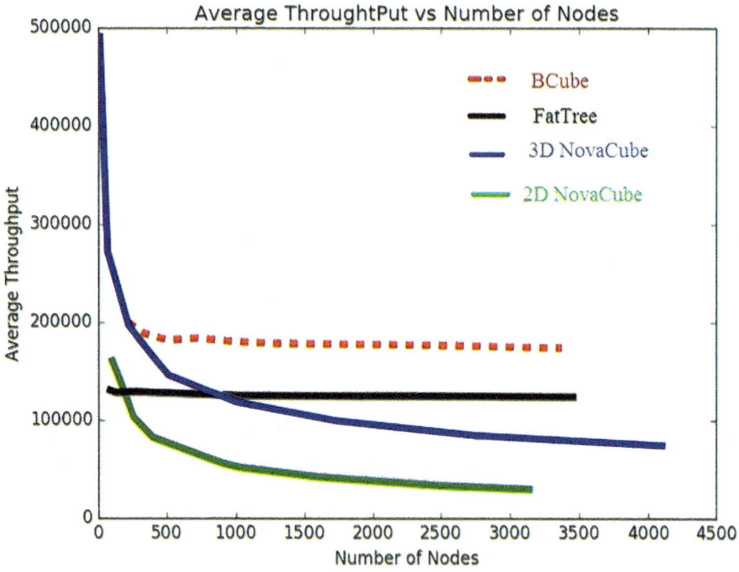

Fig. 5 Average throughput (Mbps) versus number of nodes

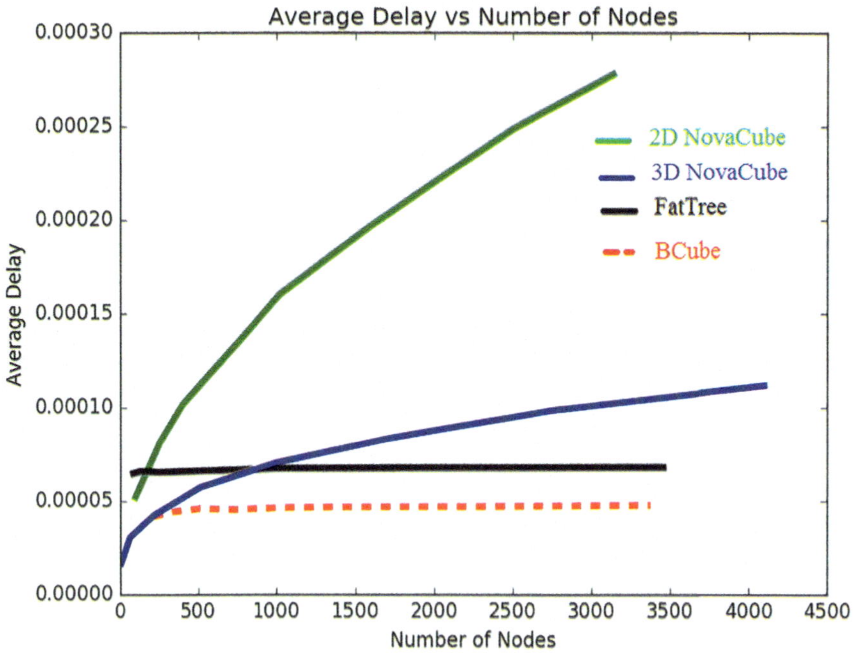

Fig. 6 Average latency (ms) versus number of nodes

As per Eq. 1, 3D NovaCube of 1000 nodes is equal to 8, whereas network diameter of FatTree is 6 and BCube1 is 4. Even though network diameter of NovaCube is high compared with FatTree for 1000, the performance of NovaCube is better because of high path diversity. But when we keep on increasing the scale of NovaCube to 4096, DNovaCube is equal to 12 whereas network diameter of FatTree and Bcube is equal to 6 and 4 respectively which is constant. Because of this huge variation in network diameter in NovaCube, its performance becomes low at large scale topology.

$$D_{\text{NovaCube}} = \frac{k|2 \times n}{2} \tag{3}$$

Table 2 shows performance study of these tree topologies with respect to scalability, number of switches per server in network, and the cost and consumption of power. By observing Table 2, with respect to number of nodes per switch, in order to realize 3375 nodes of BCube topology, it requires 675 external switches. To realize the 3456 servers of FatTree topology, it requires 720 external switches. Whereas, to realize any scale of DCN architecture, NovaCube architecture does not require any external switches. While comparing with NovaCube, BCube is taking exactly an extra 20% of external switches So it needs extra cost for these 20% of switches and that is the reason why it is cost inefficient and it will take extra power for this 20% switches. To cool these switches, it will take extra cooling devices and power

Table 2 Comparison results of NovaCube, BCube, and FatTree

	3DNova Cube	BCube	FatTree
No of switches/nodes	0/4000	675/3375	720/3456
Scalability	Flexible	Not flexible	Not flexible
Cost and power consumption	Low	High	High

consumption for these devices. Whereas, FatTree topology requires exactly an extra 20.83% of external switches, so it needs extra cost for this 20.83% of switches and that is the reason why it is cost inefficient and it will take extra power for these 20.83% switches. To cool these switches, it will take extra cooling devices and more power consumption for this extra cooling system. When comparing with scalability, the architecture complexity of BCube and FatTree makes it flexible to increase number of servers and it is also flexible to adopt the routing algorithm for this increased architecture. Whereas, NovaCube architecture is symmetrical in nature, and it does make any complexity, to increase the scalability and as well as flexibility to adapt the routing algorithm for this increased architecture of NovaCube.

5 Conclusion and Future Work

In this paper, we presented a comparative performance study between three different types of topologies, namely, NovaCube, FatTree, and BCube. Simulation results show that BCube and FatTree required 20 and 20.83% of extra external switches while comparing with 3D NovaCube of size 3500 servers. It proved that NovaCube architecture is given optimal results in cost and power consumption. Experimental results also show that NovaCube is effective architecture for small-scale Data Center Networks with respect to performance metrics, average throughput, and average Latency. Whereas, FatTree and BCube are the most effective for Large-Scale Data Center Networks with respect to Average throughput and Average Latency metrics.

In future work, by keeping this experimental result in mind, one can come up with hybrid Data Centre Network Topology, where one can use torus structure logic in recursive Torus based DCN architecture or FatTree based torus DCN architecture to reduce cost, power, and without loss of throughput.

References

1. Brin S, Page L (2012) Reprint of: The anatomy of a large-scale hypertextual web search engine. Comput Netw 56(18):3825–3833
2. Popa L, Ratnasamy S, Iannaccone G, Krishnamurthy A, Stoica I (2010) A cost comparison of datacenter network architectures. In: Proceedings of the 6th international conference. ACM, p 16

3. Greenberg A, Hamilton J, Maltz DA, Patel P (2008) The cost of a cloud: research problems in data center networks. ACM SIGCOMM Comput Commun Rev 39(1):68–73
4. Guo C, Lu G, Li D, Wu H, Zhang X, Shi Y, Tian C, Zhang Y, Lu S (2009) Bcube: a high performance, servercentric network architecture for modular data centers. ACM SIGCOMM Comput Commun Rev 39(4):63–74
5. Al-Fares M, Loukissas A, Vahdat A (2008) A scalable, commodity data center network architecture. ACM
6. Wang G, Andersen DG, Kaminsky M, Papagiannaki K, Ng T, Kozuch M, Ryan M (2010) c-through: part-time optics in data centers. ACM SIGCOMM Comput Commun Rev 40(4):327–338 (2010)
7. Farrington N, Porter G, Radhakrishnan S, Bazzaz HH, Subramanya V, Fainman Y, Papen G, Vahdat A (2010) Helios: a hybrid electrical/optical switch architecture for modular data centers. ACM SIGCOMM Comput Commun Rev 40(4):339–350
8. Lin D, Liu Y, Hamdi M, Muppala J (2012) Flatnet: towards a flatter data center network. In: Global communications conference (GLOBECOM). IEEE, pp 2499–2504
9. Hyper-bcube: a scalable data center network. In: 2012 IEEE International conference on communications (ICC). IEEE, pp 2918–2923
10. Abu-Libdeh H, Costa P, Rowstron A, O'Shea G, Donnelly A (2010) Symbiotic routing in future data centers. ACM SIGCOMM Comput Commun Rev 40(4):51–62
11. Wang T, Su Z, Xia Y, Liu Y, Muppala J, Hamdi M (2014) Sprintnet: a high performance server-centric network architecture for data centers. In: 2014 IEEE International Conference on communications (ICC). IEEE, pp 4005–4010
12. Wang T, Su Z, Xia Y, Qin B, Hamdi M (2014) Novacube: a low latency torus-based network architecture for data centers. In: 2014 IEEE Global communications conference. IEEE, pp 2252–2257
13. Riley GF, Ammar MH, Zegura EW (2001) Efficient routing using nix-vectors. In: 2001 IEEE Workshop on high performance switching and routing. IEEE, pp 390–395
14. Benson T, Anand A, Akella A, Zhang M (2010) Understanding data center traffic characteristics. ACM 40(1):92–99
15. Carneiro G, Fortuna P, Ricardo M (2009) Flowmonitor: a network monitoring framework for the network simulator 3 (ns-3). In: Proceedings of the fourth international ICST conference on performance evaluation methodologies and tools. ICST (Institute for Computer Sciences, Social-Informatics and Telecommunications Engineering), p 1

Author Index

A
Adyesha Singhdeo, 347
Ajith Sai Visal, 1
Akhil Karnam, 319
Akshatha Prabhu, 49
Arindam Sadhu, 125
Arjun, C. S., 49
Arul, N., 287
Ashok Kumar Nanda, 337
Ashok Kumar Saini, 275
Ashok Reddy Kandula, 205
Atharva Shinde, 329
Avala Raji Reddy, 371
Avantika Mahadik, 19
Aya Atiya, 241

B
Badam Prashanth, 309
Bhavishya Pachika, 319
Bhawna Ojha, 135
Bikramjit Sarkar, 125
Binitha S. Thomson, 355
Bolla Aditya, 371
Bollampelly Chandana, 401
Brahmananda Reddy, A., 319
Brindha, K., 263
Burri Naresh, 217

C
Chaitra Sai Jalda, 337
Charan Yadav Chintalacheri, 309
Ch Ravikumar, 217
Ch. Vinay Kumar, 193

D
Darisi Abhinay, 1
Deepan, P., 287
Deepthi, P., 371
Desu Devi Sri, 95
Devesh Kumar Srivastava, 275
Dhanamma Jagli, 297
Dharmendra Pathak, 225
Disha N. Girish, 73
Divya Boreda, 337
Divya, K., 205

G
Ganpat Joshi, 371
Gaurav Dubey, 135
Guntoju Kalpana Devi, 309
Gurram Sunitha, 391

H
Harika Pappala, 205
Hussien, Ahmed, 241

J
Jairam Naik Khetavath, 363
Janardhan Babu, V., 105
Jayshree D. Mallapur, 253
Jonnadula Narasimharao, 371

K
Kalaivani, K., 263
Karm Veer Singh, 39

B. R. Devi et al. (eds.), *Proceedings of Fifth International Conference on Computer and Communication Technologies*, Lecture Notes in Networks and Systems 897,
https://doi.org/10.1007/978-981-99-9704-6

Karthika Venkata Satish Reddy Jakkireddy, 205
Karuppasamy, K., 263
Khemchand Shakyawar, 135
Kirankumar B. Balavalad, 253
Kirankumar Y. Bendigeri, 253
Kirubakaran, S., 145
Konda Srikar Goud, 161
Kumar Gautam, 337
Kunal Das, 125

L
Likhitha Varsha, C., 183
Logyn Medhat, 241

M
Maganti Vasudha, 9
Mahalakshmi Jain, B. A., 9
Mahender, U., 391
Mahesh Kotha, 145
Manoj Kumar Chenna, V. S., 401
Marri Sireesha, 391
Matoori Praveen, 401
Mehernadh, K., 183
Mohan Sai Motineni, 205
Mohanty, J. R., 347
Mrunal Shinde, 329

N
Nagaraju, G., 411
Nagu Chandra Sekhar Reddy, 401
Nalla Siddhartha, 297
Nandini Sidnal, 253
Nikhil Shinde, 329
Nithin, A., 183
Nitin Dixit, 135

P
Pankaj Kunekar, 329
Parisa Monika, 95
Parthajit Roy, 65
Pavani Kotha, 105
Pavan Phani Kumar, T. V. L. N., 115
Peda Gopi Arepalli, 363
Pradeep Kumar, V., 337
Pradeep Yadav, 135
Prajakta Musale, 329
Prashanth, A., 411
Prashant Sharma, 19
Priyanka, M., 73
Puja Chavan, 329

Q
Qayyum, Mohammed, 183

R
Rahul Lal, P., 1
Rajalingam, B., 287
Rakhi Arora, 135
Ramadevi, Y., 171
Ramgopal Kashyap, 225
Ranjith, A., 411
Rashmi, D., 9
Reddy Madhavi, K., 309, 391
Roheet Bhatnagar, 241, 275
Rose Varuna, W., 355

S
Samuel Chepuri, 171
Samuel, M., 193
Sandhyarani, 401
Santosh B. Kumbalavati, 253
Santhoshkumar, R., 287
Sarada Devi, T. S. N. G., 115
Sasi Kiran, J., 391
Satyanarayana Nimmala, 217
Senthilkumar, M., 145
Shams, Mahmoud Y., 241
Shivani Pachika, 319
Shobha Rani, N., 49
Shruthi Reddy, C. R., 371
Soumyadip Sarkar, 125
Sreejyothsna Ankam, 105
Sri Rama Krishna Datta Movva, 205
Srinivasa Rao Dhanikonda, 297
Srinivasarao Dharmireddi, 145
Srinivasa Rao Giduturi, 161
Srinivas, M., 411
Srujan Raju, K., 401
Sruthi, J., 183
Subba Rao, M., 309
Subhash Chandra, N., 297
Subhash, P., 183, 193, 411
Subhasree Bhattacharjee, 125
Suji Aparna, S. V., 145
Suneetha, M., 95
Suraya Mubeen, 309
Suresh Satapathy, 347
Surya Narayana Reddy, V., 193
Suthendran, K., 145
Swathi Rudra, 391

Swati Adhikari, 65

T
Tamanna Jena Singhdeo, 347
Tazzeen Fatima, 309
Tharun, G., 193

V
Vadapally Praveen Kumar, 217
Vaibhav Narawade, 19
Versha Verma, 39
Vidhya, R., 287
Vipin Saxena, 39
Vishal Reddy, G., 193
Vishal Verma, 39
Voruganti Naresh Kumar, 145, 309, 391

SPRINGER NATURE

GPSR Compliance

The European Union's (EU) General Product Safety Regulation (GPSR) is a set of rules that requires consumer products to be safe and our obligations to ensure this.

If you have any concerns about our products, you can contact us on ProductSafety@springernature.com

In case Publisher is established outside the EU, the EU authorized representative is:

Springer Nature Customer Service Center GmbH
Europaplatz 3
69115 Heidelberg, Germany

The manufacturer's authorised representative in the EU is Springer Nature Customer Service Centre GmbH, Europaplatz 3, 69115 Heidelberg, Germany. If you have any concerns regarding our products, please contact ProductSafety@springernature.com

Printed and bound by CPI Group (UK) Ltd, Croydon, CR0 4YY
01/10/2026
02225860-0016